PLAGUE

PLAGUE

**One Scientist's Intrepid
Search for the Truth about
Human Retroviruses and
Chronic Fatigue Syndrome
(ME/CFS), Autism, and
Other Diseases**

KENT HECKENLIVELY, JD
JUDY MIKOVITS, PHD

FOREWORD BY HILLARY JOHNSON

Skyhorse Publishing

Disclaimer

The factual information presented herein is based on the experiences, documents, and recollections of the authors, as well as the various individuals who were involved in these events.

However, the conversations reported within the text are written from the memory of the authors and others involved in these events, as well as supporting documentation when available, and only represent their best recollection. These recollections are not intended to be statements of material facts, but rather the opinion of the authors and others involved in these events as to what was said and their interpretation of the meaning of those conversations.

Neither the authors nor the publisher claims that the conversations are accurately recorded, and we apologize in advance for any omissions or errors in content or meaning.

All the people and events depicted are real. Some of the names have been changed for the protection of the individuals involved.

Many of the primary documents supporting the information found in this book are available at www.plaguethebook.com.

To the researchers who labor every day to protect the integrity of their work from special interests and the patients who depend on this unbiased science.

An inefficient virus kills its host. A clever virus stays with it.

—James Lovelock

You do not become a "dissident" just because you decide one day to take up this most unusual career. You are thrown into it by your personal sense of responsibility, combined with a complex set of external circumstances. You are cast out of the existing structures and placed in a position of conflict with them. It begins as an attempt to do your work well and ends with being branded an enemy of society.

—Václav Havel

Contents

A Note from Judy Mikovits, PhD

My co-author, Kent Heckenlively, said they would come after me. It was at a dinner with him during the Autism One Conference in Chicago, Illinois, in May of 2010. How could I possibly imagine that about a year and a half later, on November 17, 2011, while I was preparing to return to Dr. Frank Ruscetti's lab in Frederick, Maryland, and participate in the Multi-Center validation study directed by Dr. Ian Lipkin, an email would be sent to Frank by none other than Dr. Tony Fauci, head of the National Institute of Allergy and Infectious Disease? In the email, Fauci stated I could participate in the study, but if I stepped foot on National Institutes of Health property, I would be immediately arrested! Is that science? The next day several police cars would descend on my home, take me away, and have me held without bail for five days in a Ventura County, California, jail. How convenient then for the journal *Science* to publish my mug shot the next week. They sent a clear message to anyone in research who dared stand by data that revealed an inconvenient truth about corruption in public health.

While I doubt that my co-author expected all of that to happen to me, I don't think he was surprised. As an autism advocate for many years, he knew the enemy. I did not. I had been a twenty-plus year government scientist and Director of the Lab of Anti-Viral Drug Mechanisms at the National Cancer Institute, and published more than fifty peer-reviewed scientific articles. My career was dedicated to trying to end the suffering of those with cancer and acquired immune deficiencies. Science might have many flaws, such as raging egos and turf wars, but surely it could not be fundamentally corrupt on whether animal viruses from cell cultures were possibly contaminating biological products, including vaccines. I was certain the research community learned from the tragedy of HIV/AIDS. I strongly believed in the scientific process. I don't believe the scientific method is being followed today in important areas of government- or industry-sponsored science. Or that any of those at the top of government science have the slightest bit of respect for the United States Constitution. They have gone rogue, and they need to be taken down. For five years I have tried to pursue this matter legally but have yet to see the inside of a courtroom or call a single witness.

My attorney, one of the country's top environmental lawyers, says he has never seen a case of such rank injustice.

Hopefully, by the time you are reading this, you will be aware of Dr. William Thompson, the CDC whistleblower who revealed that government scientists deliberately removed study participants to cover up a link between earlier measles, mumps and rubella (MMR) vaccination and rates of autism in African-American boys. The CDC also concealed data about a group it bizarrely calls "isolated autism," meaning that they were normally developing before their vaccination. I could not believe it when I saw they used the same strategy of fraud that the CDC used for more than two decades in chronic fatigue syndrome/myalgic encephalomyelitis (CFS / ME)! Remove all the real patients and make certain to use esoteric statistics to hide the real meaning of the data. We are all at risk. At the time of this writing, Thompson and those scientists who participated in these crimes against humanity remain employed by the CDC. Why are they not in jail?

I believe the sole duty of a scientist like me is to perform experiments that help us understand why some people get diseases and others don't and hopefully improve the daily life of people around the globe. What did I investigate? I wanted to get to the bottom of CFS/ME and relieve the suffering of a disease that afflicts approximately twenty million people worldwide. Most with CFS/ME lead horrific lives, often required to spend twenty-three hours of every day confined to their beds, in dark rooms, some wracked with unimaginable pain. Others will look relatively normal to the untrained eye, but they suffer with brain fog, are easily fatigued, and are plagued by the daily question of what happened to their previous vitality. These people are robbed of their lives. I wanted to give back what had been taken from them.

We found that these people were infected with a retrovirus, XMRV (xenotropic murine leukemia virus-related virus, a mouse retrovirus), which had been "discovered" in 2006 by scientists from the Cleveland Clinic and the University of California, San Francisco. In our *Science* publication announcing the isolation in 2009 (a joint publication with the Cleveland Clinic and the National Cancer Institute), we detected evidence of the retrovirus in about 67% of patients and in about 4% of healthy controls. I stand by that data. The only mistake in the article was the sequence of the viruses we isolated was not that of Robert Silverman, who originally identified the retrovirus. We did not know it at the time, but Silverman had not isolated the virus from an actual patient, as is done in traditional virology. Instead, he had created a molecular clone from three different samples.

The clone he created had many mistakes in the sequence. Silverman was a co-author on our *Science* paper but did not disclose any of this information to us. This is a severe breach of the ethics of scientific research. Yet he was not blackballed from science. He still works and gets grants to this day. Silverman was the one who refused to do the studies blinded, making it more difficult for us to discover his errors. He covered up his mistake for two years and lied to Frank and me. Others found evidence of a family of retroviruses as well, including Harvey Alter, recipient of the prestigious Lasker Award. Not just the gamma retroviruses, but the HTLV-like delta viruses that Elaine Defrietas had isolated twenty years earlier. (Elaine was also taken down by CDC fraud.)

What was my crime? What did I do to cause the furies of the scientific world to descend on me? I performed as an ethical scientist, as I had for more than 30 years. In my previous studies I had worked on HIV-AIDS. When I defended my doctoral thesis, one question the panel of experts asked me was whether the basketball star, Magic Johnson, who had recently been diagnosed with infection by the HIV retrovirus, would go on to develop AIDS (based on the data in my thesis). Our data suggested that the current paradigm of treatment for HIV would have to change. That instead of waiting to treat until the immune system was greatly damaged by HIV, treatment with the antiretroviral therapies should be started as soon as possible. That was a total change and considered a big risk to take the new highly active antiretroviral therapy so soon. Magic was courageous and took that risk and millions of lives were changed for the better. Decades later, my answer is still sound. You can go up to Magic Johnson and shake his hand because he embraced new data and trusted scientists, even though the thought was heretical at the time. Thinking about what has happened in the twenty-five years since still makes me cry. Our discovery of the retroviral association in CFS/ME, autism, and other devastating diseases should have given the victims of these devastating diseases the same opportunity that Magic Johnson had. The benefits of the therapies far outweigh the risk. Billions of tax dollars have been invested to develop these therapies and prove they are safe in children.

My real crime was saying that if this retrovirus was causing CFS/ME in adults, it might be causing autism in children. (Our research and that of others showed that the retrovirus or its closely related cousins infected anywhere from 4% to 8% of the healthy population.) Standard practice in HIV-AIDS is that when a child is born to an HIV-infected mother, that child is immediately put on anti-retroviral therapy, prior to any immunizations.

Retroviruses like to hide out in the B and T cells of the immune system, the very cells a vaccination is designed to stimulate. An HIV-infected child might develop AIDS as a result of an immunization. Everybody agrees with that. Maybe an XMRV-infected child was developing autism indirectly as a result of their immunizations. A small study that we presented at a symposium at the National Institutes of Health showed that 14 out of 17 children with autism showed evidence of XMRV infection.

You see, I gave a solid, real-science answer to why parents observed a change in their child after a vaccination. As my co-author discovered and reported in chapter 5, the first outbreak of chronic fatigue syndrome/ME, among 198 doctors and nurses at Los Angeles County Hospital in 1934-1935, was preceded by their receipt of an early polio vaccine grown in mouse tissue and given with an accompanying immune system booster preserved with the mercury preservative thimerosal. This has been a tragedy more than eighty years in the making.

Even worse, in 1994 the scientific community reported the very real possibility that growing human viruses in animal tissue and cells used every day in laboratories around the world, then re-injecting that material back into humans, could introduce new animal viruses into the human population. In fact, our research about the XMRV retroviruses in 2011 showed that this catastrophe had already taken place! It is such a simple idea, so easy to convey, that if people really begin to focus on it, the entire vaccination program would crumble. God did not intend for animal viruses to be injected directly into the human bloodstream.

And what are the current leaders in science doing about this threat to your health? I participated in a study to see if the blood supply was contaminated with these retroviruses. The simple answer? Absolutely yes! But that is not what was reported. Instead, they jailed me, called my work a fraud, and published a mug shot in *Science* to make certain that the millions who were sick would not seek restitution, and that mistakes of previous decades would remain hidden. They lied about it, while blood banks around the country started quietly spending tens of millions of dollars to purchase something called the Cerus Intercept system, which my lab proved decontaminates the retroviruses in blood supplies. To this day, millions stay desperately ill, unable to receive the therapies that may give them a better quality of life.

Then there are people like Dr. Ian Lipkin, the grand virus hunter of Columbia University, who directed that multi-center study in 2012, after I had been falsely arrested and my mug shot posted in the pages of *Science*. (Read chapter 20 in the book for the full story.) The same study in which

Fauci promised my arrest if I walked into an NIH lab! The same study that fraudulently concluded that neither XMRV nor any similar retroviruses were linked to chronic fatigue syndrome/ME. (The actual report said that 6% of patients and 6% of controls showed evidence of infection with these retroviruses.) Fauci stopped the study early because, against all odds, with me participating in retrovirus isolation by phone, we found many more positives that likely would have revealed far more than 6% and may have revealed the association with CFS/ME. They thought keeping me out of the lab and using the tried and true CDC technique that had worked in the Thompson fraud would work again! Simply use the wrong patients. The patient pool used removed any patients who had "a medical or psychiatric condition that might be associated with fatigue." The condition has been known for years as chronic fatigue syndrome and they removed any patients with a medical or psychiatric condition that might be associated with fatigue? This is what passes for science in today's corrupt America.

But the Ian Lipkin of 2012 was singing a different tune on a public conference call with the Centers for Disease Control in September of 2013 when he reported that he'd found evidence of a retrovirus in 85% of a different sample pool of chronic fatigue syndrome/ME patients. (Read chapter 21 in the book for those details.) However, he wasn't going to investigate that, but rather the abnormal immune markers associated with the retrovirus. The pattern of abnormal immune markers was the very thing that made me suspect a retrovirus. I reported on this on March 29, 2011, when I presented my data within minutes of a Lipkin presentation at the New York Academy of Sciences. That was also the very meeting where we presented data on the contaminated blood supply and the good news that the Cerus Intercept system could decontaminate it! Lipkin confirmed my initial findings, but neither he nor the medical establishment will investigate any further. He got a thirty-one million dollar grant from Fauci and the National Institute for Allergy and Infectious Diseases in March of 2014. That's 21st century American science.

My co-author has told me he knew I was in danger because I was an honest scientist. He understood the size of the storm bearing down on me and that he was powerless to stop it. But he convinced me that if we wrote this book, it might be a lifeboat for me and the millions who are suffering because of the corruption. Even though I am now bankrupt, living with my husband in a seven hundred square foot apartment in southern California, and virtually ignored by the rest of the scientific community, we have made it through the storm. We consider ourselves blessed. Frank and I are still working every day to help the victims and bring our knowledge and experience to the patients. I am so thankful for the many patients who continue to believe in me, as well as the unfailing support of my husband, David, and my long-time collaborator, Frank Ruscetti.

There is a disturbing pattern in American life that nobody takes responsibility, no solutions are proposed, but all those involved ask for more money so that in some distant future we will find answers. I believe we already know enough to start making significant change in the health of millions. If we tell the truth, we can find answers. Fighting an entrenched establishment that may be continuing these epidemics by following outmoded thinking provokes a response. However, the truth remains, the truth endures, and the truth can heal. This book is my best attempt to tell the truth about my research, the culture in science today which is hostile to new ideas, and what science can really do if is allowed to pursue promising areas of inquiry.

Foreword

A Disease Able to Affect the Economies of Nations

by Hillary Johnson

Hers was the last presentation of the day and the moderator allowed a final question. I raised my hand. "Is it true that you have discovered a *novel* pathogen in this disease?" I had heard a rumor but my expectations couldn't have been lower. Silence ensued while Judy Mikovits gripped the sides of the podium as if weighing her options. "Yes," she finally said. There was scattered laughter in the hall, as if she had just made a joke. Only an audience of people who have been ill for a very long time and who have even the slightest knowledge of the fraught history of their disease could be so devoid of hope that such a claim could be considered laughable. "It's not a *new* pathogen," Mikovits pressed on, suddenly lowering her voice to a degree barely audible. "But it *is* new to this disease. We have submitted a paper to *Science*," she added.

On that intriguing scientific riddle, the annual scientific conference held by the U.K. charity Invest in ME at One Birdcage Walk near London's Houses of Parliament ended. It was May of 2009. My friendship and professional relationship with Judy Mikovits began. She could not have imagined the inferno she was about to enter, the ups and downs and startling turns of which are described in this book.

I would watch and listen in sympathy over the next three years as Mikovits, challenged and derided by critics in a way few scientists will ever experience, seemed to leap from one circle of hell to another. Ultimately, her crime would emerge as heresy, a *sub rosa* charge reserved for anyone who dared offer evidence for a viral cause and, worse, transmissibility, in a disease governments on every continent had for three decades effectively disappeared with a contradictory mélange of explanations, none of them logical.

With her pending publication still undergoing editing, Mikovits was about to lead a battle charge in what has often been described as medicine's holy war. But then, the eleven-syllable "myalgic encephalomyelitis," understandably shorthanded to "ME," has never been regarded as an ordinary disease, nor has the public health response to it been anywhere near normal. Instead, it's been saddled with an extraordinary burden of *meaning*, a word Susan Sontag employed in her famous essay "Illness as Metaphor." "Nothing is more punitive than to give a disease meaning—that meaning invariably being a moralistic one," Sontag wrote. "Any important disease whose causality is murky, and for which treatment is ineffectual, tends to be awash in significance."

Since its emergence in the late 1970s in pandemic form, ME has been touted as a psychiatric affliction of people "with poor coping strategies or histories of unachievable ambition," as one American government scientist wrote in an influential paper in 1988. "Ultimately, any hypothesis regarding the cause of [ME] must incorporate the psychopathology that accompanies it and, in some cases, precedes it," he added. Aided by a passive lay press, government scientists have sought to dismiss the disease by labeling sufferers with all manner of deficiencies and malevolent motives. That list has included malingering and cheating welfare systems, being either sympathy-seekers or Type A personalities who one day simply fell apart, or people who read about the disease and "wanted to have it." The Centers for Disease Control (CDC) probably wins any contest as to which US government entity has done more to trash victims and belittle the disease. CDC has trotted out "hysteria," "yuppie burn out," a genetically-linked "inability to handle stress," a history of childhood sexual molestation, "doctors working themselves into a frenzy," "collusion between patients and doctors," and "an epidemic of diagnoses," to name a few of the untenable "causes" handed down from on high over the past thirty years.

Any collective memory of just how suddenly and aggressively this disease emerged in the late 1970s and early 1980s, especially in large coastal cities like New York, Boston, Los Angeles, and San Francisco, recedes with each passing year. Anyone born after, say 1985 or 1990, will be unable to recall a period in their own lives when this scourge was virtually unknown. Yet, by the mid-1980s, distressed doctors and desperate patients had turned the disease into the top category of inquiry at both the Centers for Disease Control and public health departments in the major cities of this country; eventually, their calls both to the CDC and NIH exceeded queries about AIDS at the height of the AIDS epidemic.

In 1988 a small group of worried scientists and research clinicians met to discuss the disease and its origins in Newport, Rhode Island. Charles

Carpenter, professor of medicine at Brown University Hospital noted, "We're seeing something that wasn't there in the fifties and sixties. Most of us feel this is new. If this had been going on in the fifties and sixties, even if we had discarded it as psychiatric, it would have been written about, and it's not in the literature. And that suggests there is a dominant agent that's driving the disease." Another doctor, Paul Cheney, agreed. "How could we have *possibly* missed this disease for all these years?" he asked. "Although a large number of patients are subtle and may not be that sick, there are a significant number of patients who are really quite incredible, and I just can't believe the medical profession could have watched this—*missed* this—for decades, or millennia. It's too striking." Twenty-six years after Carpenter and Cheney voiced their concerns, current estimates place the number of sufferers worldwide at twenty million, with a million or more of them in the US, exceeding the number of patients with breast and lung cancer, AIDS, and multiple sclerosis combined. Today, a doctor in New York City Skypes with patients in Uzbekistan, Scotland, and Norway. Japanese, Chinese, and Latvian investigators attend medical conferences in San Francisco. ME is now the most common chronic disease most people have never heard of until they acquire it.

By 2009, when Mikovits appeared on the scene, ME was a disease with a shocking history of neglect that rose to the level of government-sanctioned human rights violations perpetrated upon millions around the world. Patients were denied not merely medical care, pay outs on disability claims, and the emotional support that might have been forthcoming from family and friends had they suffered from a "real" disease, they had been thoroughly disenfranchised and subjected to ridicule and abuse from all quarters. If they were children, they were denied educations. Adults and children were on occasion incarcerated in psychiatric institutions against their wills, a ghastly outcome resulting, in at least some well-documented cases, their deaths. The spiral into years and decades of poverty that frequently accompanied onset of the disease was perhaps the most intolerable outcome of all.

ME was a storm gathering on the horizon in the late 1970s, a rumored malady that sounded too incredible to believe: "Like mono—except you never recover," was how the disease was first described to me by a fellow journalist who had his ear to the ground. This once rare disease was about to explode in tandem with the AIDS epidemic, but those with ME would share a very different fate. Rather than spend billions of dollars on research and reorder the medical research cosmos on its behalf, public health authorities would seek to disappear ME by any means available, whether by marginalizing its victims or, as documented in my book *Osler's Web*, conspiring to prevent Congressionally mandated research on it.

There is a magnificent treatise waiting to be penned about the divergent paths taken by governments of great nations as regards AIDS and ME, so thinly separated as to symptoms and aberrant biological signs. Both diseases have as hallmarks a deranged immune system that allows myriad viral infections to propagate—viruses that normally would remain latent, an elevated cancer risk, dementia, and more. As long ago as 1986, a top neurologist was unable to differentiate the two diseases while studying MRI brain scans of ME patients pocked with multiple small lesions and reduced gray matter and those of patients with AIDS-related dementia, or "ARD." Patients suffer from a devastating cascade of symptoms rendering them ghosts of the people they once were; more than half become completely disabled, a quarter permanently bed-bound. Recovery is rare. Morbidity studies have demonstrated that ME patients are as ill as end-stage AIDS sufferers, advanced cancer patients, and people dying from congestive heart failure.

* * *

With her publication in *Science* the following October, Mikovits would provide highly persuasive evidence for an AIDS-like viral infection existing in close to 70 percent of patients and 4 percent of healthy controls. The virus was called XMRV and had been discovered at the University of California in 2006. A member of the gammaretrovirus family, XMRV was classified as a murine leukemia virus, one which likely had jumped from its natural "reservoir"—mice—to humans at some juncture in the recent past. Although XMRV's discovery had aroused little fanfare, Mikovits's data drawing an association between the virus and ME, rare immune cancers, and, eventually, autism, raised a firestorm three years later. If her data were right, then ten million people in America alone were infected, but asymptomatic, with this virus. At one juncture during the ensuing scientific fracas an incredulous vice president of the American Red Cross, Roger Dodd, dubbed Mikovits's findings, "The Doomsday Scenario." In other words, it was too horrible to be true.

A media-stoked controversy erupted as laboratories around the world sought to replicate the findings in ME patients, a drama that played out over the next three years with a kind of fury rarely seen in science. Famous scientists shouted over each other's presentations during normally staid conferences; relationships shattered and paranoia blossomed, revealing scientists to be not atypical of the rest of humanity. Publications as diverse as *The Economist* and *Science News* followed the story faithfully and with bemused interest, as if the often down and dirty dash to confirm or deny was a tennis match instead of a pressing scientific matter, the outcome of which had the potential to affect the economies of nations. One highly conservative estimate of the annual loss to the U.S. economy from ME was $20 billion, a figure based on the unreasonable assumption that everyone disabled by the disease had been earning just $20,000 a year when they fell ill. Given the history of the long-politicized disease at the center of the controversy, the uproar was hardly surprising.

Yet, the baseball hat and flip-flop wearing scientist at the center of the controversy, who bragged of her shared ancestry with Attila the Hun, was a surprise. She was forty-six, with oftentimes unkempt, sun-streaked hair, and wide, innocent-looking blue eyes. She looked like she would be happiest at the helm of a sail boat in some far reaches of the Pacific Ocean. In reality, she was a molecular biologist and biochemist and the author of approximately fifty publications on the immunology of HIV, its cancers, and the chemistry of drugs to fight its infections. A twenty-two year veteran of the National Cancer Institute, she had been mentored in her early years by Frank Ruscetti in his lab at Ft. Detrick, Maryland. Ruscetti was himself a veteran of Robert Gallo's brutally competitive AIDS lab.

Perhaps understandably, sufferers of this disease felt that with the materialization of Mikovits and her collaborators—especially Ruscetti, codiscoverer of the first human retrovirus in 1980—the grownups had arrived. Certainly, Mikovits was not and would never be among the beleaguered ME clinicians and researchers of past decades, a stalwart if tiny fraternity in perpetual need of funds, well-known to each other and to patients. Since the early 1980s, the latter had existed as if in some kind of dystopian parallel universe, in possession of information they deemed urgent but in which few besides themselves were interested. A constant at their scientific conferences were the ashen-faced patients, mostly women, lying in fetal positions on the carpeted corridors outside the hotel ballrooms, blankets pulled up to their necks; somehow they had made their way there but the effort cost them everything.

Certainly, there was little doubt that Mikovits was a different kind of scientist, one who didn't seek the approval of the top guns at the NIH and who wasn't afraid to upbraid scientists at the CDC whether via email or in person. Indeed, she had a quality of fierceness rarely seen in science. She would call XMRV ". . . the biggest epidemic in United States history," one destined "to turn the U.S. into the equivalent of HIV-riddled sub-Sahara Africa" if it continued unabated. She labeled the Centers for Disease Control "criminal" for what she saw as the agency's failure to control the spread of XMRV. Inside her lab, the Atlanta agency's acronym stood for "Can't, Don't Care." She and her staff derided the agency's method of selecting patients— by random telephone surveys—calling the government's cohort "Publisher's Clearinghouse" patients.

Judy Mikovits was unshakable in her convictions and harsh on her critics whom she considered biased, occasionally dishonest, and often ill-informed. She was a charming advocate of patients. Startlingly for a scientist, she even consorted with them, seeking them out and befriending them, once posting her personal email address prominently to the blogosphere. Her rationale was not just humane but utterly reasonable: her understanding of this disease arose from patients and their histories and she formulated hypotheses for scientific experiments based on what she heard and observed. The ancient Greeks would have lauded Mikovits for her methods, but in the 21st Century, she was an oddball.

As far as other scientists were concerned, perhaps her greatest sin was her public conjecture about what her data might imply for other unsolved diseases, especially autism, a malady that vied with ME as a bio-political hot-button. She had identified family clusters in which parents and other close adult relatives suffered from ME and children suffered from autism, and found evidence for gammaretrovirus infection in victims of both diseases. It was one thing to pose scientific hypotheses about controversial disorders over drinks at the conference hotel bar with lab colleagues, but to explain in some detail such hypotheses on day-time television talk shows or to journalists from major American newspapers, as Mikovits did, was another matter. Federal scientists, who frequently confuse their mandate to perform research with a mandate to prevent public panic, were especially unnerved by the use of words like "infection" and "transmission" in the same sentence with words like "autism" or "lymphoma," and certainly in conjunction with what the CDC calls—in preference to ME—"chronic fatigue syndrome."

Of course, it's a rare and delicate venture in science to be first and few hazard the risk. Mikovits's primary *Science* collaborator and unabashed admirer, Ruscetti, said of Mikovits, "What I always tried to teach her is, learn the scientific method and learn it well so that you can publish something that 99 percent of the [scientific] community might say, 'You are wrong,' but which you know is right. That is the courage of a true scientist, and Judy has that." Perhaps it's unsurprising that for the next three years, Mikovits stood unbowed at the center of a crackling scientific storm that raged over several continents. If she was correct, and a highly infectious retrovirus was in fact the cause of "chronic fatigue syndrome" and ten million Americans were already infected—well, that not only changed the geometry but shattered the credibility of the nation's bulwark against infectious disease, the CDC and it's more prestigious sister agency, the National Institutes of Health.

By year three, pondering the five-foot-five-inch Mikovits and her place in the scientific cosmos brought to mind Abraham Lincoln's purported remark to Harriet Beecher Stowe, "So you're the little woman who wrote the book that started this great war!"

* * *

Mikovits had been so insulated in what could be called the HIV-AIDS research bubble, she had never heard of ME until being enlisted as scientific director of a new institute affiliated with the University of Nevada in Reno. What drew her to such a foolhardy fate? A rush of intuition, facts lining up with facts, her knowledge of AIDS immunology, a Eureka moment. "I spent my whole life working on this. I just didn't know it," she would say later.

In 2006 she sat in the audience at a conference in Barcelona and listened to a veteran of the ME wars, a doctor who had seen thousands of patients and had learned about the disease in real time right along with them, talk about a group of 300 patients he had followed for years. Dan Peterson of Nevada, an internal medicine specialist with a wait list measured in years, described what Mikovits would call "opportunistic infections" in these patients; multiple immune deficiencies; a kind of sub-acute encephalopathy that lowered IQ and destroyed even the brightest patients' ability to think straight, complemented by abnormal brain scans using several technologies. The doctor offered data for cytokine "storms," an onslaught of inflammatory proteins like interferon that leveled victims, generated in response to infections. He noted that 5 percent of patients in this carefully observed group had rare immune system cancers that would be expected to occur in the general population at rates closer to .02 percent. In all, 77 of the 300 had either blood cancers or cellular changes that were predictive of lymphoma.

Mikovits was struck to her core. "It's a retrovirus," she thought, almost saying the words out loud. HIV was one of three retrovirus families known to infect humans; maybe there were four, Mikovits wondered. Long known to infect domestic animals such as cats and cattle as well as wild animals, retroviruses caused cancer, immune deficiency, and horrendous neurodegenerative diseases. If the disease Peterson described wasn't AIDS, it was akin to AIDS or, as Mikovits would say eventually, "the other AIDS," or "non-HIV AIDS." She leapt to the microphone when Peterson finished. "I'm a cancer researcher," she said. "Number one, I look for viruses in cancer and number two, this smells like a virus." Three years later, having met hundreds of sufferers on both sides of the Atlantic, she would comment, "It's amazing to me that anyone could look at these patients and not see that this is an infectious disease that ruins lives."

* * *

There is an immense backstory to Mikovits's contemporary story; a heart-rending tale of dashed scientific careers and broken-hearted people who have made important-seeming discoveries about ME and ended up on the side-lines, incredulous that their work went unheralded and more importantly, unfunded. Not infrequently, those discoveries were evidence for retroviral infection. Experts familiar with the clinical manifestations of the disease, have recognized retroviruses as a class of pathogens with the power to cause all the symptoms and outcomes they document: immune deficiency, neu-rodegenerative disease, and greatly elevated rates of cancer. One reads again and again that the cause of the disease is "elusive" and the disease itself "mysterious." Implicit in these clichés is that some organized, concerted effort has been undertaken to solve the mystery, to no avail. Contrary to popular belief, however, searches for a causative infectious agent in the disease have been rare, limited in scope, and for the most part, either poorly funded or virtually unfunded.

As long ago as the early 1980s, Campbell Murdoch, a local doctor in Dunedin, New Zealand, began referring patients to University of Otago microbiologist and medical doctor Michael Holmes, a dark-haired, enthusiastic, and open-minded young scientist whose appearance and temperament were compared by his colleagues to Henry VIII. Interestingly, the disease was known popularly in Dunedin as "Poor Man's AIDS." The South Island of New Zealand was hit by the malady by the late 1970s perhaps most famously in the small town of Tapanui, a village less than two hours inland from coastal Dunedin. Locals there adopted the mellifluous name "Tapanui Flu." Holmes, whose primary interest was "clinical immunovirology," spent several years prior to his retirement in 2002 studying "Poor Man's Aids" for evidence of retroviruses.

In 1986, using the equivalent of $690 given to him by patients, Holmes looked at six sufferers and six healthy controls and discovered reverse transcriptase, an enzyme used by retroviruses in their replication process, in four of six patients. In addition, he found ". . . cells with convoluted nuclei comparable to those described in the ARC [AIDS-related complex] syndrome. These were not present in controls."

"We would like to propose a retrovirus etiology for CFS based not only on this pilot study but on the train of deductive observation which led us to consider it in the first place," Holmes wrote. Part of that deductive observation had been the cytokine "interferon" seen in extremely high levels in ME patients. ". . . [T]he most powerful interferon inducers are retroviruses," Holmes added.

Two years later, Holmes was awarded $7,000 from patients to continue the hunt, this time with twenty patients. Again, he saw the "convoluted nuclei" also seen in HIV disease. Four years later, he had scrounged enough money to investigate another twenty patients, with similar results. By then, 1991, American immunologist Elaine DeFreitas had published her own discovery of retroviral gene sequences in 80 percent of adult and child ME sufferers and in 4 percent of controls. Unfortunately, DeFreitas had been, in Holmes's memorable words, ". . . [S]avaged and thrown to the wolves." As a direct result, in his opinion, Holmes's interesting research reached a dead end with the scientific establishment in New Zealand and the rest of the world. After Holmes presented his findings at an ME conference in 1994 in Ft. Lauderdale, Florida—the only study on causality out of sixty presentations—a CDC epidemiologist, Keiji Fukuda, who today is the Assistant Director-General for Health, Security, and Environment at the World Health Organization, told me matter-of-factly, "To talk about etiologies is to raise false hopes. . . . It will not be agent X causing disease Y."

Fukuda's comments were representative of the denialist mindset the peppery, often sardonic DeFreitas encountered at the CDC when, after her discovery was reported in *Newsweek* and several major newspapers, scientists in Atlanta approached her. DeFreitas, who was a fast-rising star at the Wistar Institute in Philadelphia and had been mentored by its world-famous director, Hilary Koprowski, commanded about as much respect from the male retrovirologists at CDC as did the patients whose blood she was studying. In defiance of protocol, they refused to follow her methods at every turn. Ignoring her proscription against freezing blood, scientists froze ME blood samples in order to take vacations. They employed chemical reagents DeFreitas had warned against, they cast aside reagents she recommended, they regarded the ratios of particular nutrients DeFreitas used to nurture her cell cultures (and the virus) as unimportant. "There's always enough time to do it wrong, but never enough time to do it right," DeFreitas noted at the time. She also said, presciently, "If a plague were to hit this country, the CDC would be the last to know."

Eventually, DeFreitas urged the agency to send a scientist to her lab in Philadelphia to work side by side with her. She would bring the horses to water and hope they'd drink. Wistar director Koprowski offered the institute's luxurious corporate apartment to CDC scientists as lodgings. Citing a lack of money to pay the round-trip airfare of a scientist from Atlanta, CDC administrators refused both invitations. That's how much the US government either didn't care or didn't want to know what DeFreitas had found in victims of this rapidly spreading disease.

"The CDC is culpable," Mikovits told me during one conversation about DeFreitas's discovery. "They let an entire generation become infected. I think they all know there's a huge class action lawsuit here."

By the time CDC officials published their failure to confirm DeFreitas's finding in not just one but four publications and had written a letter to her boss suggesting she be fired, academic scientists were comparing DeFreitas to Joan of Arc. She had dared propose an infectious etiology for ME; she had dared to turn it into a "real" disease. I always thought a comparison with the mythological Cassandra more apt.

DeFreitas's then-seventy-five-year-old boss Koprowski was a Polish émigré of the 1930s and considered by many to be personally responsible for the European "brain drain" of the 1950s during which scientists left Europe in droves for America; many of them found a home at the Wistar Institute. Koprowski, who took a great interest in ME and characterized it as an "infectious disease of the brain," told me he believed the NIH needed an institute akin in size and scope to the National Cancer Institute to fight ME and other burgeoning central nervous system diseases. Koprowski believed all of them would turn out to have an infectious cause, much like polio and rabies.

With DeFreitas's tantalizing findings demolished by the CDC, research into the cause of ME rapidly slid into a scientific equivalent of the Mojave Desert, where the disease lay like a corpse for twenty years to be picked over by a scavenging psychiatric "lobby," as patients call the still-influential psychiatrists. The latter have turned ME into a winking euphemism they call "bodily distress syndrome," characterized by "medically unexplained symptoms," a.k.a., "M.U.P.S." They have managed to persuade much of the medical establishment that cognitive behavioral therapy is an effective treatment. Certainly, talk therapy is less expensive to administer than a full-out research effort into the cause.

Federal dollars to search for pathogenic etiologies would never be forthcoming. Nihilism settled over those rare iconoclastic clinician investigators and even rarer scientists who continued to find the malady alarming. They returned to publishing papers about symptoms, things like the pompously titled "post-exertional malaise"—the inevitable crash that occurs in patients who exert themselves—or impaired blood flow to the brain, or pathologically low blood pressure. There are today some 5,000-plus papers on abnormalities in ME that have appeared in scientific journals since the dawn of the 1980s, yet government officials continue to tell reporters "There are no biomarkers." With Swiftian logic, the same officials tell each other that the disease needs to be broken into myriad "subsets" because, they insist, it's simply unimaginable that everyone could have the same thing—a new way of saying, as did the CDC's Keiji Fukuda twenty years ago, ". . . it will not be a case of agent X causing disease Y."

After meeting Judy Mikovits in London in the spring of 2009, I wondered: had the day of reckoning arrived? Would Mikovits be the scientist who would resolve this disease or would she go down in flames? Without meaning to be cryptic, I believe the answer is a bit of "yes" on both counts.

* * *

"We have *CDC-proofed* this paper," Mikovits assured me in September of 2009 shortly before her study was published in *Science*. She might have been new to the disease, but she was well acquainted with its political history. Earlier, in July, her work had held up beautifully when two AIDS experts at the National Cancer Institute called a secret meeting of top gammaretrovirus experts to deliberate on the matter. The government's primary concerns: how to manage an unpredictable public once the news of XMRV and its possible relationship to cancer was publicized and what to do about all those infected, asymptomatic people.

"That's the piece of the data that scared everybody," Mikovits remembered. "They didn't care about 'chronic fatigue syndrome.' Ten million people infected with a retrovirus of unknown pathogenic potential? In this country, by comparison, eight-hundred-thousand people are infected with HIV."

In the tumult that followed, concern turned to scorn when several laboratories were unable to confirm the work and, in time, a persuasive argument was presented by an American scientist, initially a staunch Mikovits supporter, that XMRV was a man-made virus, a "contaminant," that had been spreading from laboratory to laboratory since the mid-1990s. Mikovits and her collaborators accepted the verdict on XMRV but drowned out in the resulting furor and even ridicule was Mikovits's voice. She insisted that evidence for gammaretrovirus infection in ME remained strong and was deserving of continued study. As with HIV, she argued, likely there were multiple strains of the pathogen and she noted that her research supported that hypothesis. She also referenced another scientist, a reclusive interferon expert named Sydney Grossberg, who had been quietly pursuing his own discovery of a retrovirus in an ME patient since the early 1990s at the University of Wisconsin. Indeed, in May 2013, once XMRV was laid to rest, Grossberg published his observation that the pathogen was a member of the gammaretrovirus family. Specifically, it was a murine leukemia virus as XMRV was believed to be. It was, he continued, distinct from XMRV. Grossberg went on to suggest that in future studies, his techniques be used to "expand the detection" of a subset of viruses "related to [murine leukemia viruses]."*

* * *

Would Mikovits have been treated as harshly had she been a man? Certainly, the discoverers of XMRV were men and none of them suffered the malignant news coverage, Internet hazing, or public drubbing she would suffer. Would journalists Jon Cohen and Martin Enserink of *Science* have tagged-teamed Mikovits, as they did in the summer of 2012, to produce an eight-page classic work of character assassination in September of 2011 in *Science*? Enserink trailed Mikovits around Brussels and the university town of Leuven, Belgium, during a scientific conference there, and Cohen later followed her from California to Reno, apparently in large part to obtain her jail mug shot, which appeared on page one of their story. I witnessed Mikovits talk herself hoarse with Enserink over a period of days in an effort to explain her scientific methods and hypotheses; I failed to recognize a word she had said to him in the article that resulted.

---

* Sidney E Grossberg et al., "Partial Molecular Cloning of the JHK Retrovirus Using Gammaretrovirus Consensus PCR Primers," *Future Virology*, Vol. 8, No. 5, April 18, 2013, 507–520.

Or was it the fact that the public health establishment largely perceives ME as a disease of women, resulting in derision and dismissal not just of patients but of the scientists who seek to identify the etiology of their disease, especially if they, too, are women? As recently as December 2013, one newly-retired NIH scientist was asked by a patient at a medical conference in New York City what the NIH leadership really thought about this disease. After a pause in which he seemed to be weighing his words, the scientist replied, unsmilingly, "They hate you." Can the rocky history of an epidemic that so far has impacted two human generations be attributed to misogyny in the way foot-dragging on AIDS during the early 1980s is attributed to homophobia?

Or does the explanation lie with the CDC and NIH imperative to save face, with the need to avoid any admission that taxpayer-supported health agencies have failed so profoundly and for so long? Or is it to do with the avoidance of the class action lawsuits Mikovits predicted? We are not primed as a society to ponder whether these agencies are conspiring, by default or purposefully, to keep the population ignorant of real and present threats to our lives and the lives of our children. The history of this disease, however, and the often devastating experiences of scientists who have tried to crack the case, force any thinking person to consider that possibility.

What cannot be disputed: Judy Mikovits changed a stale quarter-century debate about the legitimacy of ME to a conversation about its biological cause, its modes of transmission, and rational drug therapies. She reopened the Pandora's Box that was slammed shut when the CDC attempted to bury Elaine DeFreitas. She made the disease real for a while, a momentous feat in itself, and moved the focus of the scientific community away from its symptoms toward its cause. She broke the spell of nihilism. She also proposed a reasonable hypothesis about a possible infectious agent driving the autism epidemic; in a rational world, her hypothesis would be pursued aggressively. More broadly, Mikovits brought a once quiet, even esoteric conversation among molecular and evolutionary biologists out of the closet and into the mainstream: is it possible that one virus or a closely-related family of viruses might be causing the neurological diseases such as ME, autism, even possibly ALS and Parkinson's disease, as well as the epidemics of non-Hodgkin's lymphoma and leukemia?

One hopes for greater honesty and a new spirit of open-mindedness from government officials in the future. When so many are sick and the cost to the culture is so high, every discovery should be examined and explored without bias and with great urgency. Honest scientists need to know they will be supported in their endeavors to solve difficult problems rather than "burned at the stake." One hopes, too, for the salvation of the millions of people who have been disappeared by their governments for having acquired a disease that cannot be acknowledged by those governments and those who will become ill in the years ahead if there is a return to the status quo that Judy Mikovits disrupted for three brilliant years.

Prologue

The Arrest

I began comparing Judy Mikovits to Joan of Arc.
The scientists will burn her at the stake, but her
faithful following will have her canonized.

—Dr. John Coffin[1]

Friday, November 18, 2011

"Is Dr. Judy home? I'm Jamie. I'm a patient and she knows me very well. She'll remember me. She said to come by any time."

That's odd, Mikovits thought. Patients rarely showed up at her door. The only Jamie she could think of was miles across the ocean in Hawaii, hardly a place one comes from unannounced. "That's okay, David. I'll take it," she said. She swept past her husband, giving him a quick glance to indicate everything was okay as she walked to the door of her southern California beach bungalow.

Judy often wondered what David must think of her crazy life. Did he know he was signing up for a roller coaster ride when they married? She might be the world-famous rock-star scientist, but he was the rock. As a teenager growing up in Philadelphia, Judy's husband David Nolde had danced on Dick Clark's *American Bandstand* to musicians such as Sam Cooke, Neil Sedaka, and the Everly Brothers. In his professional life he had been a personnel manager for various hospitals. He was the kind of man who was good at listening, understanding people, and defusing tense situations. She was often called the brilliant one, but it was David who understood what others tried to keep hidden.

The woman standing at the door was tall and dark-haired, dressed in black. "Hi, Dr. Judy," the woman said. "Do you remember me?"

Judy Mikovits had her PhD in biochemistry and molecular biology from George Washington University and was an AIDS and cancer researcher of more than thirty years, but people often said she had a second career—a calling, in the language of her strong Christian faith—as a patient advocate. Over the years she had run volunteer cancer support groups and would often research and review treatment options for people and accompany them on doctor visits. Most people were terrified to be suddenly thrown into the medical system and were reassured by having someone along who understood the science. She also found that the majority of doctors welcomed the opinion of a researcher as they often complained that they didn't have time to keep current with the latest research.

Most people she helped referred to themselves as her "patients" even though she was not a treating physician. In the past few years she had moved from cancer research into a high-profile investigation of myalgic encephalomyelitis/chronic fatigue syndrome (ME/CFS), taking the position of research director at the start-up Whittemore Peterson Institute for Neuro-Immune Disease (WPI), housed at the University of Nevada, Reno (UNR) campus. Mikovits developed the entire research program that culminated in an article in 2009 in the highly prestigious research journal, *Science,* showing an association between a newly discovered human retrovirus, XMRV (xenotropic murine leukemia virus-related virus) and ME/CFS.[2] There had been a partial retraction of the work a month earlier,[3] but for many reasons Mikovits still believed the theory was sound and needed rigorous validation.

Over the past five years Mikovits had counseled ME/CFS patients in much the same manner as she had counseled cancer patients and felt she could tell pretty quickly if a person was suffering from the disorder. Patients were often unnaturally pale, sometimes too thin or overweight in a sickly way, and there was something about the eyes that looked different. She understood that calling what these patients suffered from "fatigue" was like calling the atomic bomb dropped on Hiroshima "fireworks." Over a spectrum of severity, many of the most severely affected spent twenty-three hours a day in bed with the shades drawn because of their utter weakness and light sensitivity. Many of the patients had been active, vital people before their affliction struck, with a good number engaging regularly in rigorous athletic pursuits, like running marathons or long-distance cycling. Their physical breakdown was often looked upon by doctors as some sort of unconscious psychological disorder, as if these people who lived life to the fullest had simply decided that life was no longer worth the trouble.

But the disease was without mercy, lasting for decades and taking decades from patients' expected lifetimes. The former chief of Viral Diseases for the Centers for Disease Control and Prevention (CDC) claimed the level of disability of many of these patients was similar to terminal AIDS patients and those in end-stage renal failure, so patient comparisons to a "living death" were apt.[4] But the years generally did not bring death, although an unusual number of patients developed rare types of cancers, salivary gland tumors or B-cell lymphomas. This fact more than any other is what drew the former cancer and AIDS researcher toward this research. Why would years of a fatiguing illness result in an elevated rate of rare types of cancer? She felt there were some intriguing avenues to explore.

Yes, Judy Mikovits had learned a great deal about ME/CFS in the past five years. Judy stared at the woman in her doorway and felt a sudden chill. She was certain the woman didn't have the disease and that she wasn't a patient she had ever seen before. "I don't know you," Mikovits said to the woman and began to push the door shut.

* * *

Regan Harris first got to know Mikovits when she called the WPI in December of 2009, after reading the *Science* article.[5] Regan was surprised and flustered to suddenly be speaking to an internationally recognized scientist, but Mikovits quickly put her at ease and asked Regan to share her story. With a deep breath, Regan began by telling Mikovits she had become sick in October of 1989, at the age of fourteen after a bout of mononucleosis. The following year she had been diagnosed with ME/CFS and from that point on, life had been a roller coaster ride.

Despite her ME/CFS, Regan had been able to graduate high school and had attended college where she received a bachelor's degree in psychology. While getting her degree, Regan researched the issue of suicide among the ME/CFS population and how these patients presented with a different psychological profile than people with depression. Regan's work eventually culminated in a poster presentation before a meeting of the American Psychology Society in 1998. After listening to Regan's tale, Mikovits told her about an ongoing research study and asked if she would like to participate. "I can never give you back the years of your childhood that were stolen from you," said Mikovits, "but I think we can prevent this from happening to other kids. Will you help me take this thing down forever?"

Galvanized by Mikovits's confidence, Regan signed the forms and went to the grand opening of the $77 million WPI and Center for Molecular Medicine at the University of Nevada, Reno in August of 2010. There she met Annette and Harvey Whittemore and their daughter Andrea, who had also been struck with ME/CFS from a young age. Regan couldn't wait to make her own contribution to this effort.

Regan moved to Nevada in September of 2010. She planned on volunteering for the WPI, hoping it would lead to a paying job. Judy and David were warm and welcoming, often taking Regan out to sample the local cuisine. When Regan first arrived, David spent some time driving her around Lake Tahoe, eventually shuttling her to Glenbrook, the exclusive gated lakefront neighborhood where the Whittemores had one of their many residences. When David approached the gatekeeper at Glenbrook the large gates opened as he said, "Whittemore."

When they got to the Whittemore home, a historical residence known as the Lakeshore House, complete with its own private dock, David motioned with a hand and said, "What do you do when your family is too big to fit in one house? You buy the one next door as well!" The Whittemores owned *two* houses on Lake Tahoe. When Regan flew home to Massachusetts that Christmas, she couldn't wait to tell her mother all about her run-in with the Nevada royalty. Regan gushed about the wealth and influence of the Whittemores, noting, "My God! They've even got a movie theater in their house. You would not believe this, Mom! Can you imagine what it's gonna be like if I can work for them? It would be so cool."

Regan's excitement was not fully celebrated by her New England mother, who said, "Regan, I never want you to be seduced by money and power. You remember one thing: anybody who is powerful enough to give you everything is also powerful enough to take it all away."

* * *

Mikovits had almost latched the door when she heard a male voice shouting, "Hold on there!" A man, identifying himself as University of Nevada, Reno campus security, stepped out from behind one of the large bushes in her yard and strode quickly to the door. Dr. Mikovits knew this man—he had investigated the robberies that had taken place at the WPI when she had been the research director. Where she *had* been research director.

That was in the past now. On September 29, 2011, she was fired, receiving the dismissal call on her cell phone from Annette Whittemore, president of the WPI, as she walked home. While the experience of being fired could shake anybody, how many could claim the news had been reported in the pages of the *Wall Street Journal*?[6] The article by the well-respected journalist Amy Dockser Marcus in her Health Blog section of the *Wall Street Journal* had given a fair account of her firing:

Whittemore told the Health Blog that she and Mikovits were not "seeing eye-to-eye" on who controlled the cells. Research on retroviruses and their possible connection to CFS as well as other diseases continues, she said. "We will keep going down that path as long as it continues to show promise," Whittemore says.

Annette Whittemore's given reasons for firing Mikovits would change several times over the ensuing months, but she detailed them in a letter sent to Dr. Mikovits on September 30, 2011, which among other things accused Dr. Mikovits of "insubordination".[7]

On October 1, 2011, Dr. Mikovits sent a response to Annette Whittemore addressing the event that had ostensibly caused her firing as well as more concerns she had about the management of the WPI. Mikovits told Annette that as the principal investigator on the National Institutes of Health (NIH) R01 grant, Mikovits alone was legally responsible for all resources on that grant and that Mikovits alone was the one who should have decided the appropriate allocation of those resources. Mikovits was pleased that Annette hoped for "a smooth transition" regarding Mikovits's departure. However, as Mikovits was the principal investigator on three grants housed at the WPI, two from the NIH and one from the Department of Defense (DOD), she told Whittemore that she fully intended to continue her research on those same grants, but at another institution—once one was found. This is common practice in the scientific community; the principal investigator takes the grants with her if she leaves the institution.[8]

Her break six weeks earlier with the Whittemores had been sudden, but Mikovits was eager to move forward with her life and research. The next day, she was scheduled to fly out to New York City to participate in the celebration of a multi-million dollar ME/CFS initiative to be run by ME/CFS physician Dr. Derek Enlander of Mount Sinai Hospital. Mikovits and Enlander were also scheduled to discuss ways in which they might collaborate after her depature from the WPI. But she would never make that trip.

* * *

A thud at her feet made Dr. Mikovits look down. She realized the woman had dropped a microphone and a recording device. "That's illegal here," said Mikovits. "You can't record me without my permission."

"We're just here to get your side of the story," replied the woman as she picked up the fallen items.

"Fine then. You can come with me to my lawyer's office. I'm on my way to meet him." Mikovits again tried to close the door when three burly Ventura County sheriff's deputies came around from the driveway. One of the deputies was brandishing a yellow piece of paper. "We have a search warrant."

The deputies came onto the landing, pushed the door open, and proceeded to enter the house, pushing Mikovits's husband along with them. "David," she called out. "Call the lawyer!"

Just that morning she had called her attorney's office to ask if there were any warrants out for her arrest. On November 4, the WPI had filed a civil case against her, claiming she left with intellectual property, specifically her notebooks and computer files. As a principal investigator on three government grants, Mikovits knew she was legally required to maintain and *protect* copies of all data under federal regulations and her UNR contract as an adjunct professor.

In addition, since her research was being challenged by the scientific community, she needed to possess this information to defend the work. The attorney had found her trepidation humorous and said he didn't see anything that serious arising out of the civil case. Just to calm her, he had checked. There were no arrest warrants.

But Mikovits still sensed something terrible afoot. She believed she had caused her former employers considerable distress. Viral Immune Pathology Diagnostic (VIP Dx)—a for-profit clinical lab loosely associated with the WPI and owned by the Whittemores and Lombardi—was selling an unvalidated diagnostic test for the XMRV retrovirus, one which they would later discontinue selling. They claimed that she had approved VIP Dx's tests, including a new serological one announced under her name, when she was not employed by VIP Dx and had not evaluated data or statements made by the clinical lab.[9]

Mikovits believed she had cut off a lucrative source of revenue for the WPI when she had vocalized all of this on September 23, 2011, at the Ottawa Conference, saying "VIP Dx lab will not continue XMRV testing because it hasn't been shown to be reproducible in the Blood Working Group [BWG]."[10]

She was fired one week later.

Others were already concluding the test was problematic after the release of the report from the BWG, the group founded to investigate whether the retrovirus posed a threat to the blood supply.[11]

Next came the replication study coordinated by Dr. Ian Lipkin of Columbia, one of the world's most famous virologists. A few days after

Mikovits was fired, Lipkin had called to ask if she had confidence in the integrity of her former employers, the Whittemores, to allow her to perform the study in Reno.[12]

Mikovits told Lipkin that she did not have confidence that the study could be performed at the WPI. It was not until November 14, 2011, that Lipkin emailed Mikovits saying he had decided not to have the WPI participate in the study, a decision which would potentially cost the institute a great deal of money.[13]

Despite these financial hardships to the Whittemores, Mikovits believed she was acting the only way she knew how—as an ethical scientist.

The woman in black took Mikovits by the arm and motioned for her to come out onto the porch. "We just want to hear your side of the story," the woman repeated. "Do you have any WPI property?"

"I do not," Dr. Mikovits answered. "Everything in this house is mine."

She knew what they were looking for. The research notebooks. The notebooks which she feared would have ended up on the bottom of Lake Tahoe, been altered, or otherwise kept from public view had she not secured them.

The open access to research, especially research funded by the government was the property of all. She didn't have the notebooks, didn't even know where they were, but she knew they were safe. She believed that her assistant, Max Pfost, had secured them. Whatever she had discovered, or the mistakes she had made, the evidence would be there for all the world to see.

"Do you have a black laptop?" the woman in black asked.

"Yes, it's sitting right on the table, but it's mine. It was a gift."

"From whom?"

"Annette Whittemore."

* * *

Mikovits remembered the extravagant 2007 Christmas party, the first WPI Christmas party, when Annette had presented her with the black laptop, a back-up disk drive, and a printer.[14] The only stipulation Annette put on her present was that Mikovits had to promise to back up the hard drive on the disk drive that stayed at the lab. Thus, as Mikovits understood it, there should be two copies of all data, one for the principal investigator, Mikovits, and one backed up on the drive at the lab. Annette even gave Mikovits the receipt for the computer in case there were any problems.

The Whittemores were political contributors to US Senator Harry Reid, a Democrat and the majority leader of the Senate, as well as many other politicians.[15] All four of Harry Reid's sons had at one time worked for the law firm where Harvey Whittemore was a senior partner.[16] In addition, Harvey Whittemore had personally helped advance the legal careers of *two* of Reid's sons—and one of the sons, Leif Reid, had become Whittemore's personal lawyer.[17]

In a 2006 article in the *Los Angeles Times*, Harvey Whittemore is quoted as saying, "You have to understand how close the Whittemore and Reid families are . . . My relationship with Sen. Reid goes back decades."[18]

Harvey Whittemore was often identified as one of the most politically influential individuals in the state of Nevada, earning nicknames such as "the 64th legislator" for his help in drafting the state's first business tax and being among a select group of four wealthy men known as the "Power Rangers,"[19] after the popular Saturday morning children's show. One reporter of Nevada state politics had quipped, "Governors come and go, but the Power Rangers stay the same."[20] Ominously, one of Harvey former associates said "Harvey Whittemore has a different moral compass than the rest of us."[21]

One of the Whittemores' children, their daughter Andrea, had been struck down with ME/CFS when she was just eleven years old. Her parents were tireless in trying to find a cure for her, and through the work of Mikovits and others, Andrea—now in her thirties—was close to recovering her health. This personal connection to the disease made Mikovits believe that she and the Whittemores would always be on the same side.

* * *

The case brought by the WPI against Mikovits was an unusual one, according to her civil attorney, Dennis Neil Jones. "The complaint alleges what I guess you could call industrial espionage. And the defense is basically a whistleblower kind of defense."[22]

It was much different than the typical cases Jones handled. Both Jones and Mikovits's bankruptcy attorney, David Follin, would be disturbed, however, by the legal maneuverings deployed against Mikovits. As attorneys, they understood the combativeness of the judicial system, but also knew there were rules and an expected logical progression of events.

But this case seemed very different from the start, in both the legal aspects, and the response of the scientific community. "It seems like the field was stacked against Judy and it's continued to be so. Any allegations she was convicted of a crime or [that] there was a successful judgment against her, is wrong," said Follin.[23] "Judy is just an amazing person. She's probably one of the most brilliant people I've ever met. All Judy wants is fairness and I can't understand how her profession can turn its back on such a talented individual who has so much to offer and could help so many people."[24]

* * *

When Mikovits thought about it later, she realized the problems had actually started soon after Mikovits and Annette Whittemore first appeared on a TV show in 2009 called *Nevada Newsmakers*,[25] shortly after the publication of the landmark article in *Science* linking a new human retrovirus to ME/CFS.

Mikovits and her team found evidence of the retrovirus in 68 out of 101 patients (67 percent) with CFS as compared to 8 out of 218 (3.7 percent) of healthy controls.[26] As if it weren't enough that they were taking on a disease which had been looked upon for more than thirty years as some form of female "hysteria," they were now planning to take on one of modern medicine's most controversial disease: autism.

"It's not in the paper and it's not reported," Mikovits said, speaking hesitantly at first, "but we've actually done some of these studies, and we found the virus present in a number, in a significant number of autistic samples that we've tested so far."

The show's host noted that this news had tremendous potential for the autism community, holding out the possibility that this might lead to treatments or even a cure. Mikovits replied by saying XMRV might be "linked to a number of neuro-immune diseases, including autism. It certainly won't be all because there are genetic defects that result in autism, but there are also the environmental effects."

Then, barely taking a breath, she crossed the Rubicon.

"There's always the hypothesis that my child was fine, then they got sick, and then they got autism. Interestingly, on that note, if I might speculate a little bit . . . This might explain why vaccines lead to autism in some children because these viruses live and divide and grow in the lymphocytes, the immune response cells, the B and T cells. So when you give a vaccine, you send your B and T cells in your immune cells into overdrive. That's its job. Well, if you're harboring one virus, and you replicate it a whole bunch, you've now broken the balance between the immune response and the virus.

So you could have had the underlying virus and then amplified it with that vaccine and then set off the disease, such that your immune system could no longer control other infections and created an immune deficiency."

If these children were harboring a retrovirus it wasn't an outlandish claim to make. It has long been established that children born to HIV-infected mothers shouldn't be immunized until they're on antiretroviral drugs and their tests show the virus to be at extremely low levels. As explained at the University of California at San Francisco web page on HIV and immunizations,

> Activation of the cellular immune system is important in the pathogenesis of HIV disease, and that fact has given rise to concerns that the activation of the immune system through vaccinations might accelerate the progression of HIV disease . . . These observations suggest that activation of the immune system through vaccinations could accelerate the progression of HIV disease through enhanced replication . . . If feasible, it is preferable to have patients on antiretroviral therapy (ART) prior to receipt of vaccination . . .[27]

Just as one wouldn't want an immunization to provoke AIDS in an HIV-positive child, one would also want to be sure a vaccination didn't trigger autism. Mikovits and Annette Whittemore had both grabbed onto the third rail of western science, the question of vaccine injury and the increasing numbers of children with neuro-developmental problems. The scientific community, often choosing comfortable-yet-unproven dogma over testing controversial ideas, made the funding of routine grant proposals even more difficult after the interview.

* * *

The saga of Harvey Whittemore's Coyote Springs development had started in 1998 when he purchased 43,000 acres of remote Nevada desert about an hour northeast of Las Vegas. The dry landscape was originally considered to be so barren that its best use was thought to be a weapons test range.[28] One reporter referred to the single outpost of civilization they'd been able to build on that God-forsaken land as "The Golf Course at the End of the World."[29]

But Harvey Whittemore had big dreams. He envisioned ten golf courses as an anchor for retirees and hard-working families who wanted the good life, but couldn't afford Vegas prices.[30] In addition to the already built Jack Nicklaus signature course, there would eventually be 159,000 housing units. If fully realized, Coyote Springs would become the second largest city in

Nevada. But it had all fallen through as the recession of 2008 started to take its toll and real estate markets across the country had bottomed out.

As of 2011, none of the housing units had been built and only the single golf course had been completed. The writer remarked that Nicklaus's single green golf oasis in the dry brush country of jagged edges and steep lines made it look like a vista out of the classic 1968 science fiction movie *Planet of the Apes*.[31]

And yet there was something audacious about Harvey Whittemore's ambitions, even in light of his troubles. The reporter who dubbed it "The Golf Course at the Edge of the World" also gave what might be considered a eulogy for many of Whittemore's projects. After first writing that normally when one sees a development gone bad you simply think the developer put his money in the wrong place and give a figurative shrug. The developer will go onto a new project. "But a golf course—at least one made with such high levels of devotion and talent as this one—is different."[32]

* * *

"You're under arrest," said the woman in black, slapping a pair of hand-cuffs on Mikovits.

"But it's my laptop!" Mikovits protested.

The police would take and hold for almost one year not only Mikovits's black laptop, but also her iPad, iPhone, the MacBook Air she had recently purchased for her Ireland trip, and the silver laptop of her stepdaughter, who had been staying with them for a few days.

"Don't say anything!" David called out.

"I won't!" she shouted back.

Four unmarked sheriff's cars immediately came around the corner from Harbor Boulevard, staging what might have looked to the casual observer like an episode of *America's Most Wanted* rather than the apprehension of a figure in a scientific controversy. Mikovits—five foot four inches of her, frizzy blonde hair, and just a shade over a hundred and forty pounds—stood on the road in her white jogging shirt and black knee-length shorts. She was shoeless, having left her flip-flops on the floor in the bathroom. One of the deputies noticed she was barefoot and asked if she had anything back at the house. "I was wearing my flip-flops," she replied.

An officer went into the house to retrieve her shoes.

"Why am I being arrested?" Mikovits asked one of the deputies.

"You are a fugitive from justice."

The arrest of Mikovits would confuse every legal expert who looked at the facts of this case for a simple reason. Nobody involved in any of these proceedings ever produced an arrest warrant. Under what law could a middle-aged scientist be taken into custody without an arrest warrant?

The question would remain unanswered.

* * *

A deputy returned with Mikovits's flip-flops and she was able to put them on her feet. A sheriff's deputy opened the back door and she was escorted into the squad car for the eight-mile drive to the Ventura police station. At the police station, she was taken to an interrogation room and read her Miranda rights by an officer. "Yes, I want an attorney and I'll remain silent," she told him.

The woman who had identified herself as "Jamie," now revealed as a member of the University of Nevada, Reno campus police, was also in the interrogation room. "We'll give you a chance to go back to Reno," she said.

One has to wonder how many times the UNR campus police have crossed the Nevada border to make an arrest of an adjunct professor in southern California.

Mikovits wondered if the whole song and dance had been an attempt to intimidate her so that she would agree to let the WPI participate in the Lipkin study, which would represent at least a quarter of a million dollars for the WPI. Arrest her in her home, drag her back to Reno, and let her stew in a jail cell until she agreed to let the WPI back into the Lipkin study? And if she didn't agree, who knew what might happen to her in a Nevada jail cell?

"I'm *never* going back to Reno," Mikovits replied, as clearly as she could.

"We'll see about that. See ya!" the campus cop sneered. After about two hours Mikovits was taken to the Ventura County Jail, booked, and told to stand for a mug shot. They gave her a thorough strip-search, including a body cavity search for drugs, took her only jewelry—her wedding band—her baseball cap, and her clothes, and issued her a standard prison orange jumpsuit. She tried to use her allotted phone call to reach David but outdated regulations disallowed calls to a cell phone. The only landline number she could remember was that of her long-time collaborator Dr. Frank Ruscetti back in Maryland. Nobody was home so the machine at his house picked up the call. Instead of allowing Mikovits to speak all that was left on the machine was a disembodied robot-like voice saying, "You have a call from inmate."

Later, Ruscetti recalled having no idea what to make of the crazy message.

Finally she called a bail bondsman and tried to post the $100,000 bond, which had been levied against her. The bondsman told her with a tone of disbelief in his voice that a "bail hold" had been placed on her case and she wouldn't be able to be released that day. "You must really have pissed off someone important," he said.

* * *

"I never had a case where somebody was charged with stealing their own research," Bill Burns of 101 Bail Bonds later recounted.[33]

When a potential client contacted Bill he usually performed a background investigation in order to get a sense of the person. Sometimes the people who found themselves arrested could be pretty smooth talkers, but their record usually told the real story. Burns talked to Mikovits's lawyer, who explained the nature of the dispute with the Whittemores and then he did his own research. He was quickly able to find out she had no criminal history, that she was a well-regarded scientist, and her husband David Nolde had also never been in trouble with the law.

A picture of his new client began to form in his mind. He had seen a similar scenario several times before—whether it was an overzealous district attorney unfairly prosecuting somebody or when a wealthy individual had influence and knew how to make another person's life miserable. The information he gathered about Mikovits in a short period of time convinced him that something was definitely out of whack.

"A lot of people suffer from this illusion of how great our legal system is," Burns later recounted, "and it really isn't great. You talk about third world countries. You could feel like you're in a third world country when you're locked up and trying to get out. You can't use the phone. You don't have the ability to mount a defense. It's amazing in a country of this size that a lot of people get screwed very badly in our system. It's very easy to end up losing everything on a case that shouldn't have even been brought."

Determining if a potential client was trustworthy was important to Burn's business. Bail bonds don't get exonerated until the case is resolved, whether that takes two months or two years. The bail for Mikovits was one hundred thousand dollars, which meant she would put up 10 percent of that money up front. Burns would normally take a lien on her house or other property as collateral for the bond, but in this case he didn't have Mikovits or David Nolde sign over anything as collateral.

"I did a hundred thousand on a signature because I thought not only was the case full of shit, but everything about it was wrong," he later said.

* * *

There were three holding cells in the basement of the Ventura County Courthouse. The cells were six-by-eight feet, with a three-foot-long steel bench, a small wall, and on the other side a steel commode, unfortunately without any toilet paper. The guards would alternate which cell a new prisoner would be put in, usually about five to a cell. When it was full or the hour was late, the group of prisoners would be taken to the new Ventura County facility down the road.

Many of the people in the holding cell were picked up that day for drug offenses or driving under the influence. For some of the prisoners it was their "appointment time" to serve all or part of their sentence. These were people whose cases had already been heard, and due to the overcrowding of the jails and the relative minor nature of their offense, would serve just a few days.

Shortly after Mikovits arrived in her cell, a woman named Karen (pseudonym), entered to serve her appointment time. She worked for a local newspaper, managing several of the vehicles, which made early morning deliveries. She had been picked up on a minor drug possession charge, was convicted, and as she told Mikovits, just wanted to put the mistake behind her and get it over with. Others were a little more frightening. One woman came in, teetering on six-inch heels, her hair eighteen different shades of the rainbow, clearly picked up for drugs. Karen and Judy exchanged thankful looks that she hadn't been put in their cell.

As the hours passed, the cells continued to fill up, with some of them apparently regulars; they would warmly greet their fellow inmates or guards as they were processed in. At some point, one of the prisoners asked if any of them were first-timers.

"I am," said Mikovits.

* * *

In the late evening, probably around ten or eleven, Ruth (pseudonym), a distraught woman in her mid-fifties, was brought into the jail. She was coughing and crying at the same time and lamenting that this was all a mistake. In the six or seven hours Mikovits had been in the holding cell she had learned a little about jail psychology: one didn't look directly at people and one kept one's head down. Everybody else was avoiding looking at Ruth as well.

"This is all wrong! This is a mistake!" cried Ruth. "I shouldn't be here! I should be home!" Mikovits knew just how she felt.

* * *

When Dr. Jamie Deckoff-Jones read the October 9, 2009, *Science* article by Mikovits and her team shortly after its publication, she looked up at her husband and said, "This is it. This is what we've got."[34]

Deckoff-Jones was a graduate of Harvard and Albert Einstein College of Medicine and a board-certified emergency physician. Her father was a brilliant man and legendary surgeon, who graduated magna cum laude from Yale and finished Harvard Medical School at the age of twenty-one. Deckoff-Jones traced the beginning of her own neurological downfall to a series of hepatitis B shots she received when she was pregnant with her third child. She also often wondered about the sugar cube polio vaccine she received in 1961.[35]

Her symptoms waxed and waned over the years and she believed the constellation of her symptoms most closely resembled some sort of combination of Lyme disease and multiple sclerosis. Her daughter came down with ME/CFS when she was thirteen years old and around the same time her husband came down with Lyme carditis, a heart condition associated with Lyme disease.

In January of 2010 she wrote to Mikovits and was amazed at the lengthy emails Mikovits wrote in response to her questions as well as her openness and inclusiveness. As their relationship grew, Deckoff-Jones took over the role of answering much of Mikovits's email questions from patients. It was Deckoff-Jones's opinion that Mikovits was spending so much time responding to patient emails that it was limiting the amount of scientific work she could accomplish in a day.

Deckoff-Jones eventually came on as the clinical director of the WPI. Her relationship with the Whittemores quickly soured. Deckoff-Jones believed the problems arose because of Annette's inability to admit what she didn't know and protect her staff. Eventually Harvey took over as the person at the WPI to whom Deckoff-Jones directly reported. She found Harvey to be a smart man and generally easy to work with but he had his breaking point.[36]

In a text she sent to Harvey, she used the word "nepotism" to describe many highly-placed individuals who worked at the WPI, like Carli West Kinne, legal counsel for the WPI, and Kellen Monick-Jones, the patient coordinator for the WPI, both Whittemore nieces. Other examples included not just relatives, but others who had long-standing personal or professional ties with the Whittemores.

"Now you've really lit my fuse," Harvey wrote back in a text after the "nepotism" comment. Shortly after that, Annette Whittemore informed Deckoff-Jones they were going to have to shelve their plans for a clinic and her services wouldn't be needed. They had had conflicts over other issues as well, such as whether the clinic should treat kids with autism. Deckoff-Jones wanted to treat them, but believed Annette saw far too many problems with such an effort.[37]

For Deckoff-Jones, Mikovits's story is important in that Mikovits was like Pandora, opening a forbidden box. "She made mistakes like everybody in the story. Me, everybody. An incredible opportunity has been lost as a result. But it's mostly Harvey and Annette's fault. Judy never had a chance. They never supported her. She didn't have what she needed to pull it off. Ever. It was a joke."[38]

* * *

Around two a.m. the day after her arrest, Mikovits was driven to the Todd Road Facility located in a lemon orchard about ten miles out of the city of Ventura. Upon being admitted to the facility, she was again required to strip, bend over, and submit to being cavity searched for drugs. Mikovits was given several pieces of paper with directions on how to be a model prisoner, but because she didn't have her reading glasses couldn't make out the words. When she complained to a guard about her need for reading glasses, the guard replied, "This isn't a resort. That's why they call it jail." Apparently a model prisoner didn't "need to read."

At one point during her processing, Mikovits was asked if she was suicidal. "No," she replied.

Even with her clear answer, Mikovits was placed in the suicide watch wing. The suicide watch wing was regularly used for people who were being arrested for the first time. It seemed that being arrested and placed in jail for the first time was such an overwhelming experience for the average person that it was presumed to make them suicidal. The light in the suicide watch cell was on the entire night, which allowed the guards to constantly monitor the prisoners for any signs of abnormal behavior. Mikovits's cellmate was a woman, Marie (pseudonym), who was undergoing treatment for a methamphetamine addiction. Because Marie was taking several powerful drugs to break her addiction and was thus at risk of falling out of bed, Mikovits was required to take the top bunk.

The cell was made of thick cinder block. The cell was about four feet wide, had a bottom and top bunk made of steel, a commode and sink attached to the wall, and a small window at the top. Instead of bars across the front entrance, there was a thick steel door with a small rectangular window. When the steel door closed, sealing her in, Mikovits felt as if she were in a tomb. The opening and closing of the heavy doors all night sent shivers through Mikovits. She could never have imagined herself in such a place. For a mattress, they were given the equivalent of an exercise mat and no pillow since they were in the suicide watch cell. Marie explained to Mikovits how to put her foot on one side of the small sink to climb into the top bunk. Upon making it to the upper bunk, Mikovits was greeted by the fluorescent, oblong light, which never went off.

Mikovits thought about one particular day in the WPI shortly after she had returned from the Invest in ME conference in England in May of 2011, when Harvey had stormed into her office. He shouted at her because he thought she had insulted Annette's efforts to reach out to another ME/CFS charity. Mikovits had done nothing of the sort but Harvey demanded, "You're going to go and apologize to Annette!"

"Okay! Okay!" Mikovits replied, hoping to defuse the situation.

Harvey's booming voice had no doubt been overheard by other staff members, but as they left Mikovits's office, he put on a big smile and slid his arm around her shoulder. But his hand didn't reach all the way to her shoulder, stopping instead at the back of her neck, where it would be concealed by her shoulder-length blonde hair. As he walked past employees of UNR, all smiles and friendliness, Mikovits felt his hand squeezing the back of her neck so hard she thought it would leave bruises. To Mikovits, the message was unmistakable: she felt like he was saying he could end her at any time he wanted and all of these people he supported wouldn't raise a voice in protest.

Harvey pulled the same little neck-squeeze trick on Mikovits in August of 2011 when they'd been leaving a restaurant with a representative of a drug company that Mikovits had introduced to the Whittemores. Harvey was hoping the company would initiate a clinical trial of a new drug therapy with the WPI and provide significant financing. Mikovits had been unusually quiet during the evening, and by the end of the meal the company had decided not to collaborate.

Since that time, Mikovits had been plagued by a recurring nightmare in which she was driving with friends of hers, having a great time, laughing and talking, when Harvey Whittemore suddenly sat up in the back seat, reached his long arm around her neck, and started strangling her. The metaphor was clear, he could do anything to her and she could not scream.

That first night in jail, Mikovits didn't worry about her own safety. She believed that Harvey's plan had been to get her back to Reno and she knew the notebooks containing evidence had been secured by Max.

Who knew what was planned for her in Nevada?

But no matter how long his arms, Mikovits doubted Harvey could reach all the way from Reno, Nevada, to her jail cell in Ventura, California. It was ironic, but she felt safer in a cell with a recovering methamphetamine addict than she had felt in months.

* * *

Mikovits let her thoughts wander to Dr. John Coffin, whom many saw as the grand old man of virology, and his quote in *Science* comparing her to Joan of Arc.

Science at the highest levels is a territorial game of power. In many cases, if a young person discovers a novel finding in someone else's turf, the self-appointed head of that domain writes the second paper and first review article and effectively squeezes the young person out. Coffin had actually written an editorial in support of her original article in the journal *Science* entitled "A New Virus for Old Diseases."[39] Now he was on the other side.

Who Mikovits wondered, compared a fellow researcher to Joan of Arc, a fourteenth century warrior saint unjustly accused of heresy, and prophesized, "The scientists will burn her at the stake": It was a ludicrous statement. Why should a scientist be burnt at the stake for publishing data that might turn out to be wrong? In the 1970s many papers were published falsely claiming the discovery of human disease-causing retroviruses. None of these people were burnt at the stake, some of them were elected to the National Academy. Was Coffin comparing the scientific community to the agents of the Inquisition? How might they feel about such a comparison? If her research turned out to be incorrect, let somebody else run the same experiments and disprove her. That was the way science went. People can be right one day and wrong the next. She could accept that. Coffin had been wrong about human retroviruses. Had he ended up in jail? Disgraced? No. There was so much more to this story.

But as much as she thought Coffin had acted inappropriately in many instances, she also felt that a great many of her problems stemmed from her former allies, the Whittemores. She believed that the recession had badly hurt the Whittemore's real estate holdings but also wondered if others with far more power might be forcing them to act against their natural inclinations.

But why would anybody not be interested in helping the millions of patients with ME/CFS and children with autism?

* * *

Even with all that had happened, as Mikovits lay in her bunk, she found herself trying to pray for the Whittemores. Mikovits had genuinely liked them. Many of her friends believed her downfall was due to her misplaced loyalty towards the Whittemores, maybe an emotional naiveté, an inability to tell when people were manipulating her. But there was no doubt that since the 1984–1985 outbreak of ME/CFS at Lake Tahoe, no other individual or group had done more to focus attention on this horrible disease than the WPI.

The poet Henry Wadsworth Longfellow once wrote, "If we could read the secret history of our enemies, we should find in each man's life sorrow and suffering enough to disarm all hostility." It was in this vein that Mikovits thought of the Whittemores as she sat in her jail cell.

Mikovits believed Annette was in over her head with the WPI, but she was a parent fighting for her child's life. She felt that so many things had conspired against them but especially the economy and not fully understanding how much the government wanted to avoid taking an honest look at ME/CFS or autism and the role vaccines might play. Mikovits tried to leave these thoughts behind and concentrate on something more elevated. She struggled to remember the words of certain Biblical verses she had heard over the years at church but couldn't recall any. It bothered her because she really wanted, needed, to pray.

Only the words to the Lord's Prayer came to her. She began reciting it over and over, almost like a mantra, and it gave her a feeling of great peace as she faced the uncertain night ahead.

> Our Father who art in Heaven,
> Hallowed be thy name;

Thy kingdom come
Thy will be done
On Earth as it is in Heaven.
Give us this day our daily bread;
And forgive us our trespasses
As we forgive those who trespass against us;
And lead us not into temptation
But deliver us from evil.

"Act, and God will act," Joan of Arc had once said. Despite all the times she had acted before and it had come to nothing, Mikovits thought she would try once again.

CHAPTER ONE

The HHV-6 Conference and the Culture of Science

In science there is—maybe—more self-interest, a little more paranoia, a little more narcissism, or else why do we go into it? You think you are good enough to solve problems of nature. Many scientists tend to keep things to themselves. If the other person does not get funded, maybe you will be funded. All these things are in play, but these are the worst elements of science or scientists.

—Dr. Robert Gallo.[1]

Barcelona, Spain—May 1, 2006

Judy Mikovits searched for a seat just barely within earshot distance of the keynote speech of Dr. Robert Gallo[2] at the 5th International Conference on HHV-6 and -7 (human herpes viruses 6 and 7). Gallo was speaking in the stately grand ballroom at the Hilton Diagonal Mar Hotel in Barcelona, Spain. She hoped to fade into the diffuse lighting and subtle European accents of the room. Mikovits knew from previous encounters that she wanted to stay far away from the famed scientist.

Gallo was there to speak about human herpes virus number 6, which had been codiscovered in his lab in 1986 by Dr. Dharam Ablashi.[3] Ablashi was also the program's committee chair of this conference dedicated to the HHV-6 virus and its possible connection to ME/CFS and other disorders.

Many Americans still consider Gallo to be the scientist who discovered the Human Immunodeficiency Virus (HIV), which causes Acquired Immune Deficiency Syndrome (AIDS). The World Health Organization estimates that since the known onset of the AIDS epidemic in the early 1980s, 70 million people have become infected with the HIV virus and about 30 million have died from the complications of AIDS,[4] making it the greatest pandemic of the modern era and ensuring a place among the Louis Pasteurs and Jonas Salks of history for those at the forefront of HIV research, hence the ferocious fight among the participants for credit. Gallo's biography at the Institute of Human Virology at the University of Maryland, which he founded and still directs, claims he is "best known for his codiscovery of HIV."[5] However, when the Nobel Prize committee awarded the Nobel Prize in 2008 for the discovery of HIV to French scientists, Luc Montagnier and Françoise Barré-Sinoussi, Gallo's name was conspicuously absent.

One can't doubt the ascendant accolades and recognition Gallo has received during his scientific career. Gallo holds an enviable twenty-nine honorary doctorates. In 1982 and 1986, he received the most prestigious American scientific award, the Lasker Prize, which is often called the American Nobel Prize for medical research. Gallo is the author of 1,200-plus scientific publications, and he authored the book, *Virus Hunting—AIDS, Cancer & the Human Retrovirus: A Story of Scientific Discovery.* According to Gallo's own account, he decided to devote his life to science after the untimely death of his younger sister at the age of six from leukemia.[6] His story is an archetypal tale in science and medicine: those personally touched by an illness often want to conquer or cure it. His career choice seemed to be his natural métier. Judy Mikovits made a similar decision to enter science after watching her beloved grandfather die of cancer, and later on her stepfather suffered the same fate.

The dispute over who actually discovered the HIV retrovirus, whether it was Gallo or a French team led by Luc Montagnier, became so heated that in 1987 it required intervention by US President Ronald Reagan and the French President Jacques Chirac.[7] The truce was struck when Gallo conceded he likely used the French HIV isolate to develop the test and allowed both Gallo and Montagnier to claim credit as "codiscoverers" of the retrovirus. These events likely marked the first time in history that credit for a scientific discovery has been decided by two heads of state.

Pulitzer-prize winning journalist John Crewdson of the *Chicago Tribune* was paramount among Gallo's early HIV critics. In several articles over a three-year period, Crewdson scrutinized many of Gallo's claims about his role in the discovery of the HIV retrovirus. Crewdson's investigative reporting culminated in a book-length special supplement to the *Tribune* of 55,000 words in November of 1988, entitled "The Great AIDS Quest".[8] In a later article in 1992, the *Chicago Tribune's* public editor Douglas Kneeland summarized Crewdon's conclusions and the unresolved controversy:[9] After noting that Gallo's lab had not discovered the AIDS virus as they long claimed, they had benefited from the resulting test.

> As a result, *the United States has collected $20 million in patent royalties over the years from an AIDS test Gallo developed by using what even he now acknowledges was the French virus* [emphasis added].

Kneeland finished his lengthy editorial by reflecting on what this investigation of high-profile science exposed about the pitfalls of money, ambition, and controversy.

> This case was not of abiding importance because it was typical. It was not. But as a worst-case example, it tells us about the treacherous quicksands greed and ambition place in the path of even professional truth-tellers in the scientific research community. And it shows too well what happens when politicians, bureaucrats, lawyers and marketers get too close to science.

Mikovits found that many of the same criticisms would be just as applicable to her investigation of the XMRV retrovirus and ME/CFS, her own version of "treacherous quicksands" that would entrap her in distortions and legal battles. But she was nothing like Robert Gallo.

She had seen the illustrious scientist at close range at the very beginning of her scientific career and did not want to imitate him.

* * *

Three years of investigation by the Federal Office of Research Integrity culminated in a report they released on December 30, 1992,[10] which found that Gallo had committed "scientific misconduct." Gallo vigorously and vocally disputed the findings. However, on July 11, 1994, the Department of Health and Human Services stated ". . . a virus provided by the Institut Pasteur was used by the National Institute of Health scientists who invented the

American HIV test kit in 1984" and promised the French $6 million dollars in restitution.[11]

Shortly after this acknowledgment, Gallo left the NIH for a position at the University of Maryland. The *Chicago Tribune* reporter, John Crewdson, would later turn the condemnatory information against Gallo into a 670-page book, *Science Fictions: A Scientific Mystery, a Massive Cover-up, and the Dark Legacy of Robert Gallo*, which would be published in March of 2002.[12]

However, it wasn't as if Gallo had contributed nothing to the field of retroviral research or had never taken risky positions. He had pushed the science further, demonstrating that the HIV virus *caused* AIDS, using a technique developed in Gallo's lab by Frank Ruscetti for growing T-cells— an immune response cell which the virus was found to infect. He had also been among the few who took the renegade position that a retrovirus could cause human cancer at a time when it had been almost laughable to believe such a thing. Right before AIDS, cancer was the most feared national disease, and he seemed willing to take on a fearsome challenge.

The scientists involved in the presidential dispute over HIV had also seemingly made peace with each other. When Montagnier was awarded the Nobel Prize in 2008, he wrote a gracious statement that Gallo was equally deserving of the award.[13] The two would later collaborate on several papers regarding the scientific history of the AIDS epidemic.

Maria Masucci, a member of the Nobel Assembly, had another take, telling the *New York Times* shortly after the awards were given, "there was no doubt as to who made the fundamental discoveries."[14]

* * *

The temporary truce between the two giants of retrovirology would deteriorate soon enough. Ironically the break would not be over HIV/AIDS, but rather over *autism*. Despite collaborating on the greatest pandemic of modern history, the winds between Gallo and Montagnier shifted dramatically around this other, underdog epidemic.

On June 4, 2012, Gallo wrote a letter to the Chantal Biya International Reference Center, an AIDS research center in the central African nation of Cameroon, calling for Montagnier to be removed from his part-time position as scientific director. Gallo was joined in this effort by several others who—like Montagnier—were Nobel Prize winners.

A June 19, 2012, article in *Nature* recounted Gallo's efforts and Montagnier's response[15]:

Montagnier deplores what he describes as "ad hominem attacks" and "plain lies", and says that there is an "ignominious campaign" against him and his group. He says that history is full of pioneers whose ideas were at first given a chilly reception by a conservative research community. "I believe this is happening again to me, and it is very sad that it involves Nobel Prize laureates attacking a fellow laureate," he says.

The *Nature* article went on to detail what seems to have been the "last straw" justifying the efforts to remove Montagnier, his appearance at Autism One, a conference of autism parents and researchers in the field. Montagnier's theory was that abnormal bacteria were causing at least some of the symptoms of autism, and that long-term treatment with antibiotics might be helpful. But it seemed his greatest crime was in listening to autism parents describe what happened to their children.

He says that he has never argued that vaccination could cause autism. "Many parents have observed a temporal association, which does not mean causation, between a vaccination and the appearance of autism symptoms," he says. "Presumably vaccination, especially against multiple antigens, could be a trigger of a pre-existing pathological situation in some children."[16]

Parents of children with autism were understandably frustrated that the scientific community disregarded that some of them noticed their children's symptoms worsened or that health problems began shortly after the child was vaccinated.

Like Montagnier, Mikovits was not into arbitrary social divisions: she didn't like being told who to shun. She felt parents tended to be honest and accurate reporters about their own children. If she wanted to talk to an autism parent, she was going to talk to an autism parent. Being an aisle-crosser just didn't always make her popular amongst spotlight-grabbers.

One of the greatest thrills of her professional life came when she delivered a speech immediately following Montagnier, and the Nobel Prize winner directed the discussion toward her presentation and its ramifications. Then in a private conversation following the presentation, Montagnier commented on the XMRV controversy, compared it to his own battle with Gallo over HIV, and told her, "Don't let the critics get you down."

* * *

Dr. Frank Ruscetti, Mikovits's mentor and long-time collaborator, was inclined to take a less charitable view of Gallo than Montagnier had offered on the morning he won the Nobel Prize. Ruscetti had seen Robert Gallo at close range, working in his lab from 1975 to 1982. Dr. Kendall Smith, a professor of medicine and immunology at Cornell University, recalled meeting Ruscetti during those early years at Gallo's lab as he worked on interleukin-2(IL-2), a protein molecule that regulates the body's immune system, saying, "I always describe him as a cross between Leonardo Da Vinci and Rocky Marciano, because he is truly a pure intellectual, probably the most well-read scientist I know, and he never pulls punches."[17]

Even before he joined her in later battles, Mikovits agreed that Ruscetti was one of the most brilliant, well-rounded men she had ever met. She saw him as both honest and incorruptible. These qualities and Mikovits's appreciation of them would form the basis of a more than thirty-year collaboration and friendship. Mikovits proclaimed that nobody on Earth knew her as well as Frank Ruscetti, as he had the soul-peering ability of an honestly-lived life.[18]

Smith's fond memories of Frank Ruscetti stand in sharp contrast to what he told the reporter Seth Roberts of *Spy* magazine in 1990 about Robert Gallo's reaction to his work:

> When, in 1988, Kendall Smith published a paper in the prestigious journal *Science* that accurately described the discovery of IL-2, Gallo was infuriated and phoned Smith from an AIDS conference in Stockholm. "Kendall, I haven't read it, but people tell me you were not nice to me." (It is a peculiar habit of Gallo's to claim he hasn't read or seen or heard whatever he is vociferously criticizing.)[19]

The article in *Spy* magazine also provided an account of the discovery of HTLV-1, the first known human retrovirus. It highlighted Gallo's reaction, and Frank Ruscetti's response to Gallo's attempt to take credit.

In 1978, Bernie Poiesz arrived at the Gallo lab for a postdoctoral fellowship and was put under the tutelage of Ruscetti. For years Gallo's lab had been looking for retroviruses that infected myeloid cells, as retroviruses were known to cause cancer in animals. Researchers wondered whether they could do the same in humans.[20] With Ruscetti's help, Poiesz found a retrovirus within months, and the two of them planned to publish an article on the discovery. Then Gallo called Ruscetti down to his office and suggested Poiesz be taken off as first author. Ruscetti refused. Gallo told him he was unlikely to get far in life with an attitude like that.

Poiesz compares the discovery of HTLV-1 to the Celtics winning an NBA championship; Poiesz was like Larry Bird; Ruscetti was like the coach; and Gallo was like the general manager. Yet in the decade that followed, Gallo, who had not done any lab work for years, received almost all the credit.

Although Mikovits would never work directly under Gallo, she found herself locking horns with Gallo and NIH officials at her very first research job, and like Ruscetti she would not back down.

* * *

"Science is a very difficult thing to do well," said Frank Ruscetti in 2013, reflecting on his research career of more than forty years and experience as the senior investigator and head of the Leukocyte Biology Section in the Laboratory of Experimental Immunology at the National Cancer Institute.[21] "And there are two ways to get ahead. You can struggle in the lab, work your tail off, and get some publications. The problem with that approach is that it's a profession, which relies on the integrity of individuals to be anonymous. Anonymous in getting funding, anonymous in reviewing manuscripts, and that anonymity depends on the integrity of the individual."[22]

In Ruscetti's opinion, Gallo's brand of ruthless jockeying is common in research, as many scientists become intoxicated by scientific celebrity and quickly throw away their integrity. As in any field, science could involve substantial social networking and favoritism. "Then there are a group of scientists who believe it's never important to be first, but to be second and third and write the first review article," said Ruscetti. Those scientists will then "use political connections to convince the world they're great. And unfortunately, our field is full of people like that. And they tend to be the more famous ones."[23]

Ruscetti recalled an incident in the early 1980s when he went to see two leading scientists about obtaining some HIV samples for research. The scientists were happy to share the samples, but one couldn't stop talking about how much he hoped this research would land him a spot on *The Tonight Show* with Johnny Carson. To Ruscetti, there seemed to be to be a lust for the limelight, a Great Man complex, when in truth science should be a collaborative process. In Ruscetti's view, this thirst for personal glory and recognition threatened the purity of science.[24]

* * *

Mikovits had entered her early professional life with enthusiastic determination. She graduated from the University of Virginia in May of 1980 with a BA in biological chemistry. She was the only one of the four children in her family to go to a four-year university.[25] Recognizing her promise, Anne Harpe Peabody, a high school English teacher at Jeb Stuart High in Falls Church, Virginia, was instrumental in getting Mikovits the financial aid to attend.

Right before Mikovits's graduation from college, *TIME Magazine* ran a cover story on March 30, 1980, about the discovery of interferon, which held the possibility of being a cure for cancer. From the day her grandfather had died several years earlier of cancer, Mikovits promised him and herself that this disease would not devastate other families as it had done hers. As with Gallo, her familial history of disease had thrust her into cancer research with a sense of calling. She also didn't want to waste any time getting started. On the Sunday following graduation, she saw an job listing in the *Washington Post* for a protein chemist to purify interferon as a biological treatment against kidney cancer in a lab contracted under the NCI in Frederick, Maryland. She applied for the job and landed it.

In 1982, Gallo contracted to Mikovits and her colleagues the task of purifying HTLV-I retrovirus from infected cells grown in 250-liter fermentors using a continuous flow centrifuge. But there were risky issues with his methodology. Mikovits and her supervisor felt that the growing conditions for the retrovirus were unsafe for the lab workers, especially since there were several pregnant young women on staff.[26] The hazards to a developing fetus were at that time unknown and they wanted more safeguards. According to Mikovits, they brought these concerns to the supervisors, who were unbending and demanded that they go ahead and grow the retrovirus or sacrifice their jobs. Mikovits and her boss decided to do the work around-the-clock by themselves and still managed to complete the project on time, not putting any pregnant workers in harm's way.

A few months later, Mikovits received a letter from her boss that the NCI no longer needed a protein chemist to purify interferon. She couldn't help but feel that the same dysfunctional environment that allowed Gallo to thrive was now turned against her as payback because she was a twenty-four-year-old lab technician who dared to challenge a lionized scientist. The guardians of public health had shown reckless disregard for the well-being of their own lab workers, even pregnant women, handling a poorly understood pathogen. What did this say about their concern for the rest of the public?

After her position was "eliminated," she attended a seminar given by Dr. Joost J. Oppenheim. After his engrossing talk, she approached Oppenheim to discuss his current research.[27] He invited her to his office, and when she mentioned her position at the NCI lab had ended he suggested she talk to Frank Ruscetti, who Oppenheim had just hired as a principal investigator in his laboratory.

Mikovits was certain after the fact that she had blown the Ruscetti interview but had actually left quite a favorable impression with the scientist. When Ruscetti went to notify the personnel director to hire Mikovits he was told no, with the rationale that Mikovits "was a troublemaker."

"How is she a troublemaker?" Ruscetti queried.

"She asks too many questions."

"But she's a scientist!" he replied with outrage. "It's her job to ask questions!" That settled it for him. Ruscetti insisted that they hire Mikovits and won his battle.

A few months later, Mikovits got a call from Robert Gallo, followed by Vince De Vita, who was head of the NCI. They wanted to take a look at a paper Ruscetti was writing confirming the isolation of HIV from blood and body fluids. Ruscetti was attending an overseas scientific convention in Europe at the time and thus could not weigh in.

Since Mikovits did not hold authorship, she told them she could not ethically give the paper to them. They threatened then to fire her for insubordination. ("Insubordination" seems to be a recurring theme in the ongoing career of Mikovits, earning respect from her supporters and angering her critics). Mikovits told Gallo and DeVita she wasn't going to hand over the paper. Mikovits challenged them to go ahead and fire her. When Frank returned from Europe and learned what she had done, he was incredulous. "You did that for me?" he asked.

The stress of the past few weeks and the fear that she could be fired for a second time for standing up to powerful men got the better of her, and she angrily lashed out. "I didn't do it for you! I did it because it was the right thing to do!"[28] Ruscetti was impressed and bemused by the brassy courage of his young protégé. This was clearly a very special young woman.

Shortly after Mikovits told Ruscetti what had happened in his absence, he got a call from Gallo. "You know, Frank, the NIH can't afford to have two different labs discovering this. You have to send me your virus to make sure it's the same as my virus."

But Ruscetti had been down too many dark alleys with Gallo to enter his circle of trust. Ruscetti replied, "Well, thanks, but no thanks.

Congratulations on confirming the isolation. But I'm going to flush my virus down the toilet rather than give it to you."

Ruscetti would later observe that there were a number of dominant people in science who believed things were true simply because they spoke with an authoritative utterance. Their underlings lived in fear of them because of how they could change the trajectory of a researcher's career. Unfortunately, a commanding phrase could take on a life of its own, outlasting even its definitive disproof, which was how some dominant figures rose to the upper echelon of the scientific hierarchy in Ruscetti's estimation.

"Gallo's a classic example," said Ruscetti. "Unfortunately, there are a lot of people like him in science."[29]

* * *

When Gallo hit a crosswind several years later regarding his claim to have discovered the HIV retrovirus before Montagnier and Barré-Sinousi, Mikovits and Ruscetti weren't exactly surprised to see him sputtering against a backlash.

However, there was no such controversy surrounding the discovery of human herpes-virus #6 (HHV-6). The dispute surrounding HHV-6 would be about its possible connection to various debilitating or deadly human diseases. Working at Gallo's lab a few years after Ruscetti had left, Drs. Syed Zaki Salahuddin and Dharam Ablashi first reported the isolation of the virus.[30] Pictures of white blood cells which had swelled (earning the nickname "juicy cells"), showed they had been infected by the virus. These findings were published in the journal *Science* in October of 1986. The infected cells appeared to be B cells, but the virus also seemed to target T-4 lymphocytes (otherwise known as CD4 cells, a key part of the immune system).

Patients with AIDS soon tested positive for the newly discovered herpesvirus, which may not have been initially groundbreaking since AIDS patients tested positive for numerous opportunistic infections including others in the herpes virus family. Then reports filtered in to Gallo, Salahuddin, and Ablashi that the so-called "juicy cells" were seen in other patients with unrelated immune system disorders including young children with seizures as well as adults and children with blood disorders, kidney problems, or ME/CFS.[31] Of course, this raised the question of what causal factors could create commonality in these seemingly disparate hits to the immune system, and which virus or retrovirus was the driver (and which ones were passengers).

Even after more than twenty years, the question remains unanswered whether the HHV-6 virus lies at the heart of many serious diseases or whether it is just another telltale indicator of an even more elusive pathogen or *type* of pathogen which is decimating the immune system of its victims and leading the reactivation of dormant bugs or invasion by new bugs.

However, with AIDS, the one-two punch of HIV and its coinfections became textbook information: HIV compromised the immune system, leading to secondary opportunistic infections. To treat AIDS, antiretroviral drugs had to be used as primary treatment, with treatment for secondary infections as a next line of course, because this was the way to deal with the one-two punch.

Treating opportunistic infections alone, in the early years of AIDS, had resulted in gravely shortened lifespans and massive carnage, even when it did prolong life.

* * *

In late 2005, Mikovits was smiling and at ease in one of her favorite spots: tending bar at the Pierpont Bay Yacht Club (PBYC) in the Ventura Harbor in Ventura, California, not far from the beach house she shared with her husband, David.[32] There were no membership fees required to join the club and annual dues were minimal, giving the club a class-straddling joviality unreachable in many similar places. However, the small dues required that the facility run on volunteerism. Members were required to partake in various duties at the clubhouse at least once a year.

Judy and David Nolde did much more than the minimal duties. In addition to being rear commodore of the yacht club, Nolde (among their friends she is known as Judy Nolde, while professionally she retains the Mikovits name) liked filling in regularly as the bartender. The job allowed her to open up the club in the late afternoon or in early evening, pour beer and wine, clear her mind with the heft of a pinot noir in one hand, and connect with a diverse group of people. She loved the relaxed banter, hearing about the struggles and obstacles people had overcome. If asked, she would freely share tales of her own life.

She regaled the patrons with stories about her more than twenty years working at the National Cancer Institute in Maryland. Or, if they were in the mood for romance, she told them about meeting David at a conference in Ventura in 1999, getting married at the age of forty-two, commuting for a few months between the NCI on the East Coast and David's home in Ventura, and finally deciding that if she wanted a real marriage she needed to be in the same time zone as her husband. It was his gentle magnetism that brought her there, to a place where an accomplished scientist might be found tending bar at an egalitarian yacht club.

To be nearer to David, she got a job as director of cancer research with a biotech start-up in Santa Barbara called EpiGenX Pharmaceuticals, which was developing drugs to regulate tumor suppressor genes, leading to more effective outcomes for cancer treatment. The drugs they were developing decreased DNA methylation (increased DNA methylation caused silencing of gene expression), which normally becomes disrupted as cancer spread through the body, thus causing further downstream damage. The intellectual property for the company was licensed out of the University of California at Santa Barbara (UCSB) and Judy was intimately involved in the construction of the lab EpiGenX built, as well as securing two SBIR grants from the NIH.[33]

The company had floundered in the wake of the sluggish economic climate after 9/11. In the spring of 2005, it was in the process of being bought out by a larger company. EpiGenX had generated a fair amount of its own intellectual property, but with no funding to pay employees, Mikovits was the only actual lab employee left. She would still go into the lab every day and run experiments, but the company had also put her in charge of handling due diligence for the upcoming sale, which took a few hours every day. Mikovits knew that when the sale went through she would in all likelihood need to look for a new job.

The sale wouldn't take place for several months, so on one Friday evening in late 2005, Judy found herself working behind the bar when then vice-commodore of PBYC, Joe Vetrano, walked in with his new girlfriend, Karen, an accountant. It would prove to be a moment of serendipity, with Judy's candor working in her favor. The three of them chatted for a while and Karen started to talk about her boss, who had a daughter tragically sick with an illness believed to be caused by a human herpes virus HHV-6. Judy was intrigued as Karen conveyed the substantial level of impairment and suffering of her boss's child. Karen's boss, Kristin Loomis, had started an organization to go after the virus, called the HHV-6 Foundation. After Karen had talked for several minutes and Judy excitedly asked a few questions, Joe initiated, off-handedly, "Judy, maybe you could help them."

"Yes, Joe, why not—I'll check it out," she replied with a grin, taking away their finished drinks.

* * *

Ken Richards joined EpiGenX in September of 2000, as chief financial officer, and recalled recruiting Mikovits from the NCI in May of 2001.[34] Ken was originally from Canada, having worked for seventeen years for a Canadian corporate investment bank before transferring with them to Los Angeles in 1997. He was surprised as a savvy money-man to find that the University of California at Santa Barbara (UCSB) had a phenomenal science and engineering program, but no systematic way to bring their research discoveries to market. Richards and two other ambitious partners founded the Santa Barbara chapter of Tech Coast Angels, the largest angel funding network in the United States.

It was at a meeting for Tech Coast Angels that Ken was introduced to EpiGenX, and through that company he would meet Mikovits. He later gushed about her: "Judy was a very well-spoken, knowledgeable, and dedicated scientist who wanted to do everything possible to find effective treatments for cancer," said Richards. On the question of why Mikovits seemed to have both strong supporters and critics, Richards said, "I tell everybody, Judy is very controversial. Many people do not like her and many people admire her. I am in the later category. She speaks her mind. When she develops a view, she is dedicated to that view and will defend it fiercely. She's combative in a positive sense, and that tends to irritate some people."

Richards believed that many of Mikovits's detractors had fallen victim to an unconscious form of sexism in which an assertive woman was "viewed as a bitch" while a man making a similarly impassioned defense of his position would be "admired and respected for his firm stance."[35] It seemed like the kind of post-feminist statement that perhaps only a man could make and be fully heard, especially regarding a disease like ME/CFS that was incorectly thought to only affect women and had been derogatorily referred to as "Yuppie flu" in its early years, with mocking press implying that women contracting the disease were overly driven.

Even though Mikovits would eventually leave EpiGenX, her tie with Richards would remain solid and he would remain a steadfast supporter. In 2011, Richards was putting together a private equity firm to invest in early stage technology and biotechnology companies. "[We] needed somebody with a strong science background, the first person I thought of was Judy Mikovits."

When a few higher-ups asked questions about bringing on this controversial figure, Richards had several cards to play on Judy's behalf. In addition to a recommendation from the respected Frank Ruscetti, Nobel Prize winner Luc Montagnier was very supportive of Judy and wrote highly of her work and her integrity when he penned a recommendation to the Yorkbridge management.

* * *

After her chance run-in at the yacht club, Mikovits did some cursory research on HHV-6, and within a few hours felt comfortable that she understood most of the issues, mainly the question of whether HHV-6 *initiated* the disease process or was simply a consequence of a poorly functioning immune system. The questions were much the same as they had been in HIV/AIDS research.

Her doctoral thesis had in fact been on the development of AIDS from HIV infection.[36] She had the curious experience of defending her thesis shortly after the basketball player Magic Johnson had announced he had tested positive for HIV in November of 1991. Because Johnson was such a beloved sports star and defied widespread stereotypes about who got AIDS and who didn't, his revelation had shaken the country and cracked open borders for marginalized groups. The focus of the committee at her thesis defense was the question: *based on your thesis do you think Magic Johnson will ultimately develop and die of AIDS?*

The soon-to-be Dr. Mikovits answered that she thought Johnson's long-term prospects looked good. His infection appeared to have been recently acquired, so if he took the recently developed antiretroviral therapy (ART) that prevented the HIV virus from replicating and integrating into tissue reservoirs of the monocyte and macrophages, it would never cause the immune deficiency. For Johnson, HIV might not be able to establish itself in hidden reservoirs, if they started him on the antiretrovirals quickly after his initial infection.

In all likelihood, Mikovits told her thesis committee then, Johnson could look forward to an average life-span. Nearly fourteen years later, as Mikovits sat down to write an email to Loomis in November 2005, she reflected on Johnson's continuing good health, which many would have thought magical thinking in 1991. She included her résumé in the email and details of the upcoming sale of EpiGenX. Loomis emailed Mikovits back within a few hours and they made plans to meet at Loomis's house in Montecito. They talked for much of the afternoon. Loomis said she was interested in bringing Mikovits on staff in some sort of research capacity. Ablashi had the title of research director, so that would remain with him.

But some position would be found for Mikovits.

Mikovits wrote an excited email to Ruscetti about the job offer, noting how much she disliked the turmoil of EpiGenX's pending sale, being put in charge of due diligence for the sale, and that—more than anything else—she wanted to get back to working with patients and doing research. Ruscetti was a little more cautious, hesitant about the involvement of Ablashi. He thought Ablashi was cut from the same cloth as Gallo, and part of what had gone wrong in retrovirology in the past thirty years, turning what should have been a collaborative search for truth about the greatest scourge of recent history into a cruel and brutish competition for individual glory.[37]

Against the advice of Ruscetti, Mikovits took the job.

* * *

Within the first couple of weeks she worked as a consultant for the HHV-6 Foundation, Mikovits took Loomis to her lab at EpiGenX. Loomis was excited over the large freezer they had at the facility and explained that she worked with Dr. Daniel Peterson, a physician in Incline Village, who—along with Dr. Paul Cheney—stumbled upon and then started treating the "walking wounded" from the first modern outbreak of what was called at the time, chronic fatigue syndrome (CFS) at Lake Tahoe in 1984–1985.

The name would later be changed to myalgic encephalomyelitis/chronic fatigue syndrome (ME/CFS). It was a surreal reality for these small-town doctors in the placid setting nestled in the mountains, like being in a remote area of the Alps unaware that it was wartime and finding an office filled with never-before-seen injuries.

The patients kept stumbling into the practice in Incline Village, as baffled as the doctors, while Peterson and Cheney tried to take detailed testing and records with little certainty about what to offer beyond symptomatic treatments. Peterson reputedly had a repository of blood samples dating back to that first outbreak, which would be ideal for research. Mikovits dreamed of the diagnostic tests she could run on the blood. Loomis told Mikovits that she and Peterson should write papers together since that was not one of Peterson's strengths, whereas Mikovits had more than forty published articles to her name.

After the initial burst of excitement of the first couple weeks, though, there was a disconnect between what Mikovits thought she would be doing and what she was assigned. Mikovits figured a bright spot might be preparing for the 5th International Conference on HHV-6 and -7, to be held later that year in Barcelona, Spain. Loomis wanted her to find companies and individuals to sponsor the meeting. She knew that Mikovits had many contacts from her years at the NCI and her years in the biotech field.

Mikovits reviewed the abstracts for the conference, knowing the rigorous standards the pharmaceutical companies and scientific experts would expect. After reading the abstracts she grew more despondent. There was no real uniformity to the papers and Mikovits felt she could not present the idea of sponsoring the conference to any of her colleagues due to the fact there was so little actual science in any of them. It wasn't necessarily a criticism of the scientists. It was largely a reflection of the paltry level of funding given to study the virus. Without serious money given to study a problem, nobody could be expected to produce high-quality science.

Loomis was upset when Mikovits said she couldn't find any sponsors for the conference among the scientists and organizations she knew. After those aimless early weeks and after talking it over with her husband, Mikovits made a decision. She had promised to go to the meeting in Barcelona, but when the conference was over she would leave the HHV-6 Foundation and look for another job.

* * *

Barcelona, Spain—May 1, 2006

Gallo finished his dinner talk and Mikovits was glad she had been able to avoid listening too closely. The pharmaceutical reps at her table had been lively company. To the public, Gallo was still a respected figurehead, but some viewed him as somebody who had been "saved" because a true accounting of his misconduct would have been a stain on the prestige of American science.[38]

After Gallo's talk there was an award being given to one of the members of the board of directors of the HHV-6 Foundation, Annette Whittemore. Mikovits knew the name, but had not otherwise been aware of the Whittemores until the award announcement and the letter written by her husband Harvey as a prelude. Mikovits learned that Annette's father had been a doctor for small rural communities in eastern Nevada. Later Mikovits would learn that Annette's father had delivered all of Senator Reid's children. Annette also had a degree in special education and had worked for several years with children who had autism. She appeared a bit overwhelmed by the tribute, kindly thanking the group before quickly exiting the stage.

As Annette stepped down from the podium, Mikovits couldn't help but reflect on the contents of Harvey's letter, detailing what a loving, virtuous, person Annette seemed to be: just the kind of person Mikovits wanted to be around.

* * *

Barcelona, Spain—May 3, 2006

Dan Peterson gave the morning presentation on the final day of the conference. He was listed in the program as "Principal Investigator, Chronic Fatigue Syndrome and Immune Dysfunction Syndrome"[39] out of Incline Village, Nevada. It was only fitting he should have such an exalted title given his association with ME/CFS ever since it had appeared amongst those snow-capped peaks in 1984–1985.

Mikovits sat in the back of the room, thinking she had only a few more hours left of the conference and her attention was waning a bit, when Peterson came to his last slide. The slide contained data from sixteen people with ME/CFS and the various cancers they had developed over time, which obviously piqued her interest after so many years at the National Cancer Institute.[40] In addition to the cancers, the slide showed the results of their immune cell testing.

The T cells showed some unusual abnormalities known as clonal rearrangements. This meant instead of making many different generalized T cells to target all of the pathogens that might appear, these patients' cells seemed to be focused on a singular insult, leaving their immune systems vulnerable to other invaders as if the cells were obsessively staring down the crosshairs of a gun at one target while other invaders snuck in from behind.

This piqued Mikovits's attention because T cells eliminate virus and cancer cells and abnormalities meant the body would have a reduced ability to fight off a viral pathogen. A chronic viral infection lying in wait for many years might end up causing a cancer. Mikovits also saw on the slide a few mantle cell lymphoma cases. At the time she was participating in a cancer support group in Ventura, and a few of those women also had mantle cell lymphoma. Mikovits's years at the NCI working with Ruscetti had drilled into her the idea that whenever one saw a cluster occurrence of illnesses, one should think about pathogenic causes. This dictum wasn't always true, but it was a logical place to start.

Peterson told the assembled group that he didn't know what the unusual T-cell abnormality meant and stated that if anybody had any idea they should come up and talk to him. Mikovits nearly sprinted up to the podium to grab his attention. They talked for several minutes and Peterson was enthusiastic about her expertise and invited her up to Incline Village to look over his files and talk with patients. She would meet with the Whittemores and with Peterson, who was treating the Whittemores' aforementioned daughter, Andrea, to see if there were any ways she might assist their efforts. Mikovits also thought she might be able to link the mystery of the T-cell abnormality to the pathogenesis of ME/CFS.

Exciting collaborations seemed to be afoot and Mikovits was unexpectedly looking toward Nevada, a state that had used the slogans, "Wide Open" and "Battle Born." Both slogans would prove prescient for her as she ventured into what seemed to be a wide-open field of possibility and found herself battling bullying opposition and intangible demons.

Reno had hosted its own slogans and nicknames over the years too, such as "A Little West of Center" and "Far From Expected," catch phrases that seemed to connote a climate of broad-minded risk taking that would welcome an insubordinate scientist who stuck to her guns.

CHAPTER TWO

The Move to Nevada

*Throughout the 1990s and early 2000s . . . Whittemore . . .
was the supreme legislative lobbyist. He represented the gaming,
tobacco and liquor industries . . . He could get legislators to pass
the legislation his clients wanted and then help the same legislators
finance their next political campaigns.*[1]

—*Las Vegas Review*, February 26, 2012

Sparks, Nevada—early June 2006
Judy Mikovits first met with Harvey and Annette Whittemore at the Red
Hawk Resort and Country Club in Sparks, Nevada. Red Hawk was a place
with preternaturally green golf courses rimmed with mountain views,
dining areas with tasteful stonework and towering wood pillars, and a luxe-
yet-earthy aesthetic: even the restaurant's house wines were from *La Terre*
(the earth) winery, a reminder that patrons were in a verdant but obviously
man-made paradise.[2]

The juxtaposition of green fairways and the arid desert was one Harvey
seemed to love as a Nevada developer. Other regions of the country had
transitioned from dust bowl to arable plains, from marshy tall grass to
farmland, and this process of revamping unruly land often involved both
visionary thinkers and sometimes a heavy human hand. Sparks sat about
thirty minutes east of Reno, and many initially thought it was a crazy

location for a city. It was plunked right in the middle of the desert, where average rainfall was less than eight inches per year. Economic growth in Reno in the 1950s had created a rising demand for low cost housing, and Nevada was known for risk-takers seeking unexpected payoffs. Developers successfully grabbed water rights to divert streams from the Sierra Nevadas that towered to the West, thus allowing the city to prosper. The approximately 90,000 people of Sparks were housed in a neat layout around the centrally-located tower for John Ascuaga's Nugget Casino, which was understatedly flashy by Nevada terms. Nevadans often joked, "Reno is so close to hell you can see Sparks!"

Water rights, being so central to life and property and a protected commodity in the West, could be harder to secure than cattle in modern-day Nevada. The Silver State might as well have been named after the battle over silvery streams and rivers. When her car glided closer to the Red Hawk resort it was difficult to be unimpressed by the sheer audacity of what Harvey Whittemore had pulled off in the high desert. The Resort at Red Hawk sat on land that had originally belonged to George Wingfield, who during his sixty-three years in Nevada had been a miner, banker, rancher, investor, and—like Harvey Whittemore—a developer and political power broker.[3]

Harvey and his fellow developer David Loeb chose to preserve the original ranch house of the man nicknamed "King George" when they built their resort among the sage, reeds, and rushes where Wingfield had once set his duck blinds and raised Labrador retrievers in the shadow of the nearby Pah Rah Mountains. In the spring of 1997, Harvey and his partner opened the first of their two golf courses, the Lakes Course, designed by the famed golf architect, Robert Trent Jones II. Sitting upon Audubon Certified protected natural wetlands and featuring natural lakes, murmuring springs, and cottonwood trees, the course was designed to embrace the idyllic landscape.

In the ensuing years the Lakes Course was named "Best Golf Course in Reno" seven times by the *Reno Gazette Journal*.[4] In 2001, Harvey's group built a second course designed by player-turned-golf-architect Hale Irwin. Then they broke ground on a permanent clubhouse and a variety of residential options ranging from large homes to condominiums.

Mikovits immediately noticed the respect bordering on reverence that the employees seemed to have for Harvey and Annette. Later she noticed glimmers of nepotism in the same Whittemore mirage, as she learned that many of the employees at Red Hawk were the children of close friends or even family members. In person, Harvey could be considered a physically intimidating man, standing well over six foot four, his reddish hair cropped and thinning, almost military style. He also had a well-trimmed goatee, giving him a rebellious look for a businessman and adding to the waft of celebrity that followed him around Red Hawk and beyond. Harvey's office was mightily extravagant. An enormous bay window overlooked the Lakes Course and autographed sports memorabilia ornamented the walls along with flat screen TVs.

In many meetings to come, Mikovits would typically join the Whittemores in a small conference room. Before the meetings started, a waiter typically slipped in to ask if they needed drinks or food. This first meeting, however, they met in the evening in a conference room right off of the restaurant where they could dine properly on the restaurant's New American cuisine. On the heels of the recent Barcelona conference, they were joined by Dan Peterson and Dr. Greg Pari, a professor at the Department of Microbiology and Immunology for the University of Nevada, Reno (UNR). In 2010, Pari would become chair of that department.[5]

During the meeting, the Whittemores first proposed to Mikovits their idea of starting a research institute in collaboration with UNR. They believed an institute dedicated to all aspects of neuroimmune disease, one that could encompass the work of the HHV-6 Foundation under the auspices of a common academic home, was the best approach to streamline the research and bring scientific powerhouses together.

Pari thought it was a terrible idea, even though the Whittemores had envisioned him as perfect for the job of research director. Pari told the Whittemores if he was involved in a significant capacity his salary alone would cost somewhere north of two hundred fifty thousand dollars a year. But Pari didn't want the job anyway.[6] Even with his protests, he did remain interested enough to show up at WPI parties, and he would later join its Scientific Advisory Board after Mikovits was fired.[7]

With Pari declining the offer to be research director of a still-not-yet-existent WPI, the others agreed that Mikovits should spend the summer commuting to Peterson's office to learn about the disease and begin laying out plans for the WPI. Her routine quickly fell into place. She took a Southwest flight early Monday mornings, spent the week working at Peterson's office at Incline Village, then flew home to be with David on Friday night. This would remain her routine before the WPI acquired laboratory space from the university.

* * *

One Friday, as Mikovits flew back to southern California, she thought of what she had learned over the week about Andrea Whittemore. Annette later wrote a long and thorough article, which gave much of her daughter's history as well as the rationale for the founding of the institute. It was published in the June 2010 issue of *Molecular Interventions*[8]. She wrote of Andrea coming down with a mono-like illness in 1989, months of flu-like symptoms, and symptoms including tachycardia (abnormal heart rate), swollen lymph glands, muscle pain, and night sweats. The doctor diagnosed the condition as being psychological in nature, which seemed ridiculous to Annette.

In November of 2009, shortly after the publication of the article by Mikovits and her team in *Science* that showed an association between the XMRV retrovirus and ME/CFS Andrea Whittemore-Goad went public with her story on a Facebook posting for the WPI.[9]

"My name is Andrea Whittemore-Goad," she wrote. "Until last year I was uncomfortable telling my story to complete strangers, but now if this is what it takes for all to understand the severity of this disease, I will." She recounted how as a young girl she'd had a poor reaction to a DPT booster and that in fourth grade she came down with a mono-like illness after a tonsillectomy. After the surgery she had gastrointestinal problems and tachycardia and none of the doctors could tell her parents what was wrong. But the medical professionals were scared. One psychiatrist told her to get out of her office because she didn't want to catch what Andrea had. Another told her parents that Andrea was "school-phobic."

Annette continued the story in her *Molecular Interventions* article.[10] A neighbor suggested they visit a Peterson who had treated patients from the Incline Village outbreak of 1984–1985. Peterson turned out to have a good understanding of the disease, and although his treatments seemed to provide mostly symptomatic relief, Andrea made slow but steady progress under his care. Her mother elaborated:

> She continued this modest improvement until she decided to enroll at the University of Nevada, Reno. The admission policy required the measles, mumps, and rubella (MMR) vaccination prior to starting classes. Within five days of the MMR vaccination, Andrea had a severe relapse and never regained her previous level of health.[11]

Andrea's Facebook post recounted that her reaction to the MMR vaccination left her confined to a wheelchair.[12] Over the next several years, the family tried numerous strategies to improve Andrea's health, without much luck until Andrea was twenty-one and they stumbled upon Ampligen, a substance that acts to stimulate the body's antiviral defenses.

Compared to other treatments, for Andrea it was a godsend, a gift she didn't take for granted since she was one of the very few patients with access to the drug under controlled trials. She took Ampligen off and on for eight years:

> While taking Ampligen, Andrea improved to 75% of her previous levels of energy and stamina, but despite many of the positive outcomes, she continued to fall ill with opportunistic infections. For unknown reasons, Andrea began to develop reactions to Ampligen, making her too sick to continue.

In 2006, when Mikovits first started working in Nevada, Andrea was doing relatively well, although there were clear indications that the Ampligen was producing less than a full recovery. This had also been true of early monotherapy for AIDS, where stumbles and severe relapses often followed improvements.

* * *

When Mikovits first arrived at Peterson's office in Incline Village to begin her investigation she encountered her new "staff,"—three summer student interns—waiting for her.[13] The crew included David Pomeranz, who later went on to USC Medical School, and Byron Hsu, a young man from Berkeley: both hired from a job listing on Monster.com. The third member was Katy Hagen, whose mother was a friend of Annette's.

After talking with David, Byron, and Katy for a few minutes it became clear that none of them knew much about science in a research setting. Yet they all seemed like motivated, brainy, greenhorns and Mikovits had often worked with similarly raw students at the National Cancer Institute. She enjoyed the challenge and energy associated with new learning and often found that their inexperience could be a positive as they would work hard to master the right techniques, entering with what Buddhists call a "beginner's mind."

In their previous discussions, Mikovits remembered Peterson talking about his "repository" of samples. She had already compiled a mental list of investigative studies which might be conducted with them. Things were moving forward. There had been a meeting with Dr. John McDonald, who at the time was dean of the medical school, and Harvey and Annette had already started meeting with state and national politicians about the feasibility of putting an institute together at UNR.

Mikovits asked Byron to find a few patient samples using dates and patients from the original table Peterson had showed in Barcelona, especially the ones with Mantle Cell Lymphoma (MCL). Then she started discussing with David and Katy what they would need to do to build an Excel data sheet to match the critical records from each of the patients to their samples in the repository. It seemed to Mikovits that Byron had only been gone for just a minute or two when he was back.

"I think you need to see this," Byron said. Mikovits followed Byron down the hallway to a small room which held a large deep freezer about six feet tall by three or four feet deep and just as wide. Byron opened the freezer and Mikovits saw it was filled from top to bottom with plastic bags containing tubes with written names and birthdates scribbled on the side of each tube.

Oh my God! This is the great repository? Mikovits thought. Many of these samples had in all likelihood been logged prior to the more stringent privacy laws. From a scientific standpoint there were more significant problems. Freezing whole blood destroyed evidence of RNA viruses and the even more delicate RNA messenger molecules whose importance was just beginning to be understood, leaving behind only DNA.

Mikovits and her new team drove down to the local grocery store and bought a block of dry ice, a few bottles of isopropynol, and cotton swabs. The interns would start the summer by rubbing the names off of thousands of tubes, putting dates and identification numbers on them as required under the more stringent privacy laws, and building the information into Excel files.

Mikovits would start by contacting the patients, making arrangements to draw new, fresh blood, properly harvested. On one of her weekends back in southern California she stopped by EpiGenX. She got permission to take some extra pipettes, bottles of trizol, vials that could withstand being cryogenically frozen, and reagents which would allow the samples to be frozen without destroying the nucleic acids.

The repository may not have been what she expected, but Mikovits would make sure that this new repository was a resource they could trust to answer their most challenging questions.

* * *

The first hypothesis Mikovits came up with was that they might be able to find a biomarker of a viral infection in the sampled patients by measuring the ability to produce alpha interferon from plasmacytoid dendritic cells (PDCs) introduced into a patient culture sample. PDCs are immune cells circulating in the blood. A healthy individual will produce large amounts of interferon from their PDCs when exposed to retroviruses like HIV.

These cells can be harvested from blood and the amount of interferon they produce can be measured.

Frank Ruscetti had done some of the original work on PDCs and human retroviruses and had a good store of the cells. Mikovits was able to get fresh primary PDCs and the interferon-producing PDC cell line, CAL-1, and use them to test patient samples. To her surprise, none of the samples were pumping out interferon. It was an interesting finding as PDCs infected with HTLV-1 did not produce interferon either. A key part of the immune response was disabled in ME/CFS patients, just as in HTLV-1, which caused cancer and neuroinflammatory disease. Mikovits and Ruscetti were both intrigued.

Mikovits spent a good deal of time taking data from patients about their maladies. During this time she also did consulting work for a biotech company that had developed a unique chemokine/cytokine multiplexed assay. She used this test that measured inflammatory markers (usually a sign of infection) on the patients. In addition, she checked their levels of natural killer (NK) cell activity and RNase L (other indicators of an abnormality in the innate immune systems of ME/CFS patients). When using the cytokine/chemokine profiling test Mikovits later recalled that:

> it was consistent with a viral infection. It didn't have to be a specific virus, it's just what we found when we ran the original viral expression micro-array experiments that same summer. There was a highly dysregulated expression of RNase L, there were various levels of significant expression for many different kinds of virus. It really didn't matter. Lots of HHV-6, lots of enteroviruses. So much noise that I did remark that these people looked like AIDS patients.[14]

She was not the only person to make the clinical observation that ME/CFS patients resembled people with AIDS. Before HAART therapy for AIDS patients, when AIDS dementia complex (ADC) often went unchecked, Dr. Anthony Komaroff of Harvard University published a vivid paper that compared the brain SPECT scans (to measure blood flow) of patients with ADC, ME/CFS, and unipolar depression to controls.

Komaroff's images showed well-lighted brains in depression and the control group (with the illuminated red, orange, and yellow tones representing blood flow to various areas of the brain), but an optically-stunning near "lights out" in ADC and ME/CFS that was shockingly similar, as if a violent storm had knocked out the power grid in two neighboring states.[15]

That summer, Mikovits called Ruscetti and told him that among Peterson's patients she had encountered a fifteen year old and a thirty year old with shingles.

"That's ridiculous," Ruscetti replied. "They'd be AIDS patients."

"Yeah. Exactly what I was thinking."

Since the patients weren't dying en masse as they did with AIDS, Mikovits knew it couldn't be the HIV retrovirus. But what if it was another retrovirus, leading not to the predictable degeneration and death of untreated HIV/AIDS, but to a chronic, long-term disease, characterized by a loss of cellular energy and a lowering of the defenses of the body's immune system, as in the slower-moving retrovirus HTLV-I? That would explain why patients presented with pathogenic coinfections, just as AIDS patients did, as well as with cancers that often seemed to develop after decades of being sick.

Cancers were, of course, associated with both the retroviruses HIV and HTLV-I, though with AIDS in the years before aggressive therapies they took a swifter course, and AIDS was even once called the "gay cancer" due to the unlikely appearance of Kaposi's Sarcoma (KS) in younger gay men before AIDS was defined—when KS at that time was thought to primarily target older Jewish men.

Retroviruses are a family of ribonucleic acid (RNA) viruses, which contain the enzyme reverse transcriptase. The virus enters a cell, uses the reverse transcriptase to direct the cell to create viral deoxyribonucleic acid (DNA), which then integrates into the DNA of the host cell. The virus thus becomes part of the DNA of many of the cells of an infected person. Retroviruses had documented transmission through sexual contact, exposure to infected blood or blood products, or passage from an infected mother to her newborn child during gestation, delivery, or even breastfeeding. Retroviruses are transferred via body fluids, rather than by a vector such as a mosquito or a tick.

Mikovits thought about the infections, conditions, and cancers associated with HIV infection. HIV causes problems by damaging the immune system, leaving it vulnerable to other organisms that can cause disease. The majority of people infected by HIV first develop a flu-like illness within a month or two of infection, often suffering fever, muscle soreness, headaches, swollen glands, and chronic diarrhea—similar to the relatively common "flu-like" onset of ME/CFS. Latent infection can last from eight to ten years, giving some patients in the early years of the pandemic the illusory impression that something they were doing was staving off the progressive decline.

Similar misattribution has at times impacted ME/CFS patients due to lack of viable treatments—or, more commonly, led to many claims from outsiders of pseudo-cures, which can thrive when an illness's course is unknown. Some with HIV develop full-blown AIDS sooner, and others later. Usually after one has been infected for ten years, the patients suffer from night sweats, fevers which last for weeks, weight loss, headaches, and chronic diarrhea.

In poor nations, those with HIV/AIDS commonly develop tuberculosis.[16] HIV/AIDS patients are also more susceptible to salmonella, cytomegalovirus, candida, cryptococcal meningitis (an inflammation of the membranes and fluid surrounding the brain and spinal cord), toxoplasma gondii (a parasite spread by cats), and cryptosporidium (a parasite which lives in the intestines and bile ducts, leading to chronic diarrhea).[17]

One of the common neurological disorders is AIDS dementia complex, which usually leads to behavioral changes and diminished mental functioning. Mikovits couldn't help but think about the mental and cognitive changes associated with ME/CFS, such as memory loss, drop in IQ, problems on neuropsychological testing, word-finding difficulties, computation problems, and many others.

Considering all the clues and in light of early observations that ME/CFS had overlap with AIDS, it seemed they were on the trail of a retrovirus.

* * *

Early in her work at Peterson's office, he advised her to be extremely discreet about whom she saw in the office.[18] She was starting to learn the politics of ME/CFS.

Mainly, that if you had it, you didn't want anybody to know.

Peterson, given his prominent role during the first modern outbreak of 1984–1985 in Incline Village, was known as the doctor to the rich and famous. Mikovits was amazed at the small but steady flow of Hollywood actors and professional athletes who came through Peterson's office to get some relief from the disease.

If one saw somebody famous in the waiting room, the protocol was to act like the person was like any other patient there. That came easily to Mikovits, as she never fancied herself a fan of anybody, and even if she found herself in an elevator with a favorite baseball player, she would let him have his privacy. When she wasn't talking about science or doing her volunteer work, she was actually a quiet person. She knew this would no doubt surprise many of her colleagues, who often referred to her as "Tsunami Judy" for the torrent of words that could pour forth from her when she was excited about a subject.

Since it was summer, the Whittemores were typically in residence at their home in the Glenbrook neighborhood at Lake Tahoe. Past the double-gated security checkpoint, residents often drove around in golf carts and the houses all had names. The historic name of the Whittemore home was the Lake Shore House and resembled a two-story southern mansion, with an enormous wrap-around porch, a beach where one could swim out into the cove, and a private dock out back where Harvey had several boats.

Mikovits enjoyed the life at Glenbrook, often walking out on the dock just as Harvey was pulling up in his new thirty-foot power boat, while an employee grabbed the line. Harvey would tell her and anybody else who might be on the dock to go ahead and jump in, and they'd speed off across the enormous Alpine lake. Sometimes they would get all the way to the California side, tie up the boat at Jake's on the Lake restaurant, have dinner and drinks (for which Mikovits never saw the bill), and then hop back in the boat and race home. It was a lifestyle to which anyone could become accustomed.

There were some townhomes located next door, and the Whittemores bought one to host their friends and their children. The townhouse had a high loft with large windows and a reading cove where Andrea would often sit with a book while her family's guests enjoyed the beach. Mikovits couldn't help but think it was due to Andrea's condition and not wanting to provoke what many ME/CFS patients called "a crash" caused by overexerting her limited supplies of energy.

Down the road about a mile and an easy walk was "The Barn" which was owned by somebody else but was where the Whittemores held most of their social events and political fundraisers. It was once an actual barn but had been completely renovated with nice wooden floors, a bar, picnic tables, pinball machines, basketball hoops like the ones at a county fair, antique Coke machines which would dispense the soda in their classic hourglass shaped bottles, and horse saddles and other farming implements on the walls displayed like pieces of fine art.[19]

The Whittemores were also crazy about sports, a passion they shared with Mikovits. They had a skybox on the fifty-yard line for the University of Nevada, Reno, football games, seats on the floor for basketball, and season tickets for the Reno Aces, a minor league baseball team. Mikovits couldn't help but notice that Harvey's seats at Aces Stadium were in front of those for Dean Heller, the Republican Senator for the state who served alongside Senator Harry Reid in Washington, DC. Not far away from those seats were those for Congresswoman Shelley Berkeley, who would unsuccessfully challenge Heller for his US Senate seat.

Nevada's circle of power was small, and now, working for the Whittemores, Mikovits had a front row seat to it all.

* * *

Over the Fourth of July holiday in 2006 Judy and David went to Hawaii, but Judy wasn't participating in the Aloha spirit. Instead she spent the whole time parked in a cozy cabana chair with her silver laptop from EpiGenX writing grant proposals. Loomis, Whittemore, and Peterson would then determine if they could fund the research or if they could present it to one of their wealthier patients who might be interested in funding. It was an abnormal life for a researcher, but one that was becoming common for many scientists. As government sponsorship of science declined, the public was often left to fill the gap.

As Mikovits settled further into her weekly routine, the Whittemores arranged for her to get a free dorm room at Sierra Nevada College, which was close to Peterson's office. Ocasionally, the Whittemores would treat Mikovits to a few days in a luxurious condominium at Red Hawk. The condos were large with lavish marble appointments and beautiful faucets and sinks. There were always big soft robes and towels as well as fancy hand soaps and lotions. The attention to detail and elegance was almost intoxicating. Mikovits could go down to the restaurant or swimming pool and know she would be waited on hand and foot.

Mikovits had a friend, a coworker from the National Cancer Institute in Frederick, Maryland, who lived in Reno and was a professor at the university. On a couple of occasions, the two would meet at Red Hawk and they'd spend time at the swimming pool.[20] While Mikovits friend didn't personally know the Whittemores, she knew of them and told Mikovits that the Whittemores had a good reputation among her colleagues. Several times that summer the Whittemores took Mikovits to dinner at David's Grill at Red Hawk. They might start off with crab cakes with California pepper aioli or just dive right in to a fettucine alfredo with a gluten-free pasta option. Many prominent business people and local politicians would often drop by the table to say hello to the Whittemores. Harvey would always introduce Mikovits as their "dear friend" who was helping them with Andrea's disease and his wife's institution. And a few times they'd jokingly say they were "trying to make her family."

Their family was so admirably close-knit and accomplished that it was hard not to be warmed by those words, and Judy Mikovits was already starting to feel like part of the Whittemore clan.

* * *

The summer was ending and Mikovits wasn't sure what she should do, but she had to transition from the luxuriant summer of power boats and her young assistants into something more stable. She had an offer from a bio-tech company for whom she had been consulting recently. The company had some natural product drugs which were close to clinical trial for cancer treatment and they offered her a lucrative compensation package to become the vice president of research. And while she had loved her "summer staff" and the patients, she didn't exactly know if this institute would ever get off the ground. Then there was the problem of funding. The research community didn't seem to believe ME/CFS was a priority, so there were lots of appeals to wealthy individuals or families who might have a loved one with the disease. The begging for philanthropy is a constant curse of a maligned illness, but ME/CFS had the additional curse of patients being too sick to pound on doors or to march through the streets and shout funding demands through a bullhorn.

"Harvey, you can't let her leave!" Annette had pouted when Mikovits had told Annette she was considering another opportunity. "Make her a good offer so she'll stay!"

Harvey asked Mikovits what it would take to keep her.

Mikovits said she wanted a five-year guaranteed contract (as would one see for a valued player in baseball) and the ability to keep their beach home in Oxnard because that's where they planned to retire. If she did take the job in Reno, they'd want to become full members of the community by purchasing a small home or condominium. Harvey asked if she'd withhold making a decision for a while to see if he could use his influence and make some things happen.

Mikovits agreed to give Harvey some time.

By a happy coincidence Frank and Sandy Ruscetti had been invited to UNR by one of their former colleagues from the NCI to give a seminar at the end of the month. Mikovits thought it would be a perfect opportunity for him to meet the Whittemores and weigh in on the long-term prospect of her continuing to work for them.

Not that she had taken his advice in the past. When she'd met and married David, Frank Ruscetti had told her it was a bad time to leave the NCI and try to get into the biotech industry. He had advised against her signing with EpiGenX, and he'd also advised against going to work for the HHV-6 Foundation.

* * *

Reno, Nevada—October 31, 2006

As Frank Ruscetti and his wife, Sandy, drove to the Whittemore's house in Glenbrook, Frank considered how far his dear friend had come in the twenty-three years since he had hired the protein chemist who could not find a cell under a microscope.

But even though she'd left the East Coast for the West, Frank and Judy still talked on the phone several times a week. In the years they'd worked together Judy had become like family, occasionally baby sitting their only son so that Frank and Sandy could have a night out at the Shakespeare theater. Now in this momentous decision of her life, whether to take a job as the research director of an institute which hadn't even been built, she wanted Frank and Sandy to come and check it out.

Frank liked the Whittemores and felt the offer to Judy was a fair one. He was impressed with the Whittemore's friendliness, how unpretentious they were, and their dedication to finding answers for the ME/CFS that plagued their daughter.

It all culminated in a small, simple dinner with the Whittemores at the Lake Shore house along with Mikovits and her husband and Dr. Peterson and his wife, Mary. The Whittemores cooked spaghetti and garlic bread and tossed a salad while the guests hovered about them in the kitchen and played Trivial Pursuit. Mary was an educator and an aspiring playwright like Frank. Everyone was enjoying a casual moment so comfortable it seemed they'd been friends for years instead of just meeting a few hours earlier.

After the meal was over, Frank took Harvey aside and gave him the third degree: What were Harvey's motivations? What did he want to get out of this? Harvey replied that his daughter was now in her mid-twenties and had been ill since she was twelve years old. Money could do a lot of things, but if your child was sick, how much did that matter? His taut face belied a father's anguish that his daughter had lost her most carefree years to illness, with no end in sight.

Harvey seemed to know a great deal about local history and Frank was pleased that Harvey had interests beyond business. Frank thought he glimpsed a man who relished intellectual combat when he asked Harvey if he knew where the namesake of the city of Reno was buried. Harvey knew the city was named after Union General Jesse Reno who died in the Civil War, but didn't know where he was buried.

"His grave is in Frederick, Maryland, near the National Cancer Institute," Frank told him, a little smugly. "He died at the Battle of South Mountain. I've been there a couple times."

Frank could see Harvey was slightly unnerved that he had known a bit of Nevada trivia that Harvey didn't. Harvey was clearly a man who didn't like to lose. That could be a good thing. They talked for a while longer and then Frank abruptly said, "Harvey, this has been a lovely evening, but I must confess to a sense of disappointment."

"Why's that?" Harvey asked, slightly taken aback.

"I've always believed like the French philosopher Balzac that 'behind every fortune lies a great crime.' And during this time you've made a lot of progress in disabusing me of that notion."

They had a good laugh over that. When Frank and Sandy were at the airport the next day, Judy called him on his cell phone. "What should I do?" she asked.

"Take the damned job!" he grumbled.

She followed his advice, as she said she would.

Frank later lamented it was the worst advice he had ever given anyone.[21]

CHAPTER THREE

Day Two in Jail

Judy, there's something in this house.

—Mikovits's stepfather as he lay dying of aggressive prostate cancer

Saturday, November 19, 2011

At a quarter to six in the morning, the tower guards pealed over the loud-speakers, "Breakfast will be served in fifteen minutes." A speaker in each cell linked the prisoners to the main tower, a squawking, ever-present reminder of the institutional setting. Messages could be dispatched to all cells or to an individual one. Each prisoner quickly learned a Pavlovian response to the noise: she was supposed to be fully dressed, have her bunk made, wrap her small towel around her neck, make her booking notice visible on her arm, and be facing the far wall opposite the door with her nose actually touching the wall.[1]

Mikovits couldn't recall what had happened in the moments before the guard jostled her out of the dreamy state where she could forget where she was for a minute. Maybe she had drifted off to sleep again. Marie was preparing for the roll call so Mikovits hurriedly swung herself down from her bunk to do the same. The previous night she had been issued a bar of soap slightly larger than a silver dollar as well as a single aluminum foil packet of toothpaste. She quickly used both of them, not realizing these were the only toiletries they would provide to her for the next five days.

Not realizing the protocol, Judy faced the door as she waited and noticed the thickness of the steel entry to their cell and the small window, no more than eight inches long by six inches wide, which allowed the guard to peer in and see if they were ready. She felt a tap on her shoulder and turned to see her cellmate motioning her to the back wall. "You need to do this," Marie said, turning to face the wall with her nose touching. "Trouble for all today if you do not do this." She exaggerated her face and gestures.

The guards came by, eyeballing her booking number and attire. She passed muster, and she and Marie were perfunctorily led out of their cell and lined up against the wall until the guards were finished checking everybody. When all of the prisoners were assembled, they were marched to breakfast. Mikovits took a moment to observe the strategic layout of the facility. Each cell block was like a triangular-shaped petal attached to a center, with the cells at the far wide end narrowing down to a single exit which led into a large circular room that was the center of the "flower" of these cellblock petals around it. The circle served as their "free time" room as well as where they got their meals before returning to their cells to eat. The "tower" was in the middle of the four "petals" that circled it, allowing the guards to see the individual cell-blocks as well as the free time areas.

The breakfast that morning was powdered eggs and either French toast or pancakes, which looked like it had come out of the freezer and had been thrown in a huge oven on a sheet tray. She took the filled plate, but did not plan to eat, especially not the eggs, which looked about as appetizing as decades-old food in a bomb shelter. She expected to be able to post bail by the end of the day and be released. As she looked at the reconstituted eggs on the sad-looking paper plate all she could think of was how happy she was that her last meal before incarceration had been a wonderful breakfast burrito at Mrs. Olson's Coffee Hut. Mrs. Olson's was renowned for burritos so big you could slowly excavate them for a week (or she had joked), with fresh ingredients and homestyle flair: heavenly by comparison.

When the prisoners were done masticating the monotonous food, they pushed their trays out the door as directed. Other prisoners on their "free time" would come by to collect the trays and dump the paper plates in the trash.

At around eleven in the morning the speaker squawked, "Mikovits! You have a visitor! Prepare yourself!" Mikovits started to head toward the back wall in order to put her nose against it when her cellmate stopped her. "No, you stand at door, hands behind your back," Marie said, modeling the proper posture. "And you walk to visiting area, same way. Hands behind back, no handcuffs."

"Oh, okay." Mikovits assumed the position and looked to her cellmate for confirmation.

"Okay." Marie said with a validating nod. The female guard opened the door, did a cursory check of Mikovits, and, gesturing for Judy to walk in front of her, motioned her forward to the visiting area. As Mikovits walked in front of her, the guard pointed to the signet ring on the middle finger of Judy's right hand and snapped. "No jewelry, inmate!"

"It's been on there a long time. Probably longer than you've been alive," said Judy. It was her signet ring from the University of Virginia at Charlottesville. One day in 1981, after hours in a cold room purifying natural products for cancer therapies, it had fallen off again, and in a fit of pique Judy had slipped it over her middle finger where it had remained snugly attached to that day. The guard was one of many people who claimed they had a "trick" for getting rings off, normally involving twisting the metal in a certain way or using a special cream or lotion. The guard left for a moment and returned with one of her "magic" creams. She put it on, and tried to twist and yank the ring off of Mikovits's finger. Mikovits's knuckle became red and inflamed as the ring tightened like gag Chinese finger handcuffs. "It has to come off," said the guard after several minutes of trying. "This is for your own protection. No telling what a prisoner might do to try and get it."

Mikovits understood the threat of harm coming to prisoners wearing jewelry, but there was no getting that ring off her finger. For Judy it was a small triumph; a way to retain her own identity as an honest scientist with hard-working roots. Finally, the guard capitulated and led her in haste to the visitors, area. As Mikovits walked down the cellblock in front of the guard she could see a few other prisoners getting ready. Visiting times were Friday, Saturday, and Sunday from 7:30 a.m. to 5:00 p.m. with cellblocks available for rotating in two-hour blocks and a half hour between each block. The routine enforced both compliance and a sense of drudgery.

Family members wanting to visit a loved one were advised to call before showing up at the facility due to the fact that prisoners were often reassigned to different cellblocks with different visiting hours. Inmates were allowed two scant half hour visits per week and at each visit were limited to having two adults or one adult and one minor child. Any person entering the facility to visit a prisoner was subject to a search.[2]

The rigidity of the visiting area shocked Mikovits: it seemed as obdurate as some of the prisoners. The walls were obsidian-black stainless steel and in the visiting booths there was a single chair with a phone on the wall next to it. Mikovits went to the booth where David was waiting. She tried to give him a brave smile as she lifted the phone. "I'm doing okay," she quickly mouthed into the receiver. There was no audible reply from the other end, although she could see David's lips earnestly moving. Mikovits turned to one of the guards. "The phone isn't working," she complained.

"Read the directions," the guard replied in a surly voice.

Mikovits leaned close but found that without her reading glasses it was impossible. Like a lot of people gradually losing eyesight, she made do by picking up drugstore reading glasses. Her prescription was +3.25 and she knew that soon she'd need something stronger from an optometrist. All those years of peering through microscopes had taken their toll. "I can't read it," said Mikovits. The guard exhaled in frustration, as if he heard illiteracy complaints several times a day. "If you can't read then your cellmate should be able to read the directions for you and tell you what to do."

"No, I can *read*, it's just that I don't have my glasses."

"You just pick up the phone, dial in the four digit number they gave you when you came in here, followed by your booking number, and the call will go through."

"What four digit number?" Mikovits asked.

"The one you got when you came in here."

Mikovits remembered the small stack of papers she'd been given when she was processed at two o'clock the previous morning. It must have been in there. "I think I left it back in my cell."

"Then I guess you'll just have to have your visit tomorrow."

The moment was too much for Mikovits to endure. She started to cry. How could she be in such close proximity to David, be able to see him, and yet not be able to talk to her husband and comfort him? He looked haggard, his face ashen: the situation was obviously taking its toll. She knew he was probably worried sick that she'd been abused in jail, but that was the least of her woes. She also wanted to find out how he was faring through the shock and stress. He might not even know that she wasn't going to be carted off in the early morning hours to Reno, that she was safe for now in Ventura. This was an upright, law-abiding man, and now he had a wife in jail.

What had she done wrong?

Nothing. She had tried to help the patients and protect the government's money.

The guard motioned for her to stand and put her hands behind her for the walk back to her cell. David looked befuddled and helpless behind the glass.

She walked out of the visiting area, down the cellblock, into her cell, and heard the steel door cut through the air behind her, sounding menacing and final in the way that only steel can.

* * *

David was determined he wasn't going to leave the jail without talking to his wife. After Mikovits had been taken from the visiting area, David went to talk to one of the guards. He explained in his disarming manner that she didn't have her reading glasses and was unfamiliar with the jail procedures. Couldn't they bring her back one more time so she could actually speak to him? Meanwhile, Marie watched Mikovits come in and saw the distraught look on her face. When asked, Mikovits told her cellmate what had happened.

"You cannot see?" Marie asked.

"Maybe if the numbers were big enough."

Marie looked through Mikovits's papers, found the four digit number and her booking number from her wrist band, and wrote them on a piece of paper large enough for Mikovits to read. "Can you read that now?"

"Yes. Yes, I can," said Mikovits. She thought that tomorrow she would get it right.

A few minutes later, though, the guard in the tower made an announcement over the loudspeaker: "Mikovits, you have a visitor."

"Okay, let's try it again," said the guard when he arrived. Mikovits put her hands behind her back, the guard entered, and they made the walk down the cellblock again. When she entered the room again and saw David, she felt triumphant. She sat down, picked up the phone, held the paper to her face, and dialed the numbers. The phone rang on David's side of the glass and he picked up. "How are you?" he asked gently.

"I'm fine, honey. I can handle this. I've done nothing wrong." She really was resolute and at peace. She had been shocked by the arrest but felt the real danger was in going back to Nevada. She worried that if she was under Harvey's almost-despotic control over his lackeys in the Silver State, there was no telling what might happen. He had extended his far-reaching influence into California to detain her, but folks in the Golden

State weren't his flunkies. They would take the orders he had padded with official-sounding legalese, but they wouldn't do his bidding.

"Good. Because I'm trying to make things work out here. I got the bail, but I don't know what's going on. It's very strange. I don't understand it. I'm talking to Harvey, and Frank is helping as well. Frank is going to talk to Harvey. It's all going to be okay." They spoke hurriedly for a few more minutes and then their time was up. Fifteen minutes passed in a flash after one was locked up in jail for nearly a day. When the guard came this time, Mikovits felt serene. She was back in control of a destabilizing situation. David was working on her behalf and Frank also had her back. They were trying to reason with Harvey. They would all figure out a solution to get her out.

Back in the cell, Marie was sleeping heavily. All that Mikovits could do was crawl onto her bunk and gaze at the ceiling. Since she didn't have her glasses she couldn't read anything. She could only lie there, making her think of the ME/CFS patients who had to stay supine and reduce neurological stimulation, trapped in a liminal world of rest. At around four o'clock prisoners were taken out of their cells for "free time," which usually lasted about four hours. The circular area, which contained the "view tower," also had a room with concrete floors, steel picnic tables bolted to the ground, carts with books and magazines, and a small television mounted high on the wall, about ten feet up to protect it from the inmates. Without her glasses, she couldn't read or follow the football game that was playing on the TV: it was as if she was in a foreign land.

She tried pacing around the room, hoping the movement would wear her out a little bit for better sleep, but she still wasn't clear on all the protocols and knew that she had to watch the others for cues. When the door leading from the tower opened up and a guard walked in, all of the prisoners were supposed to immediately find a seat. A few times Mikovits found herself the only person standing, not realizing why everybody else seemed to be playing an incomprehensible game of musical chairs. Prisoners were also taken out of their cells for lunch and for dinner. Although she didn't eat any of the lunch, she figured she should try to have some dinner, even just a slab of mystery protein to keep her strength up. Dinner was a hamburger and she took it, sliding the meat off of the bun and eating just the patty. It tasted like the cardboard end of an old oatmeal cylinder, but she had to eat something.

* * *

After dinner, and with her cellmate sleeping again, Mikovits found her thoughts drifting back to the chain of events that had caused this calamity. She tried to think positively about Harvey Whittemore but her anger bubbled to the surface. How could he do this when they had so recently been allied for the same cause? How could he treat her like family and then do this? Hadn't they learned who she was over the past five years? She felt the walls of the cell closing in again. Then she prayed for Harvey and Annette to come to their senses and stop this madness.

Mikovits remembered her last discussion with Annette Whittemore a few months before at the Ottawa Conference in September of 2011. "I'm not going to be a part of this," Mikovits had said after explaining again that the XMRV diagnostic test wasn't clinically validated and thus they shouldn't be selling it prematurely.[3]

Mikovits told Annette that as the Principal Investigator on their grants, she was more than willing to take responsibility for the previous mistakes. "But that's in the past. I won't take responsibility for failure to take direction," she continued. "We need to do the right thing, stop everything until we figure this out!" Those were the last words she said in person to Annette.

Mikovits realized her words and tone to Annette had been harsh, but she knew they were honest. If she could go back and soften them she probably would have done so, but her underlying message would remain unchanged: They could not in good conscience sell that ineffective test.[4] And the question that remained unanswered, which the scientific community had little interest in investigating, was why all these people were sick with diseases like cancer, ME/CFS, and autism.

To Mikovits, the path was a straight line. A scientist should investigate human disease, try to unlock its secrets, and when she does, figure out the quickest course to bring relief to those afflicted. All other issues were of secondary concern.

The important question was how to deal with those who were sick now.

Anything which kept patients with ME/CFS a single day longer in their darkened rooms, which kept children with autism from being able to speak the thoughts that flashed through their overexcited brains, or that put an honest scientist in jail could only go by a simple name: evil.

CHAPTER FOUR

A Retrovirus in Chronic Fatigue Syndrome?

It's a classic gamma retrovirus, but it's totally new. Nobody's ever seen it before. Its closest relative is, in fact, from mice, and so we would call this a xentotropic retrovirus, because it's infecting a species other than mice. . . We've done it for many patients now, and we can say they're all independent infections.

—Dr. Joseph DeRisi: "Hunting the Next Killer Virus," February 2006, Monterey, CA, TED Talks[1]

Incline Village, Nevada—Summer and Fall 2006
The days in Nevada fell into an efficient routine. Mikovits saw an average of five to six of Peterson's patients a day at Sierra Internal Medicine.[2] She and her team randomly numbered and entered every patient into the database to blind the clinical information they took next and then they drew about 30 to 40 milliliters of blood. They separated and aliquoted (portioned) the blood into plasma, sera, and pellets for DNA analysis. After that, they mixed the samples with trizol, a chemical solution that preserved the nucleic acids and proteins when they were frozen.

Trizol had been in use since 1987 and for many researchers it became the preferred method for RNA/DNA/protein extraction. Even though trizol had an extremely potent and foul smell, Mikovits preferred it because it maintained RNA and protein integrity while ripping apart cells and cell components. One could feel comfortable going back to a sample preserved with trizol years and even decades later, and when it was unfrozen one would see little, if any, degradation of the nucleic acids or proteins. To do longitudinal research on an epidemic such as ME/CFS, this meant that patient's blood could be captured like a snapshot in pathogenic time.

In addition to preserving the samples in trizol, Mikovits's team would take some cells and preserve them in DMSO (Dimethyl sulfoxide, used as a cryprotectant to prevent ice formation that would cause cell death) and place them in liquid nitrogen so they could be thawed at a later date for cell growth in cultures. Peterson's patients were scheduled to return to see him every three months and this gave Mikovits the opportunity to collect more clinical data on each patient with additional blood. She hoped to eventually have five or six samples from each person as his or her symptoms waxed and waned, spanning seasonal patterns, life stresses, and additional variants. This was important in pathogen hunting since it helped override the natural inclination of pathogens to evade detection as well as providing samples to reproduce any results.

In addition to taking the clinical information and placing samples in the new and improved repository, Mikovits had them tested for cytokines (cell signaling molecules that can create a disease "fingerprint" of sorts), put them on the flow cytometer to determine types of immune cells, and looked at white blood cells under a microscope. Mikovits also used donations to purchase an expensive database also used in major hospitals like Sloan-Kettering Cancer Center in New York City (the world's oldest and largest private cancer center). The database would track the master sample and then the aliquots (a portion of the original sample) that were utilized in every study. This allowed samples to be traced back to the exact aliquot from the master sample. Again, because preservation was everything, the master sample would remain untouched by anything that might contaminate it.

Mikovits was determined that nobody would ever look at her repository and question its organization as she had done when Byron Hsu had led her to Peterson's original repository.

* * *

"It was like the great plague was moving from town to town, striking every six months," Paul Cheney later recalled. "First, there was Lake Tahoe in

1984–1985, then the small town of Yerington, Nevada, forty miles to the southeast, then six months after that, Placerville, California, forty miles to the southwest."[3]

This new illness was a freak show of physical abnormalities, parading its strange, markedly odd symptoms along a slow route of quiet towns around Lake Tahoe and the Sierra Nevadas near where the famous Donner party met in 1846 with horrific catastrophe including starvation, exposure, disease, hypothermia, and cannabilism, as they were cut off from both the East and the West.

Eleven years after these outbreak, journalist Hillary Johnson published the definitive narrative of this modern calamity in a 700-plus page book called *Osler's Web: Inside the Labyrinth of the Chronic Fatigue Epidemic*. Johnson chronicled the government's lackluster, oftentimes nefarious, response to the then-emerging disease, beginning with a near day-by-day recounting of the outbreak in Incline Village, Nevada. Johnson's description of the extraordinary lengths to which Cheney and Peterson went to unravel the mysteries of the disease is worth revisiting because the doctors ultimately arrived—much like Mikovits two decades later—at a retroviral hypothesis.

As Johnson reports, Peterson was a well-trained internal medicine specialist who set up a private practice in Incline Village after paying for his medical education by serving as a clinician to an impoverished population of migrants and farm workers in Idaho.[4] Cheney, at the time, was nearing the end of his Air Force enlistment and serving as the chief of medicine at Mountain Home Air Force Base Hospital.[5]

After learning that Peterson was looking for a partner, Cheney decided to visit the Peterson clinic for a week, discovered that he liked the picturesque area and the surprisingly sophisticated practice, and joined Peterson in October 1983.

The exceptional natural beauty of the region was a given and the economic opportunities were abundant, especially since Cheney and Peterson were the only two board-certified physicians in Lake Tahoe.[6] The doctors ran what many locals considered the "powerhouse practice" in town, according to Johnson. Rapidly, they became the doctors of choice in the exclusive community where most people had the ability to see any doctor in the world.

In October of 1984, however, Cheney and Peterson began to see an unusual group of patients, starting with the girls' basketball team of a local high school, several of whom complained of what seemed at first to be a severe case of mononucleosis.[7] In March of 1985, Johnson writes, the doctors began to witness large numbers of adults coming down with similar complaints, including teachers who shared a faculty lounge.[8] Most of these patients were in their thirties, which was atypical, as mono generally strikes adolescents and young adults (hence why it was nicknamed the "kissing disease" due to salivary transmission, though any contact with infected saliva can cause the disease).[9]

The numbers started to rise exponentially in May to June of 1985, and the two young physicians were astonished by what they saw. In a series of late-evening conversations, as described in *Osler's Web*, the two were forced to face the obvious: they were witnessing a very large epidemic among formerly healthy adults, with an average age of about thirty-eight years old, occurring within a small geographical area.[10] Initially, the disease *resembled* mononucleosis. Patients had sore throats, swollen glands, fever, large spleens that could be felt, and atypical lymphocytes (lymphocytes are immune cells that fight foreign invaders) on peripheral smear tests.[11]

In 2013 Cheney recalled, "Some cases looked like classic mono and in others somewhat different than mono. Something of an encephalitic onset, with severe pressure headaches, light sensitivity, disorientation, and vestibular problems."[12]

While Cheney observed that the cases would initially resemble the flu or mono, when the acute phase ended there came "this extraordinary fatigue, causing them to be unable to function, and in addition, lots of cognitive issues. Trouble with word search, finding their way through traffic in a one horse town, forgetting things, having to take notes, not being able to watch TV programs or read books because they couldn't follow the plot."[13]

Eventually, Johnson continues, the doctors concluded that whatever had been in their patients' bloodstream "had now invaded their brains." But what was it? Nothing in their medical training or clinical experience, Johnson noted, had prepared the doctors for anything like this devastating disease.

* * *

A critical turning point in Johnson's narrative occurred when Cheney and Peterson sent their blood samples to Susan Wormsley, a biochemist and flow cytometry expert at Cytometrics Laboratory in San Diego. Flow cytometry quantifies and qualifies the condition of the immune system cells.[14] Wormsley told Johnson:

"Right from the time we separated and stained the cells, we saw a lot of debris," she said. "Just broken-apart cells, *pieces* of cells and platlets. And we don't see that in anything else that gets sent to us. Now, naturally, with everything that is sent to us, the people are sick, and most of them have cancer-leukemia, lymphoma—but we didn't see this kind of debris except in these patients."[15]

As Wormsley explained to Johnson, the answer was clear. Some type of virus or toxin had to be killing the cells of these patients.[16]

But there was another issue. In order to perform one of the tests, the kappa/lambda test, the researchers needed a good portion of B cells in their samples.[17]

"Right from the beginning," she tells Johnson, "these people seemed to have extremely low percentages [of B-cells], sometimes only one or two percent of their white blood cell population instead of the eight to twelve percent that we normally see. I noticed it because with a normal person, ten milliliters of blood gives you plenty of cells to do the entire assay. But I wasn't able to get enough B-cells to feel comfortable with Paul's patients."[18]

Another abnormality found in these unusual patients involved the ratio of T-cell subsets, Johnson writes.[19] T-cells are an immune system cell which regulates production of disease fighting antibodies. Scientists recognize two distinct types of T-cells, "helper" T-cells which boost antibody production, and "suppressor" cells which suppress antibody prdocution.

The ratio for a healthy person is one helper T-cell to every two or three suppressor T-cells. The helper cells are like new police recruits anxious for action, while the suppressor T-cells are like the seasoned captains who would be more cautious in the use of force. With AIDS, HIV piggybacks onto the CD4 receptor on T-cells and works its way into the cells, turning them into virus factories to reproduce itself and render the cells useless against infection.

From *Osler's Web*:

"One of the most striking immunological aberrations Wormsely observed, however, was abnormal ratios of T-cell subsets. T-cells are a major category of immune system cell: they regulate production of disease-fighting antibodies. Two primary T-cell subsets are "helper" and "suppressor" T-cells, which boost and suppress antibody production, respectively. In AIDS the normal ratio tends to be dramatically skewed in favor of suppressors. Since this finding is virtually diagnostic of AIDS, Cheney and Peterson were curious to know the T-cell subset profile in the Tahoe malady.

Wormsley's results showed that four of five Tahoe patients did have abnormal helper suppressor ratios. But unlike the ratios in AIDS sufferers, they were low in the number of suppressor cells. Instead of one-to-two or one-to-three, which are typical of healthy people, the Incline patients had helper-suppressor ratios of five-to-one, ten-to-one, and higher. *It was the mirror image of AIDS.* (italics by authors)"[20]

Could there be a more frightening description of a disease? In AIDS the immune system was told to essentially "stand-down," allowing all sorts of pathogens to run wild. By contrast, in ME/CFS it seemed that the immune system was instructed to go into "full-attack mode," indiscriminately wiping out the good with the bad.

The delicate balance of the immune system had been broken.

* * *

Cheney and Peterson investigated whether this low suppressor/high helper T-cell ratio had been observed before and learned it was extremely rare.[21]

"Peterson and Cheney began using the curious helper-suppressor ratio as yet another laboratory abnormality—in addition to the abnormal Epstein-Barr virus antibody profile—to support their diagnosis of the disease. It was a fragile stand, they knew, and they were uncertain of its significance. Yet, it was real, and it was not normal."[22]

The elevated rate of another rare lymphoma, Mantle Cell Lymphoma (MCL), increased Mikovits's curiosity about ME/CFS when she saw Peterson's presentation at the HHV-6 Conference in Barcelona, Spain. Mikovits would come to believe that the development of lymphomas represented a long-term progression and final outcome of the cellular damage wreaked by a retrovirus.

The same idea had occurred to Cheney decades earlier, when HIV/AIDS was still a largely untreated pandemic. The saying, "If it looks like a duck, quacks like a duck" surely applied to what they were witnessing. Both of them could see a retroviral duck, one that quacked too loudly for a dissimilar decoy: now they needed to find the location of the pond.

* * *

Cheney wasted no time pursuing the retroviral hypothesis. The doctor knew that there was a commercial test for one other human retrovirus besides HIV, and that was HTLV-I, the human T-cell leukemia-lymphoma virus I, discovered by Frank Ruscetti and Bernie Poiesz in Gallo's lab.

Cheney sent five samples to a Specialty Lab in Los Angeles, Johnson writes, noting that the samples came from five patients who could not be classified as among the worst or the least affected, and two of them were teachers from the outbreak at an elementary school. The prevalence rate in the North American population for HTLV-I was predicted to be about .031 percent, but *four* of the five samples (80 percent) turned out to be positive.

Cheney and Peterson thought they had landed their retrovirus. But as reported in *Osler's* Web, a second and a third test of the samples turned out negative, raising the possibility that their first results were false positives. Cheney was undeterred, certain he was on the trail of what caused the disease. One of his colleagues referred him to Dr. Elaine DeFreitas of the Wistar Institute, the nation's first independent biomedical research facility, founded in Philadelphia in 1892.

DeFreitas was initially reluctant to enter into a new collaboration. During an interview with Johnson, she described her early interactions with Cheney:

> "What he was telling me about these patients was fascinating, and his theories and his experiments were very interesting. And he seemed absolutely convinced that there was a retrovirus involved in this. And I would tell him, 'I'm [working] on multiple sclerosis. I can only handle one disease at a time.' But he was so persistent. Finally I decided that the quickest way of shutting him up was to take five or six patients, run the samples—to do what I could with them here. And they were all going to be negative—I was absolutely convinced of it."[23]

It took several months before Cheney heard back from DeFreitas. He learned that his samples had been passed along to the Gallo lab at the National Cancer Institute. Gallo's lab was puzzled because they had been finding antibodies to a new virus in a handful of different illnesses. When they got to Cheney's samples, they found all of them were infected by this new virus.

The new virus was human herpes-virus #6 (HHV-6) and Gallo claimed its discovery in the journal *Science* on October 31, 1986.[24]

As mentioned before, HHV-6 confused scientists for the next twenty years as to whether it *caused* disease or was just one more pathogen that appeared like wildfire in the system of these immune-compromised patients.

It might be like the causative virus of Kaposis's sarcoma (Human herpes virus-8) was discovered in men with HIV infections who also had KS, just as it previously had been found in the aging, dysfunctional immune systems of certain old men.

HHV-6 was the virus Mikovits would hear about in a yacht club in southern California in 2006, forcing her to abruptly change course like a mariner heading into a perfect storm.

* * *

Paul Cheney first met Judy Mikovits on a mountaintop Buddhist retreat about two and a half hours north of San Francisco.[25] Nestled in the California wine country, the Ratna Ling (Buddhist for "health") Retreat Center is located on 120 acres of redwood forests in the coastal range, about a mile away from the Pacific Ocean. Among the towering redwoods and fog-shrouded ridges of Sonoma County, Ratna Ling plays host several times a year to a unique gathering of scientists and medical researchers.

The driving force behind this series of ongoing conferences under the gorgeous post and beam space of Ratna Ling was a wealthy San Francisco businessman who converted to Tibetan Buddhism and then poured money into this small heath retreat. When his wife came down with ME/CFS, he formed a consortium of the best minds in related medical and research fields and invited them to Ratna Ling. As a business big shot who had created a successful algorithm for the world's economy, his guiding principle was to get smart people with diverse ideas into the same room, and let those ideas percolate.

Cheney recalled, "The first conference was thirty people and it was one of the greatest intellectual experiences of my life. I got to see in front of me all the best thinking in one room. It was very collegial. It was a Buddhist temple: you had to take your shoes off, and we ate vegetarian meals. There were no cell phones, televisions, radios, or anything. Each of us presented for about twenty minutes. Then people around the room would make comments and have a discussion. It was very respectful of other views, even if they were the exact opposite of yours.

"And I learned a lot. I became comfortable with certain aspects of what I believed, and saw flaws in my own thinking. It was wonderful. This was repeated every few months. The next one was on Lyme disease, the one after that was autism, and then there was one on autoimmunity. If you were invited to any single conference you could attend any of the other ones.

I attended the one on chronic fatigue syndrome and the one on autism because I had findings on autism to share. Judy spoke at the chronic fatigue syndrome conference, and that's where I met her."

Cheney went onto say he was respectful of Mikovits's passion, intelligence, and knowledge of the field.[26] They would continue to stay in touch and he would be one of the first people Mikovits would tell when something very exciting had been found in the disease which he had studied for so long.[27]

* * *

At the Ratna Ling Conference on chronic fatigue syndrome, Cheney also met Dr. Joseph Burrascano, a prominent Lyme disease expert who co-authored the International Lyme and Associated Diseases Society (ILADS) guidelines for treatment and diagnosis of Lyme and related coinfections. Cheney rapidly bonded with Burrascano, respecting the physician's intellect as well as his passionate but even temperament.[28]

Burrascano told Cheney that he had treated acute Lyme disease on Long Island where the disease is highly endemic, for a decade between 1970 and 1980. The patients presented with the standard known tick bite, bullseye rash, and acute illness that resolved with the medication Doxycycline. Not one of these cases advanced into chronic Lyme disease in those days. But around 1980, something changed in his practice. Patients started getting a new disease, a *chronic* Lyme disease that did not go away.

It reminded Cheney of a time in 1987 when he attended a conference in Portland, Oregon, with about fifteen other clinicians treating ME/CFS patients.[29] The moderator asked the clinicians to raise hands to note the year when they first began to see large numbers of ME/CFS cases in their practices. The moderator called out 1978 and nobody raised a hand. The same thing with 1979: not one hand. For 1980, two hands popped up. One was from a clinician in the San Francisco Bay Area and the other was from New York City, the two areas that were the epicenters of the AIDS epidemic.

This revelation made Cheney think more deeply about the pathogenesis of ME/CFS. What did it mean that San Francisco and New York City presented with the first increases in these illnesses? Aside from high rents, coastal proximity, international airports, large diverse émigré populations, and foodie cuisine, what did these two cities have in common? It was mysterious.

Most viruses strike the young or the old, but whatever was at the heart of this disease was striking those in adolescence and early adulthood. Few over fifty years of age came down with it, and it certainly wasn't rampaging through nursing homes. Kids in elementary schools were not sneezing on each other's milk cartons and causing large outbreaks in their schools. Whatever was initiating the problem seemed to require a robust immune system to counterattack it, which was similar to the profile of those struck down by mononucleosis (caused by the herpesvirus, Epstein Barr), hence perhaps they were witnessing a herpesvirus or the destruction of an otherwise-intact adult immunity. Cheney felt it was likely that ME/CFS was an epi-phenomenon of the AIDS epidemic. He considered HHV-6, type A, a likely causative candidate, but also thought HHV-6A may have recombined to some degree with the HIV retrovirus. Both HHV-6A and HIV could readily infect the same type of cell.

Cheney was also intrigued by a paper he read from Dr. Patricia Coyle of the State University of New York at Stony Brook in which she compared many of the features of chronic Lyme disease and ME/CFS and found that a large number of the clinical markers of the two were indistinguishable.[30]

Many autism doctors later asserted that Lyme played a significant role in autism[31] and Cheney soon saw an overlap of both of these conditions with ME/CFS.

In February of 2011, however, Dr. Steven E. Schultzer's research team of New Jersey along with Dr. Richard D. Smith of the Pacific Northwest National Laboratory published in the journal *PLOS ONE* that chronic Lyme disease (which they called Neurologic Post Treatment Lyme disease) and ME/CFS patients presented with unique, measurable proteins in their spinal fluids—different from each other, and different from controls.[32] Was it possible, then, that Lyme (caused by a spirochete) was not unlike syphilis (caused by a spirochete) in untreated AIDS—a coinfection that furthered the retroviral damage?

Of what transpired with Mikovits and the controversy over XMRV, Cheney would say, "I truly do believe with the results she got from the spleen-focus forming virus antibody test [designed to detect known and undiscovered mouse xenotropic retroviruses], that there's something there. We just don't know what it is. I suppose it's possible that it's a human endogenous retrovirus that's active, making an envelope protein and a gag protein. It could be a novel gamma-retrovirus, yet to be discovered. It could be a piece of HIV stuck in a herpesvirus, making it a cross between a retrovirus and a herpesvirus. The most likely virus for that would be HHV-6, since it integrates into the human DNA, just like retroviruses do. So we could have a cross of some kind."[33]

"I don't know what this is," Cheney said. "But I sure as hell know what this disease is. It's very coherent. You can measure it by objective standards that are not found in control groups and cannot be explained by normal medicine."[34]

* * *

Even before Annette Whittemore met Mikovits, she was interested in Dr. Joseph DeRisi's invention called the ViroChip[35] (which had first identified XMRV in the tumors of prostate cancer patients in 2006[36]) for the ME/CFS patients, including her daughter.[37]

The ViroChip was a single diagnostic assay that contained a genetic sequence for every known virus and that also had the ability to identify novel viruses. It had evolved out of a marriage of virology and technology and was expected to more rapidly advance the science of viral detection. DeRisi had discovered XMRV along with Dr. Robert Silverman of the prestigious Cleveland Clinic, who would later come to play a crucial role in their work.[38]

Annette told Judy she had written a letter to DeRisi inquiring about using the chip, but she never received a reply. After Mikovits started working with Peterson, Mikovits revived the ViroChip idea, but the two groups were unable to come to an agreement.

At the National Cancer Institute, there was a similar chip developed by the molecular technologies program. Ruscetti had access to it and the NCI was fortuitously looking for investigators willing to try it. This option seemed like a good opportunity, but one the research team would later come to regret.

There were entire virus families that were not represented on the chip, including some in the XMRV family. A retrovirus virus might have something like 9,000 base nucleotide pairs in its genetic makeup, some of which were quickly evolving and others which were not. Among animals, even small genetic variations create wildly different species. For example, a human and a chimp vary in their genetic makeup by approximately 2 percent, but nobody would ever identify them as the same species.[39]

However, an RNA virus can vary in its genetic makeup by a much greater percentage and still be considered the same species. The different subtypes of HIV differ by 25–30 percent while the viruses in a specific subtype themselves can typically differ by 10–15 percent.[40] A typical assay might have only 50–200 base nucleotide pairs in its sequence to determine the identity of a certain virus. It was therefore critical in creating an accurate assay to determine which stretches of the viral genome were not evolving, or—in the parlance of virologists—which were *conserved.*

About a year after they settled on the NCI chip that first summer of 2006, Mikovits did some additional research, talking with Dr. Ian Lipkin, the head of Columbia University's Institute of Infection and Immunity.[41] Lipkin thought that both the NCI and DeRisi chips were flawed. Lipkin articulated at length about the technology he had developed and why it was better than either chip.

Even though Mikovits did not understand every technical detail, she was able to discern that Lipkin did indeed have the superior technology.[42] But their experiments were already well underway and it would have been unethical for Lipkin to intervene in the research at such a time. In retrospect, Mikovits wished she had learned about Lipkin's technology before she used the NCI chip with Peterson's top 70 samples and as a component of the Integrative Neural Immune Program's (INIP) Intramural Research award investigation. This knowledge might have radically altered the course of events.

Dr. Dennis Taub and Frank Ruscetti cowrote a grant proposal for the INIP program, looking at the role of chronic inflammation stimulation by an active herpesvirus infection, and what role this might have in the development of immune dysfunction and mantle cell lymphoma in ME/CFS patients. The work would be done with the samples Mikovits had prepared for Peterson the summer of 2006. The proposal was awarded three years of funding in September of 2007—six months before she would discuss chip technology with Lipkin.

Less than a year since Mikovits set forth her plan, a new era of research into ME/CFS was barreling ahead.

* * *

As Mikovits worked on building a program to unravel the mystery of ME/CFS, she became acutely aware of the influence wielded by the Whittemores in Nevada. In the fall of 2006 the Whittemores threw a themed "I Hope You Dance Fundraiser" (since few ME/CFS patients could dance any more) at the Peppermill Casino in Reno, a coffee shop turned gaming center when the owners partnered with the Seeno family in 1979. The event featured the legendary singer Joan Baez. The event was populated by Nevada royalty: most of the state's congressional delegation, the current and former governor of Nevada, and US Senate Majority Leader, Harry Reid, with his wife and four sons. Senator Reid received a special award for his support of the institute and Joan Baez gave him an autographed guitar from her collection.[43] Future honorees included Governor Kenny Guinn, State Senator Bill Raggio, and Ron Parraguirre, who would be appointed Navada Supreme Court Chief Justice in 2010.[44]

In late fall of 2007, Mikovits organized the WPI's first Scientific Retreat hosted by the Whittemores at Red Hawk for a number of notable scientists in the field, including Drs. Nancy Klimas and Suzanne Vernon, as well as others from the UNR, many of whom possessed appropriate skill sets to figure out how to attack this problem. The group included immunologists, virologists, microbiologists, and any other field that Mikovits felt might conceivably be able to offer some assistance.

Afterward, everybody drove out to the Whittemore's Lakeshore house. There at Glenbrook, Mikovits was first formally introduced to Harry Reid, who seemed to be an unpretentious man. He didn't overpower a room the way Harvey did, and it wasn't just because Harvey towered over the Senator in height.

Mikovits was not interested in politics. Her husband David did like rousing conversation about politics and history, and thus was thrilled to engage with the Senator. Over the years that Mikovits worked in Nevada, the Reids were a common fixture at Whittemore events.

* * *

In 2005, prior to the arrival of Mikovits, the Whittemores had persuaded the Nevada Legislature to unanimously pass Senate Bill 105, which appropriated ten million dollars to the University of Nevada, School of Medicine, the Nevada Cancer Institute, and the Whittemore Peterson Institute.[45] Harvey pledged to contribute two million dollars to this effort as well as raise an additional two million from friends and associates. These donations comprised the initial seed money for what would later become the Center for Molecular Medicine at UNR, to house the WPI.

In 2007 Mikovits joined Harvey and Annette in advocating for and securing passage of Senate Bill 443, which set aside two million dollars for the construction of a facility dedicated to the research and treatment of neuroimmune disorders, as well as providing money for equipment and furnishings.[46] The legislature also provided funds for continued operational funding for the WPI in their legislation.

On January 14, 2008, Annette Whittemore wrote a glowing letter to Mikovits on what had transpired in the past year and a half, noting that Judy had taken the program from a small, one-room office to a first class research laboratory.[47]

Annette's tone belied her genuine warmth for her friend and colleague and—as she would have put it—her new family member. Mikovits also received a check for twenty thousand dollars, her annual bonus. She felt she had earned every penny of it.

* * *

Another intriguing clue to unraveling the riddle of ME/CFS came in a paper published in 2005 by Belgian researcher, Dr. Kenny de Meirleir.[48] De Meirleir reported that the RNase L enzyme in ME/CFS patients wasn't working effectively. This meant that patients would have a reduced ability to defend against an RNA virus or retrovirus.

It was a simple, yet crippling defect in the immune system.

Mikovits and her team worked in 2007 and 2008 on developing tests to monitor cytokine and chemokine (a family of small cytokines) levels of the patients. Unlike in HIV, T cells from ME/CFS patients were not dying in culture. It was more like HTLV-1 which could be latent in T cells for decades without killing them and still cause a neurological disease, HTLV-1 associated myelopathy/tropical spastic parapersis (HAM/TSP.) Only rarely (<5 percent of the cases) did HTLV-1 infection *transform* the T-cells, causing leukemia or result in HAM/TSP. Now they had found two ways in which ME/CFS patients' immune systems were dysregulated similar to patients with HAM/TSP, the dysfunction of Ruscetti's plasmacytoid dendritic cells and the inflammatory cytokine signatures.

Her perspective began to solidfy in October of 2007 when she and a postdoctoral researcher who had recently graduated from the UNR—Dr. Vincent Lombardi—attended the Michael Milken Prostate Cancer meeting in Incline Village.[49]

Lombardi had already received a fifty thousand dollar grant to study RNase L and prostate cancer from the Nevada Cancer Institute. At the Milken meeting in October 2007, he first met Robert Silverman of the Cleveland Clinic (the scientist who is credited with discovering XMRV in prostate cancer cells) and discussed the research with him. In March of 2008 Silverman sent him some reagents including the plasmid containing XMRV clone to test his material. Mikovits had helped Lombardi with some of the design of his experiments for the prostate cancer grant, but they had been so busy working on ME/CFS that Lombardi hadn't actually been able to get caught up and do the work. Silverman's reagents sat in the freezer for close to six months.[50]

Mikovits was familiar with Silverman as they had both worked in prostate cancer research and also attended meetings of the American Association of Cancer Researcher prostate cancer meetings. He was an immunologist, with an expertise in the type I interferon pathway in cancer. Silverman's poster presentation at the Milken event was on the RNASEL genetic defects in prostate cancer patients who tested positive for the newly discovered XMRV retrovirus. Mikovits attended the conference, displaying a poster on behalf of the biotech company for whom she had consulted, showing how their drugs affected inflammation and cytokine signatures of disease. They had some early research on how these drugs worked on ME/CFS patients not shown on the poster but was the topic of their private conversations.

Little interest was displayed in Silverman's and Mikovits's posters so the two of them along with Lombardi had some time to talk then, and she just happened to be displaying her poster next to Silverman's. It was a fortuitous collision of thinking. A quick perusal of the two posters showed some intriguing similarities.

Mikovits's work showed high levels of pro-inflammatory cytokines and chemokines, such as IL6, chemokine IL-8, and they showed disregulation on interferon alpha.[51] Mikovits, Silverman, and Lombardi were all intrigued by the abnormalities in RNase L pathway because it might explain why the patients had so many chronic viral infections. Maybe they were dealing with a retrovirus that had some similarities to HIV and HTLV-I. Could Silverman's XMRV explain the differing results?

Mikovits was somewhat skeptical that Silverman's reagents would find this newly discovered retrovirus in her patient population. Mikovits recalled that the microarray was showing an elevated expression of just about every virus, as if something had thrown the immune system out of control.[52] But the subjects weren't obviously dying like AIDS patients, and aside from the increased incidence of certain types of lymphomas, they weren't developing cancer, although Hillary Johnson had documented other cancers—such as salivary gland cancers—in the patient populations elsewhere.[53]

When they got back to their lab, Mikovits and Lombardi instructed a new graduate student, Max Pfost, to do some PCR testing using Silverman's reagents. Mikovits had a dual role at the WPI, also serving as an adjunct professor in several departments at UNR, working extensively with students and teaching them the proper and thorough way to do a scientific investigation so that they would come away with foundational knowledge.

Max Pfost would go onto become one of Mikovits's closest colleagues and they would develop almost a mother-son type relationship.[54] Pfost was not quite cut from the same cloth as many of the other graduate students, sporting bike racing tattoos on his right arm, along with a quote in Latin from the 1980s film, *American Flyers*, which read *res firma mitescere nescit*, or "a firm resolve is not easily broken."[55]

On Max's left arm the subject was science. He had illustrations of DNA proteins being translated, immune cells, B cells, anti-bodies being secreted, and veins with viruses exploding out of them.[56] When the debates between Mikovits and Coffin over XMRV became heated, Max had a small coffin tattooed on his middle right finger, so that if he ever met Coffin and went to shake his hand, the veteran researcher would know just what Max thought of him.[57]

When Mikovits worked with Max, they picked out samples from twenty of the sickest patients, including several diagnosed with lymphoma, and she asked him to do PCR on the samples.[58] Most of the samples came back negative.

But two or three were positive.

However, the bands were the wrong sizes, meaning there could possibly be a related virus. If she had been at a typical lab with a well-trained staff, she might have simply thrown those samples out, figuring something went wrong in the experiment.[59] She might have returned to the cytokines, chemokines, RNase L, natural killer cells, or something else entirely, and conclude that there was not a novel virus in the samples.

But because she was teaching a young researcher she followed the most rigorous practices and asked Max to sequence the bands. Another nagging doubt in her mind was that some of the bands were quite bright, a strong positive signal. When the results came back they contained sequences of Silverman's XMRV retrovirus. There were some deletions at curious places and other inconsistencies, but if one aligned the sequences in just a certain way the pattern looked like XMRV gag—that is, sequences translated into a structural protein of XMRV (gag polyproteins are used in the viral replication cycle of a retrovirus).[60] It was like a Christmas tree lighting up.

Mikovits showed Pfost how to optimize the PCR, how to ever so slightly vary the annealing temperature so that he could find everything that was closely related. Mikovits was not looking for a stringent match, but more of a loose association that might suggest they were looking at a taxonomic family member to XMRV.

Lowering the stringency of the PCR changed everything.[61]

They found a lot of samples in the group of twenty ME/CFS patients with sequences highly similar to XMRV gag. They pulled an additional thirty samples and tested them under the loosened PCR standard. A number of those were positive as well. They tested samples that had been taken at different intervals from a single individual.

In doing this, they frequently found patients who tested negative in one sample, but positive in another.[62] Both viral latency and methylation issues could (at least temporarily) conceal the presence of a virus from PCR tests.

This interesting implication that ME/CFS patients could be positive for XMRV was something they felt compelled to discuss with Silverman and Ruscetti. She hoped she could convince Frank to come for a holiday in San Diego and show him the preliminary data. If Ruscetti thought the preliminary data was compelling, they had a green light.

* * *

They met in January of 2009 in San Diego during a meeting of the American Association of Cancer Research (AACR) special focus Prostate Cancer meeting, first to sign a confidential disclosure agreement regarding XMRV and the recent data generated by Pfost and Lombardi in the Mikovits lab.[63]

After signing the agreement, Lombardi and Mikovits showed the preliminary data to Silverman and Ruscetti to see if the two experts felt the data warranted a collaboration. The four of them—Mikovits, Ruscetti, Lombardi, and Silverman—agreed early on to authorship of the paper. Lombardi would be first author and Mikovits would be senior author, as is customary for a post doctoral fellow developing the mentor's research hypothesis under her direction.

The agreement they ended up signing on January 20 stated that the Cleveland Clinic had "novel assays to detect xenotropic murine leukemia virus-related virus (XMRV) infections in humans" and that both the National Cancer Institute and the Whittemore Peterson Institute had "certain confidential information relating to detection of XMRV in patients with chronic fatigue syndrome."⁶⁴ The esteemed Cleveland Clinic, the iconic NCI, and the still-formative WPI (which did not yet have a building, just a borrowed lab) were going to commence a pivotal study on whether the XMRV virus was associated with ME/CFS.

It had been a wild, two-and-a-half year ride for Mikovits. Now she and her colleagues were on a virology big game hunt.

They were on the hunt for a retrovirus.

CHAPTER FIVE

The Appearance of Chronic Fatigue Syndrome and Autism in the Medical Literature

From these experiments, it appears that of all the ordinary laboratory animals, the mouse should be the best in attempting to produce poliomyelitis, for the virus survives in its brain for a longer time than in that of the guinea pig, rabbit, or rat.[1]

—Dr. Maurice Brodie of the New York Department of Health, in 1935, reporting on experiments with his experimental polio vaccine.

Accordingly the virus in the form of an injected suspension of mouse brain was introduced into several of these cultures of human tissue which were then handled exactly as in the experiment with the mumps agent.[2]

—Nobel Prize Lecture, December 11, 1954

Attempting to identify the first outbreak or even the first patient with a new disease (often referred to as "patient zero") can be as frustrating as it is for a homicide investigator trying to solve a cold case. Defying the staid image of science as a measured progression of small steps, the exercise may look like the antithesis of scientific method, involving wild leaps of specula-tion, departure from evidentiary reason, and even a radical reinterpretation of subjects once thought to be well-understood.

Early AIDS investigation is a good example. New cases of the disease started cropping up in the early 1980s among gay men. These men were often dying of Kaposi's sarcoma (KS), a cancer that had, up to that point, typically struck elderly Jewish or Italian men.[3] In December of 1981, a gay man named Bobbi Campbell posted Polaroid snapshots of the purplish KS lesions in his mouth and throat and on his torso in the window of Star Pharmacy in San Francisco's gay mecca, the Castro District. He warned other gay men that something ominous was spreading around, and those who witnessed his stark admonition reported being chilled to the bone, sensing the community was tipping into a terrifying new era. A few years later, as the "gay cancer" spread through the district like wildfire, Star Pharmacy hung a plaque on their wall noting that they had reached a prescription sales record.[4]

The Jewish and Italian men who contracted this normally benign cancer would usually die *with* the unsightly but not fatal Kaposi's sarcoma lesions but be killed by other causes. These new atypical KS patients, on the other hand, also suffered from a staggering array of atypical ailments such as pneumocystis pneumonia that were generally not so virulent but were rapidly fatal in this population. The investigators tried to see if they could uncover any common links between the earliest cases of KS in gay men. Eventually they were able to find a "patient zero," a charming, hypersexual French-Canadian airline steward and athlete named Gaëtan Dugas who frequented gay bathhouses in New York and San Francisco, and had also traveled overseas.[5] The "little black book" of Dugas held the names and numbers of many of the first cluster of AIDS cases in the United States, whom researchers were then able to link to other cases.

* * *

Shortly after the publication of her October 2009 article in the journal *Science,* Mikovits attended a conference in January, 2010 of Defeat Autism Now! (DAN) doctors. These medical professionals held the opinion that a trifecta of genetics, infection, and toxins caused autism. Some of the DAN doctors focused on genetics, others on bacterial or viral infections, and others on ubiquitous toxins like mercury or lead. They shared the belief that autism spectrum diseases (ASD) resulted from some combination of these elements, creating a perfect storm. Mikovits had long been curious about a connection between ME/CFS and autism, as her team had preliminary data for what appeared to be a higher than expected number of children with autism among ME/CFS patients and their families.

* * *

On June 14, 1934, the train carrying Dr. John R. Paul—a small, wiry Yale Medical School professor with a fierce intellect—and Dr. Leslie Webster—a tall, good-looking physician from the Rockefeller Institute with a Clark Gable mustache—was met at the Pasadena rail station by Dr. George Parrish, City Health Officer for Los Angeles. The men were immediately whisked to the Mayor's office.[6] High officials of the beleaguered, polio-stricken city of Los Angeles welcomed them as saviors.

In a letter written to his wife on the day after his arrival, Dr. Paul recounted the journey and the pandemonium that greeted their arrival.[7]

> As we commenced to get Los Angeles papers, we realized we had newspaper troubles ahead. A telegram was received requesting that we disembark at Pasadena. There we were met by the City Health Commissioner, Dr. Parrish and his aides. We were told they had sidetracked us at Pasadena because of the size of the crowd at the terminal. Then they said the Mayor was waiting to receive us and award us the Freedom of the City. I said "Positively we will not give out statements, we came here to work: leave us alone."

The scientists were treated like heroes before they had done a single lab test and they must have felt like the stars of their own movie.

> After dinner, the Police Commissioner of Los Angeles gave us his private car and we were driven at breakneck speed with red lights, policeman's radio, and siren going full blast. We skipped all traffic stops and just roared 18 miles to Hollywood's Greatest Premier Move Theater, where the great owner, Sam Grauman, gave us complimentary seats. He held up the main show 3 minutes to meet us. My feelings cannot be committed to paper!!![8]

The *Los Angeles Times* gave a clear account of what the two men planned to accomplish on their journey to the west.

It is the plan of the two physicians, with the assistance of Dr. James F. Trask, of Yale Medical Institute, who will arrive Sunday, to undertake extensive laboratory tests and research in an effort to isolate the poliomyelitis virus and through inoculation experiments with monkeys seek prevention of the disease in the interest of humanity.[9]

They were men of science in the City of Dreams. They didn't know it, but in addition to polio, they were about to witness the first recorded outbreak of ME/CFS.

* * *

Paul and Webster were part of a second wave to arrive in Los Angeles from New York City. They were there to observe a polio epidemic as it swept the city and, if possible, gain valuable information about it. One physician, however, was interested in protecting the doctors and nurses who were on the front lines of the terrifying epidemic.

Dr. Maurice Brodie, a young Canadian physician working at the New York City Health Department and at New York University,[10] had developed a polio vaccine. According to an account in the popular magazine *The Literary Digest*, the funds for Brodie's work came "partly from the Warm Springs Foundation, in which President Roosevelt, himself a sufferer from infantile paralysis at the age of forty, is the moving spirit. The rest is being contributed by the Rockefeller Foundation and the New York Foundation."[11]

Brodie had reputable support from the president's own charity in addition to the Rockefeller and New York Foundation. Talk about a trifecta: the researcher could not have had more powerful patrons than the president and the country's wealthiest charities. Brodie later detailed the process of creating his polio vaccine, which consisted of up to forty-five passages of the virus through mice, and noted:

The passage material was a mixture of 1 part of active virus and 4 parts of a suspension of mouse brain and brain stem. The material from the 24th and 45th passages injected into a series of mice in multiple inoculations, produced no effect in the mice nor did the virus in either of these passages survive for a longer time than in the preliminary experiment, when it was demonstrated in the mouse brain, 3 but not 5 days after intracerebral injection.[12]

Trials of the Brodie vaccine were subsequently conducted in North Carolina, Virginia, and California, and included 7,000 children[13] in addition to the "300 nurses and physicians" from the Los Angeles County Hospital.[14]

* * *

In a later account of the 1934 Los Angeles epidemic, Paul seemed conflicted about the polio outbreak as well as the undefined new disease he witnessed. He wrote:

> To sum up various opinions about the Los Angeles epidemic, there seems to be little question that Los Angeles County was visited by an epidemic of poliomyelitis in the summer of 1934, although even this fact has been denied in some quarters. But a major question has been whether there was not some other, unrecognized illness combined with poliomyelitis cases, as Dr. Webster had suspected almost at the start of our virologic studies, and perhaps he was correct.[15]

In his account of the new disease which affected 198 nurses and physicians, Paul relied on a later account by Dr. A. G. Gilliam, who observed that "the majority of atypical symptoms could be described as 'rheumatoid or influenza' in character."[16] Paul noted from the beginning that the vagueness of many of the symptoms caused some researchers to believe they were dealing with an outbreak of mass hysteria.

Paul noted that there were later outbreaks of this fatiguing condition he observed, namely in the winter of 1948 in Iceland and at the Royal Free Hospital in London in 1954, where the condition was first referred to as a myalgic encephalomyelitis (ME), which technically translates into muscle pain (myalgic) accompanied by inflammation of the brain and spinal cord (encephalomyelitis). The similarity of the second word to polio*myelitis* should not be overlooked, as the new disease was quickly seen as polio-*like*.

In addressing the question of whether Paul and Dr. Webster failed to properly observe and classify a new disease arising alongside to the Los Angeles polio epidemic of 1934 Paul wrote:

> Nonetheless the Los Angeles episode is a reminder that even those who believe themselves to be experts occasionally ride for a fall, although they may be extremely loathe to admit it, especially to their patients. It is sometimes the bitterest pill they have to swallow . . . As a weak excuse, it may be said that we had our hearts so set on isolating the poliovirus that we could think of little else.[17]

In their haste to isolate the polio virus, Paul and his team believed they may have overlooked a new disease in their midst.

* * *

What might an investigation by scientists into a *virus* (polio is caused by an enterovirus, it would become a model for studying the biology of RNA viruses) look like if it involved mice, monkeys, and men in the 1930s? While Brodie did not publish extensive notes of his experiments, future generations are fortunate to have a well-documented example of the play-by-play of a similar investigation.

This effort predates the use of the Brodie vaccine on the nurses and physicians at Los Angeles County Hospital. It was performed by scientists working at the Rockefeller Institute; scientists with whom Brodie would have had close association. It was an investigation into the harrowing disease, yellow fever, a viral fever that often induces a hemorrhagic response that causes patients to bleed to death.

A 1932 article by the Rockefeller Institute stated exactly what they hoped to achieve with their research: "The method here presented for vaccination against yellow fever was devised primarily to interrupt the long series of accidental infections of persons making laboratory investigations."[18] Lab workers had become inadvertent soldiers in the pursuit of science and needed the "body armor" of an effective vaccine.

The scientists were dying from the disease they were studying. In the four and a half years since researchers had started using rhesus monkeys as experimental animals to investigate yellow fever, there had been thirty-two infections among lab personnel and five scientists had died.[19]

The concept of how to weaken a virus so that a human will produce an effective immune response to protect the host rather than injure or kill the host in cooperation with the virus is relatively simple. When viruses jump between species, their genetic structure undergoes changes that often make the virus more benign. The alteration of just a few base pairs as the virus jumps can create an adjustment that causes a previously terrifying virus to become a pathogen the body can handle with relatively little difficulty. Modern day Ebola virus infection is a perfect example for first human infections that are rapidly 100 percent lethal but after several passages in humans loses its lethality.[20] However, the process can just as easily go in the opposite direction, turning a previously harmless virus into a killer. Non-endemic pathogens can also become virulently lethal in a new geographical location—which is why the pathogens carried by Europeans when they came to the Americas was so predictably devastating to the native population.

From early-to mid-twentieth century, directing the evolution of the world's most dangerous viruses through various animal species and gentling them as a rancher would a wild horse, was the holy grail of medicine. The question that would haunt researchers at the time and echo through research halls over decades was whether in the attempt to conquer one disease a researcher might inadvertently create another. This question became worth billions of dollars: failing to adequately answer it could have a devastating impact on the entire medical and scientific community.

In their own words, this is what the scientists of the Rockefeller Institute did three years before the Brodie vaccine was used by 300 nurses and physicians at Los Angeles County Hospital:

> The preliminary experiments with various ways of vaccinating against yellow fever seemed to show the most dependable and effective of the methods tried was the injection of living yellow fever virus with a simultaneous or preceding infection of potent immune serum.
>
> Moreover, it appeared that yellow fever virus, after many passages through mice, had lost most, but not all of its virulence for monkeys, and probably for man, although retaining its power to immunize. It was proposed, therefore, to test in monkeys the safety and the immunizing power of a vaccine composed of living virus fixed for mice and human immune serum, and if these tests gave satisfactory results, to commence immunizing persons exposed to yellow fever.[21]

While it may surprise people today, early experiments with vaccinations were often coupled with attempts to boost the immune system of an individual by use of an "immune serum." Vaccines carried the risk of overwhelming a person's immune system, so steps were taken to increase the strength of the immune system as a precaution.

The Rockefeller researchers saw great promise in the work of their fellow colleague in the Institute, Max Theiler, (winner of the Nobel Prize in 1951 for the development of a vaccine against yellow fever), who had succeeded in lowering the virulence of the yellow fever virus by passaging it through mouse brain tissue:

> When Theiler showed that yellow fever virus which had been adapted to mice had lost much of its virulence for monkeys, the hope was raised that the untreated virus in mouse-brain tissue, after enough passages in mice, would lose its virulence for man and would be safe to use as a vaccine.
>
> At the time of our experiments the French strain of yellow fever virus had been through over 100 successive passages in mice and was nevertheless still able to produce fever in monkeys, although it had apparently long lost its power to kill them.[22]

The mice seemed to be the key to weakening the virus so that it would cause a fever in the rhesus monkeys when exposed to the virus (a sign that the immune system was attacking a pathogen), but not kill them. The generated fever showed the immune system had been provided enough help to fight off the virus and attain immunity to future exposures of the virus by encoding it as a threat. Human testing was the next step:

> Sixteen persons were vaccinated against yellow fever in this laboratory, the first on May 13, 1931, by the methods developed with monkeys . . . The first person to be vaccinated was admitted to the Hospital of the Rockefeller Institute for Medical Research through the courtesy of Dr. Rufus Cole and Dr. T. M. Rivers, and was closely observed for us by Dr. G. P. Berry of the hospital staff . . . There were no subjective symptoms and no abnormalities of temperature, pulse, blood pressure, heart action as shown by electrocardiogram, or urine, during 10 days in the hospital, 2 days before vaccination and 8 after. The other persons vaccinated remained on duty in the laboratory.[23]

Researchers at the Rockefeller Institute seemed to have established a sensible protocol for the yellow fever vaccine. It involved passaging the virus through mouse tissue more than 100 times, then testing the vaccine on monkeys, and if that worked out without safety concerns, vaccinating humans. While this vaccine was used for some time among researchers in African countries, eventually it was replaced by other vaccines that were not grown in mice.

The reason?

When the vaccine was injected into the *brains* of mice and monkeys, their brains quickly swelled with encephalitis and they rapidly died. This was not initially considered a safety concern since vaccines were not directly injected into brain tissue.

The problem with the yellow fever vaccine grown in mouse brains by the scientists of the Rockefeller Institute was the subject of a presentation Dr. G. Stuart to the World Health Organization in 1953.

> The neurotropic strain used in this vaccine is a derivative of the pantropic French strain of yellow-fever virus which, during prolonged serial brain-to-brain passage in mice as shown by Theiler to have lost its ability to produce visceral yellow fever in rhesus monkeys—animals which, when inoculated extraneurally with the modified virus, ordinarily developed only a mild non-fatal infection and consequent solid immunity.
>
> At the same time, however, the modified strain had acquired an enhanced neurotropic virulence for both mice and monkeys, producing in these

animals, when injected intracerebrally, a rapidly developing fatal encephalitis.[24]

To put it in plain language, if the vaccine was injected into the bloodstream it would provoke a mild immune reaction, after which the subject would be immune to yellow fever. But if the vaccine components somehow migrated into the brain, they would quickly produce a fatal swelling. The problems with this vaccine grown in mouse brain tissue were clear and Stuart addressed them in his presentation:

> [T]wo main objections to this vaccine have been voiced, because of the possibility that: (i) the mouse brains employed in its preparation may be contaminated with a virus pathogenic for man although latent in mice . . . or may be the cause of demyelinating encephalomyelitis; (ii) the use, as antigen, of a virus with enhanced neurotropic properties may be followed by serious reactions involving the central nervous system.[25]

It was all there in stark black and white (plus there was that word: "encephalomyelitis"). There were serious problems with using a vaccine developed by the passage of a human virus through animal tissues. The vaccine might pick up a virus from the mouse that was deadly to humans.

This newly acquired virus could then cause the myelin sheaths, which coat neurons like the protective plastic on an electric cord, to fall apart and turn the nerve endings to brittle threads. And the virus might make the body overreact, provoking autoimmune disease in which the body attacks not an invader, but rather turns on itself. Mikovits would make the same arguments nearly seventy years later.

Mikovits would also add the possibility that a recombination event might morph the human and mouse viruses into another entity altogether.

* * *

Brodie's once-promising polio vaccine never made it into general usage. A twenty-year-old man vaccinated with Brodie's polio vaccine developed paralysis in his inoculated arm and died four days later.[26] According to vaccine expert Dr. Paul Offit, in his account of the Brodie vaccine trials, "Two children, five and fifteen months old, developed polio within two weeks of receiving inoculations."[27] The Brodie vaccine may have also induced the death of two-year-old Jackie Baldwin of Healdsburg, California (in Sonoma County, north of San Francisco) in July of 1934, who died after receiving a polio vaccine administered by his father, a local physician.[28] This death in California is especially interesting as it took place around the mid-point of the first known ME/CFS outbreak among the doctors and nurses at Los Angeles County Hospital.

By December of 1935, many members of the medical community—such as Dr. James Leake of the US Public Health Service—were recommending discontinuance of the Brodie vaccine.[29] Despite what had been reported in *The Literary Digest* that the Warm Springs Foundation of President Roosevelt had contributed funding for the development of the Brodie vaccine, in late 1935 the foundation distanced itself from the vaccine and claimed only that they contributed money towards the building of a laboratory for Brodie.[30]

According to Offit, after 1935 Brodie found it difficult to get a new job, eventually accepting a minor, demoting position at Providence Hospital in Detroit. In May of 1939, Maurice Brodie died at the age of thirty-six of a suspected suicide.[31]

While Brodie's work with an ill-fated polio vaccine may have prompted his untimely death in 1939, his use of mouse brain suspension to cultivate the polio virus would result in three other researchers winning the Nobel Prize in Medicine in 1954.[32]

* * *

The first outbreak of what would later be rebranded "chronic fatigue syndrome" started at Los Angeles County General Hospital in the late spring of 1934. At the time, General was a giant: the largest hospital in the world. This ME/CFS outbreak followed a polio outbreak on the West Coast in which 2,449 suspected cases were seen at Los Angeles County General Hospital. It was considered an unusually mild polio epidemic because so few deaths were reported.

One hundred-ninety-eight doctors, nurses, and hospital staff were eventually afflicted with ME/CFS from May through December 15 of 1934. (Other accounts report even higher numbers.) The disease outbreak was first thought to be polio contracted from patients.

But it was something distinct.

Dr. A. G. Gilliam, an assistant surgeon for the US Public Health Service, was dispatched to Los Angeles to investigate the large cluster outbreak. Prior to this mission, Gilliam just *happened* to be investigating the efficacy of the Brodie vaccine in North Carolina.[33] Gilliam's boss was Leake, who eventually spoke out so strongly against the Brodie vaccine. Gilliam later had a distinguished career as a professor of epidemiology at Johns Hopkins University. He wrote of the Los Angeles epidemic:

> These cases represent an attack rate of approximately 4.4 percent for all employees of the General Hospital; the rate in the medical units being 4.5 percent and in the osteopathic unit 1.0 percent. The personnel most severely affected were nurses and physicians, who suffered attack rates of 10.7 percent and 5.4 percent respectively.[34]

Gilliam struggled to describe the constellation of symptoms and particularly the pain of the sufferers:

> "The character of pain experienced varied within wide limits. It was frequently described as rheumatoid or influenza in character—an ache in the muscles or bones—but often of sufficient intensity to awaken the patient from sleep. In most instances it was aggravated by exercise. It was confined to no definite body area and characteristically was variable from day to day in its distribution."[35]

Headaches experienced by the patients were described as being "of a character and severity never previously experienced by the patients." The patients also suffered from muscle twitching, nausea, and vomiting. Other complaints included irritability, drowsiness, stiffness in the neck or back, photophobia, constipation, and tremors. The distribution of standard neurological symptoms (twitching, irritability, photophobia) along with flu-like symptoms (nausea and vomiting) certainly matched reports by Peterson and Cheney in Incline Village, decades later (along with encephalitis and meningitis-like symptoms such as stiff neck, crushing headache, even "drowsiness" which is a dangerous sign in known encephalitis-causing pathogens, as it can signal an approaching coma).

Perhaps the most telling correlative to the more-recent Incline Village outbreak of ME/CFS, however, was the description of the condition being "aggravated by exercise." This odd characteristic is so descriptive of ME/CFS that it has a name, "post-exertional malaise" or PEM.

But the patients at the LA hospital did not recover, instead falling into a pattern of lingering muscle weakness and fatigue that proved disabling. A majority of the cluster group who tried to return to duty "subsequently found themselves unable to continue work and in consequence were sent off duty again," again a portrait of modern-day ME/CFS survivors. As the Centers for Disease Control and Prevention's ME/CFS site states: "The majority of people with ME/CFS are affected by post-exertion malaise, which is defined as intensifying of symptoms following physical or mental exertion, with symptoms typically worsening 12–48 hours after activity and lasting for days or even weeks." The site conclusively adds, "While vigorous aerobic exercise is beneficial for many chronic illnesses, CFS patients can't tolerate traditional exercise routines."[36] At the time Gilliam compiled his findings, he reported that 55 percent of the patients were still unable to reenter the labor force.

In summarizing his report, Gilliam observed that the disease appeared to show a preference for "females rather than males, and females under 30 years of age rather than those over 30 years." Again, this parallels the more recent ME/CFS outbreaks. He also straddled the idea that this outbreak could be a new variation of polio or a unique entity:

> The observed distribution of the disease is therefore what might have been expected had it been an epidemic of a disease, such as scarlet fever, spread by direct personal contact with cases and carriers. It is quite different from what one would have expected in an epidemic such as typhoid fever, spread by contamination of the hospital water, milk, or food supply . . .

In his report Gilliam struggles with the question of whether the disease is the polio which is affecting the public or something completely different. Eventually he comes down on the side of believing something new has entered the world.

> Irrespective of the actual mechanism of spread and of the identity of the disease, the outbreak has no parallel in the history of poliomyelitis or other central nervous system infections. There is nothing in past experience with this class of diseases, which would have permitted anticipation of this institutional epidemic.[37]

It is curious that Gilliam refers to this as an "institutional epidemic" because it appears to have affected solely the employees of the hospital. If it was something in the food, the water, or any of the other myriad items in general use in the hospital it would have affected both staff and patients. But it didn't. If affected only the staff.

The critical question for any researcher is, to what was the staff exposed that was uniquely different from that of their patients?

* * *

In *Osler's* Web, Hillary Johnson recounted the story told to her by Canadian physician and researcher, Dr. Byron Hyde, who had studied Gilliam's report and in 1988 was subsequently able to locate and interview some of the doctors who had fallen ill during the Los Angeles outbreakof 1934–1935.

According to Johnson, "Hyde added that there had been no medical follow-up of any of the cases, which he suspected was the result of a legal settlement between the epidemic victims and the hospital. A condition of the settlement was that the victims refrain from discussing their ordeal publicly. The one hundred ninety-eight staff members sued the hospital," Hyde told Johnson, "and eventually settled for six million dollars in 1939, which would have purchased three houses for each victim in the best section of Los Angeles. Contingent on receiving the payment was non-publicity of the epidemic.[38]

The value of six million dollars in 1939 would be just over one hundred million dollars in 2014. Johnson was unable to determine whether this amount had been paid in full by the Los Angeles County General Hospital, or whether other groups contributed to the settlement.

Either way, it was a good chunk of money.

What set of facts might have caused an organization to settle for such an amount, and what other incentives might there be for the victims of this disease to remain silent?

In 1992 Hyde would publish a 724-page textbook, *The Clinical and Scientific Basis of Myalgic Encephalomyelitis/Chronic Fatigue Syndrome*[39] under the Nightingale Research Foundation that he founded to better study the disease. He was assisted in this effort by Dr. Jay Goldstein of the Chronic Fatigue Syndrome Institute of Anaheim, California, as well as Dr. Paul Levine of the National Institutes of Health.

While Hyde generally applauded the job Gilliam had done in characterizing the epidemic, he believed there were conspicuous omissions from Gilliam's report, such as any sustained attempt to identify the cause of the outbreak, which would be common practice in this type of investigation.

> Not only did Gilliam give the first clear description of the acute and subacute characteristics of M.E./CFS and its epidemiological factors, but his book also implicitly posed a question that has been avoided and unanswered to this day. "(A) Did the 198 health care workers fall ill, or remain ill, solely as a result of the "prophylactic serum" that Gilliam persistently documents as having been given, often prior to any symptom, to the majority of patients in this epidemic?[40]

Could the Brodie vaccine have been the "something different" between the patients and the staff at the hospital? From his discussion with the surviving doctors of the 1934–1935 epidemic, Hyde believes it is the most likely possibility. Hyde recounts what the doctors told him.

> . . . Gilliam told Dr. Shelokov that he had been "stymied by his superiors from telling the whole story" . . .
>
> If the question concerning the pooled adult serum and convalescent prophylactic serum as a potential cause of some of the pathology is not clearly stated in the text of his book, there is ample reason for this. We know of the battle that Gilliam fought to even have the epidemic findings published. It was obvious that he was not going to rock any more boats if he was going to see the document in print . . .[41]

Hyde was interviewed in 2013 by the authors and had the following to say about Gilliam's work. "In the original document he [Gilliam] said it was caused by this immunization and it was a human transfer of infectious material, so it was actually a great experiment. He had a huge fight with the head of the public health system of the United States and the fight went on for almost six months.

"Finally, it was agreed that he would be allowed to publish it, as long as they took out the immunization section—because if you put the immunization in, it would set back all Americans on immunizations for years and cause the death of many, many people. I think it was the theory then and it still is today. As a physician you cannot say anything negative about immunizations, even if it's causing problems—and they do."[42]

* * *

In October of 1938, Dr. Leo Kanner of Johns Hopkins University met Donald Triplett, a five-year-old boy who would be known for decades only as Donald T., the first child in Kanner's 1943 article describing a new condition known as autism.[43] The Rockefeller Foundation supported Kanner's work. The Triplett family lived in the small lumber town of Forest, Mississippi, where yellow fever outbreaks had occurred.

According to his father, Donald always had always been fussy with food and was never interested in candy or ice cream. At just one year old, he could hum and sing many tunes, and by the age of two had developed an exceptional memory for names and faces. As he grew older there was something peculiar about him, an abnormal insularity and aloneness.

Donald's father was a successful, meticulous, and hard-working lawyer who, when ill, always followed, "doctors' orders punctiliously, even for the slightest cold." The boy's doting mother was calm, capable, and a college graduate.

In their book on the autism epidemic, *The Age of Autism* (2010), authors Dan Olmsted and Mark Blaxill uncovered many interesting facts about the families of the first eleven children diagnosed with autism.[44] Autism patient number two was Frederick Creighton Wellman III. He was referred on May 27, 1942, at the age of six years. His father, Dr. Frederick L. Wellman, was a plant pathologist who worked in tropical areas of the world where yellow fever was common.

The father of patient number three was a forestry professor, who worked at the University of North Carolina. North Carolina was one of the states where the Brodie polio vaccine underwent early trials and where Gilliam conducted his study on the vaccine. Again, there is the shadowy presence of vaccines, which had been passaged through mice. The father of patient number four was a mining engineer. The fathers of patients five through seven were all psychiatrists and the mothers of patients five and seven were respectively a nurse and a pediatrician.

The mother of patient number seven, Dr. Elizabeth Peabody, cocreator of the "Well-Baby" visit (an early wellness concept that caught on through subsequent decades of post-war hygienic concern), which generally consisted of an examination of the child and an increasing number of immunizations designed to curb the infectious diseases of childhood.

It stands to reason that Dr. Peabody might have been among the first people that a researcher like Brodie would have contacted to assist in his vaccination campaign. In an article about a talk Peabody gave in April of 1947, the Annapolis *Capitol* reported:

> Too many parents, said Dr. Peabody, have the proper shots given and then relax, forgetting that booster shots are needed and that immunization does wear off. Speaking specifically of some of the most prevalent ailments, she stated that a child cannot be vaccinated against smallpox too often and it should be done for the first time when a baby is between three months and one year of age.[45]

The father of patient number eight was a patent examiner and his wife was a psychologist.

Both of patient number ten's parents were in the medical field. His father was a psychiatrist and his mother worked in a pathology lab. Would the offspring of these parents have been among the 7,000 children who received the Brodie vaccine, which was claimed to offer protection against the scourge of polio? Of the twenty known occupations of the parents of Kanner's first eleven cases, ten were in scientific or medical pursuits. Four were psychiatrists, two were plant scientists, one was a psychologist, a nurse, a pediatrician, and the last worked in a pathology lab. Three mothers were teachers, and the remaining professions broke down into lawyer, mining engineer, patent examiner, clothing merchant, advertising copywriter, magazine employee, and theatrical booking agent.

In their book on autism, Olmsted and Blaxill found the evidence pointed to three distinct clusters among the first eleven autism cases.

The first was the mercury fungicide cluster centered at the University of Wisconsin and Frederick Wellman's work. Wellman's son was case number two in Kanner's study.

In addition to Wellman, this cluster connected with case number three, the son of William Miller, who was a forestry professor at North Carolina State University. The school had recently acquired an eighty thousand acre tract of land known as Hofmann Forest. Ethylmercury fungicides—including Ceresan and Lignansan—developed by Wellman and others were being used at North Carolina's research station in the Hoffmann Forest.

Of the two compounds, Lignasan, showed the most promise. Besides the students and faculty using the research station in the Hoffman Forest, three lumber companies also shared the research station. One of the three companies, Eastman-Gardiner, owned land that extended north to that of the territory of Bienville Lumber, headquartered in Forest, Mississpippi. This was the home of Donald Triplett, patient number one of Kanner's eleven. Could the mercury from these agricultural products or the mercury derivative, thimerosal, used in Brodie's vaccine which was passaged through mice set off a chain reaction in susceptible children?

While there was no direct evidence, Olmsted and Blaxill argued that it seemed logical to conclude that the smaller mills located in Forest also used Lignisan in the years before the birth of Donald Triplett. Thus, the first three cases in Kanner's group all had links to agriculture and forestry, and specifically to the Caribbean and southern parts of the United States, as well as to the use of mercury-containing fungicides. A question can reasonably be extrapolated from this knowledge. Might researchers traveling to undeveloped areas of the southern United States, or to the Caribbean, fear tropical diseases like yellow fever and have been among the first to volunteer for the Rockefeller Foundation's yellow fever vaccine?

The 1933 Annual Report for the Rockefeller Foundation, just seven months after the first successful human trial, does report just such an approach.

> Vaccination against yellow fever by the injection of human immune serum and living yellow fever virus fixed for mice was begun in 1931. The number of persons vaccinated in the laboratories of the Rockefeller Foundation has now reached fifty-six. These persons have been principally members of the staff of the Foundation assigned to yellow fever work, and government officials, missionaries, scientists, and educators about to leave for countries in which yellow fever was present.[46]

The second cluster was referred to as the medical cluster, embracing the fields of pediatrics, psychiatry, psychology, nursing, and pathology. This group contained cases five, six, seven, eight, and ten. Olmsted and Blaxill focused on the work of Peabody and the effort in the 1930s to vaccinate children with a new diphtheria toxoid shot containing thimerosal, a mercury derivative. They asserted the likelihood that any parent involved in medicine would have been an early adopter of vaccines to protect their children. Would they also have been among the parents volunteering their children for an early Brodie polio vaccine? Could they have been among the 7,000 children who received that vaccine?

Blaxill and Olmsted argued that the clustering of some of the early cases of autism around Baltimore raised questions about whether the Johns Hopkins anti-diptheria campaign of the early 1930s may have contributed to these first cases as well. Could the new diphtheria vaccines have also been the product of mouse-derived biological products? Or might these children have also received a Brodie polio vaccine? Could autism have spread among the members of the medical and scientific community, as it did in the Los Angeles County General Hospital outbreak of ME/CFS. If so, might it have interrupted development only among those families whose parents had some acquired pathogen perhaps transmitted to their offspring, who were also exposed to the mercury-containing products, and whose children were at a critical stage of development?

If one considers the infection rate of 10.7 percent among the Los Angeles nurses in the ME/CFS outbreak (presumably female), and 5.4 percent among the doctors (overwhelmingly male), then it's easy to see how even widespread exposure among medical professionals would not necessarily translate into a perceived outbreak of autism among their children, as the secondary outbreak would have looked like a small trickle by comparison.

The third cluster, containing only patients nine and eleven seemed to only overlap in that they resided in major cities, specifically, Boston and New York. These would have been natural places to conduct trials of either the Brodie polio vaccine, or other early vaccines which were at least partially derived from mouse products. And what of the "thimerosal," the mercury derivative used in Brodie's vaccine?

The trifecta of genes, infections, and toxins suggested by the DAN doctors did seem to parallel much of the historical information uncovered about the earliest outbreaks of both ME/CFS and autism. While Blaxill and Olmsted were looking for evidence of thimerosal exposure in agricultural products, there was another potential source.Brodie's experimental polio vaccine was given in combination with an immune serum, specifically designed to increase the function of the immune system. This accompanying immune serum was preserved with "merthiolate," a trade name for thimerosal.[47]

* * *

In looking at the *birthdates* of the initial eleven autism cases from Kanner's report (rather than when they were seen at Johns Hopkins), it is clear that the first known child to have autism was born on September 13, 1931. Human testing of the first mouse-monkey-human yellow fever vaccine took place exactly four months earlier, starting on May 13 of 1931. The second child, Elaine C., was born on February 3, 1932 in Boston. The third child, Alfred L., had the birthdate June 20, 1932. The last child in Kanner's first eleven children, Herbert B., came into the world on November 18, 1937.

In a little more than six years, eleven children came down with this previously unrecorded mystery affliction and found their way to Johns Hopkins Hospital in Baltimore, Maryland, where they were seen by Dr. Leo Kanner, a Rockefeller Foundation Fellow. The parents of two of these children even knew each other. A country away in 1934–1935, 198 nurses and doctors at Los Angeles County General Hospital had fallen ill with the mystery affliction that made them even sicker the more they struggled against it, like Charlie Chaplin caught in a giant industrial cogwheel.

The tight chronological association between these outbreaks is unmistakable, as is the occupational connectedness of the victims' families, who were mostly well-educated people in the medical or scientific professions. What else might these two diseases share?

After nearly eight decades, it may be difficult if not impossible to nail down an answer.

* * *

But someone may have known further information.

If Hyde's claims are accurate, a settlement of six million dollars was distributed to the nearly two hundred staff members of the Los Angeles County General Hospital in 1939. In the last official year of the Great Depression, when people were still sewing dresses from flour sacks, it seems beyond belief that the hospital would pay out an amount that would equal just over a hundred million dollars today, simply out of good conscience. It's plausible the pay-out was meant to compensate the staff for their injuries without assigning blame. But then if there was no actual or perceived blame, why were the afflicted not supposed to discuss the epidemic?

Wouldn't that have been the standard scientific investigation for such an event?

Perhaps it's a further stretch, but could the mystery entity have any link to the parents of the first autism cases? Given what we know today about the ability of viruses to recombine, is it possible that experiments at the Rockefeller Institute in the 1930s to create a yellow fever vaccine, and in later attempts by Brodie and others to create a polio vaccine, might have accidentally created a mouse-human hybrid virus that caused its own unique set of symptoms?

Would scientists today even be allowed to conduct such experiments or were they a sign of more lax attitudes toward protecting human subjects? In the 2006 Pulitzer Prize winning book, *Polio—An American Story*, author David M. Oshinsky recounts the 1930s race to develop a polio vaccine by Brodie, his mentor William Park (a professor of bacteriology at New York University Medical School), and their rival John Kolmber—who was a Philadelphia pathologist supported by a group of Philadelphia hospitals and medical schools. Oshinsky also addressed the fall-out from this race.

After first testing the vaccine successfully on monkeys:

> Park and Brodie next tested the vaccine on themselves. There were no problems beyond some muscle discomfort near the injection site. Then a dozen children were vaccinated, all supposedly "volunteered by their parents" . . . Inoculation of this material into several human volunteers having shown that it was probably safe for human administration, it was used in children."
>
> "Probably safe?" "Used in children?" These words, so chilling today, were hailed as progress by a public just awakening to the threat of polio.[48]

The effort fell apart when several scientists, including those from the Rockefeller Institute (who had pioneered the passaging of human tissue through mouse brains) and those in public health like Gilliam's boss—James Leake—stated that not only was the vaccine ineffective, but questioned whether the vaccine may have conversely triggered the disease in some patients and caused injury or death in others.

In their discovery of XMRV, Silverman and DeRisi found that the retrovirus was closely related to a known mouse retrovirus. But mice and men have shared the planet for millions of years, so how did the retrovirus suddenly jump to humans?

If we believe Gilliam's opinion that ME/CFS had a pathogenesis around 1934, we might have an explanation. We know that the Brodie polio vaccine, passaged through mouse tissue, was administered to nurses and physicians at Los Angeles County Hospital in 1934.

Might this explain why *only* the medical staff of the hospital (and not the more vulnerable patients) came down with this new disease? Just as the scientific community would come to believe that XMRV was derived from prostate tissue passaged through mice, the question is whether something similar had happened decades earlier.

* * *

What is most curious about the 1934 investigation of the polio outbreak in Los Angeles by the Rockefeller Foundation is their failure to make any mention of it in their 1934 Annual Report.[49] It seems that an effort to observe a polio outbreak unfolding would have been one of the premier public health projects of the year, and heralded to donors in the public relations materials. Yet there is no mention of it in the 1934 Annual Report. The Rockefeller Foundation was only too happy to report its success against yellow fever, however:

> Before the partial perfection of a vaccination method, first applied to human beings on May 13, 1931, one of the tragedies of yellow fever work was the comparatively great danger to laboratory workers and men engaged directly in the fight against yellow fever . . . Fortunately, this danger seems to have been ended by the timely application of a method of vaccination which consists in administering modified virus accompanied by immune serum . . .[50]

Likewise, the Rockefeller Foundation had no difficulty trumpeting their support of Kanner (who nine years later would publish the first description of autism) as he attempted to develop a program of research and training at Johns Hopkins University School of Medicine:

> The departments of Psychiatry and Pediatrics of the Johns Hopkins University School of Medicine are cooperating in the development of a program of research and teaching in child psychiatry in the Pediatric Clinic under the direction of Dr. Leo Kanner, a former Rockefeller Foundation fellow.[51]

The annual reports of the Rockefeller Foundation also offer a window into the financial state of the foundation. Upon closer inspection something curious appears along with this conspicuous omission: the disappearance of more than seven million dollars.

The Rockefeller Foundation entered the year 1935 with a little more than $153,600,000 in their principal fund.[52] In today's money that is roughly equivalent to 2.5 billion dollars. At the end of 1935, they had realized a fifty thousand dollar increase in the principal fund for a little more than $153,650,000 in that fund.

For a foundation with more than a hundred and fifty million dollars in principal, this principal increase was an expected amount. The principal throws off enough money to fund annual projects and research and if there's a small yearly loss or gain it doesn't make much of a difference to the ongoing work of the foundation.

But starting in 1936, a different pattern shows up. From the years 1936–1939 the trustees of the foundation invade principal and transfer $7,800,000 to the "Appropriations Account," resulting in a decrease in the value of the principal fund to a little over $146,000,000.[53] Where did this substantial amount of money go?

Could it have gone to the 198 nurses and doctors of the Los Angeles County General Hospital with the mystery affliction, a.k.a. ME/CFS? Besides the more than six million dollars cited by Hyde going to the patients, there would also have been administrative and legal fees. The numbers are suggestive, but in no way conclusive.

They deserve a deeper explanation.

Between 1940 and 1941 the trustees withdrew an additional $1,750,000 but this was balanced by the return of $1,700,000 in 1941 and 1942 from a cancelled "contingent project." When the account was reconciled, from the years 1940 to 1945, there was a total loss to the fund of just $50,000.

The 50K figure is a far cry from the $7,800,000 withdrawn from principal between 1936 and 1939. Why was there no mention of Rockefeller Foundation's investigation into the 1934 Los Angles epidemic where—in addition to polio—they observed the first chronic fatigue syndrome epidemic?

The team from the Rockefeller Institute were welcomed as heroes by the citizens of Los Angeles. Why then do the exploits of the small, wiry Yale professor Paul and the tall, handsome Webster disappear from the historical records of their own institution?

Paul and Webster have in essence become the "Invisible Men" of science.

* * *

Nearly eighty years have passed since those events of the 1930s.

It may be impossible to recreate the precise chain of events that led to the creation of the Los Angeles General Hospital ME/CFS outbreak and those first cases of autism. But that doesn't mean that we might not be able to make some educated guesses.

In January of 2011, an article was published in the journal *Frontiers in Microbiology* that tried to make sense of the evolution of this type of retrovirus and the possible connection to human disease. The article was entitled, "Of Mice and Men: On the origin of XMRV."[54] The authors considered two possible scenarios. First, that the virus could have spread through direct virus transmission from mouse to human, and second, that it may have come from mouse biological products, specifically vaccines:

> One of the most widely distributed biological products that frequently involved mice or mouse tissue, at least up to recent years, are vaccines, especially vaccines against viruses . . . It is possible that XMRV particles were present in virus stocks cultured in mice or mouse cells for vaccine production, and that the virus was transferred to the human population by vaccination.[55]

What if like a character in some Greek tragedy who seeks to bring about something good for humanity, the scientists who sought to end terrible diseases like polio and yellow fever inadvertently created other diseases? Could the Brodie vaccine have contained unsuspected new viruses caused by the mixing of animal and human tissue? Could other researchers who mixed human and mouse tissue in cultures for vaccines and other biological products have created other entirely new strains of viruses?

What if in the attempt to solve the riddles of one plague, the scientists had created two more?

CHAPTER SIX

Day Three in Jail

A good head and a good heart are always a formidable combination.

—Nelson Mandela

Sunday, November 20, 2011

By Sunday morning's breakfast call at 5:45 a.m., the rain outside was a steady white noise and Mikovits knew she was not getting out that day.[1] The arrest at home on Friday had blindsided everyone as there was no arrest warrant. By Saturday in her jail cell, she felt angry, mystified, and increasingly sad. By Sunday morning, these had muted into a dull resignation.

Mikovits gazed out the small window at the top of her cell. She saw tops of rows of lemon trees outside, discerned a hazy rim of the mountains in the distance, and watched the sheets of rain coming down. The guard led them out, and after grabbing their breakfast trays she and Marie went back to the cell. Mikovits picked at her food. Marie ate quickly and then dozed off. Mikovits could only flash back to recent events.

Months after her October 2009 publication in *Science*, the staff at WPI had chatted anxiously about how to respond to the increasing controversy over the association between XMRV and ME/CFS. Beyond the discovery of HTLV-1/2 and HIV, the field of retrovirology had been a quiet player in human disease investigations, but those retroviruses—particularly HIV— were scientific bombshells. The press exploded after the *Science* publication and some officials seemingly did covert damage control, with Dr. William Reeves from the Centers for Disease Control's ME/CFS program proclaiming to the *New York Times* about her research "If we validate it, great. My expectation is that we will not."[2]

Annette was a fan of the 2009 movie *Invictus*, which detailed how in his post-prison years Nelson Mandela had launched a charm offensive, using white South Africans' rugby fanaticism as a way to bring factions of the country together after apartheid had ended. With Morgan Freeman as Mandela and Matt Damon playing the captain of the South African rugby team during the World Cup, the movie had moved Annette to the point that she had encouraged Judy (who wasn't a film buff) to recruit the entire staff to watch it.

Mikovits saw the movie on a flight shortly after a conversation early in 2010 with Annette "We know that groups like the CDC and the NIH have *in the past* screwed up their investigations into this disease, but we want to bring them along," Annette had said. "We want them as our partners. If the CDC has a probe we can use, then we take it. If the NIH wants to help, we'll let them. Let's try to make everybody a winner." It seemed possible to quilt together ideas, as had been done in AIDS, once the initial resistance and prejudice was overcome.

In a field where discoveries were often side-tracked by ego grabs, the two women wanted to show collaborative potential for the common good. Though ME/CFS was often dismissively called a "wastebasket diagnosis," early AIDS might have been similarly pegged if it wasn't bogged down in other insulting misnomers such as "gay related immune deficiency" (GRID), since its initial presentation seemed beyond belief.

The so-called "wastebasket" mislabeling was, in fact, the basis for a lot of the prejudice against ME/CFS patients, leading to the false impression that patients could not possibly have so many diverse symptoms in one body. As patient-writer Jodi Bassett stated in her article "ME—The Shocking Disease," "People say it's too severe and there are too many symptoms. The entirely unique way we respond to even trivial exertion and are so disabled by it, instead of inspiring sympathy, seems to actually inspire disbelief. People seem to (bizarrely) believe that there must be some limit on how bad a disease could be, and that such severe illness couldn't be possible long-term."[3]

A retrovirus theory for ME/CFS—as with AIDS—created a unifying thread, since it made essentially every coinfection hypothesis workable as these could now be viewed as secondary infections. AIDS (as opposed to just HIV infection) was in fact defined as HIV *plus* one of twenty odd coinfections and CD4 counts below 200, with the bugs attacking a patient after the immune system was wrecked by HIV.

The resemblance of ME/CFS to AIDS was unmistakable: AIDS patients often had active coinfections in the herpes virus family, mycoplasma, candida—all found to be overrunning the body in ME/CFS by doctor-pioneers such as Dr. Martin Lerner, who biopsied heart tissue of ME/CFS patients and found opportunistic infections such as cytomegalovirus (CMV) and HHV-6, as well as Lyme and Babesia.

Babesiosis, CMV, and HHV-6 are also known to run rampant in AIDS, and Lyme is caused by a spirochete just like syphilis, which occurs in AIDS. Lerner's story was like that of Chiron, the mythological "wounded healer," as his eureka moment about ME/CFS and heart infections came after his own heart was infected and he treated himself with antiviral drugs.[4]

Other tick-borne coinfections such as bartonella are also found in both AIDS and ME/CFS. In AIDS, an association was even found between bartonella henselae and the progression to neuropsychological symptoms and dementia.[5] Bartonella in AIDS patients was studied fairly early on in the pandemic because of the tragic fact that some people with AIDS, seeking comfort in a pet, had contracted it from cat scratches.

Immune abnormalities in natural killer cells, RNase L and clonal expansions of T cells in ME/CFS also paralleled the immune dysregulation recorded in AIDS, and could be retroviral in origin. Even John Coffin from Tufts University, discussed the potential new linkage between retroviruses and ME/CFS in his early commentary on the *Science* paper.

Coffin temporarily saw they were looking through a sharper lens at old history, and in retrospect the connections between ME/CFS and a retrovirus were obvious.

* * *

Around eleven-thirty that familiar call came over the cell's speaker: "Mikovits, prepare for a visitor." As if to add to the feeling that *the jail* was some kind of wastebasket—a place for the socially forgotten—nobody else on the cellblock had had a single visitor but her. She snapped into the routine this time. She had the booking number sheet in her pocket and put her hands behind her back, and covered her signet ring with a finger.

The door to the cell opened and Mikovits knew to wait at the entrance for the guard now. From where she stood at attention, she could see the circular free time room and some of the inmates sitting at the steel picnic tables. Although the faces were fuzzy without her glasses, she saw the woman at the table nearest to her take notice of her. "Can you help me?" the woman asked imploringly. Mikovits recognized Ruth, the woman who had entered the holding cell sobbing and coughing on Friday night. Mikovits surveyed the scene to see if any of the guards were witnessing this. She had noticed that they didn't seem to mind prisoners communicating, so she chanced it. "What do you need?" she replied.

"You have a visitor?" asked Ruth. "You're going to see someone now?"

"Yeah in a minute."

"I have to get word to my family and to my work. They expect me to be there or I'll be fired. I have to keep my job."

"Your family only has cell phones?" Mikovits asked, remembering how difficult it was for her to make contact with David from the holding cell initially due to the restriction from calling cell phones. But the rules were different at the Todd Road Facility, allowing cell phone calls.

"Yes. My daughter has a landline phone, but I only got the answering machine."

Mikovits imagined the phone ringing at the daughter's home, the message machine picking up, and the robotic voice that David would later parrot saying, "Message from inmate—Ventura County Jail." Ruth did not seem like a repeat offender—like Mikovits, she looked shocked to be there, a little too pliable for the atmosphere, traumatized. Her daughter probably had no idea her mother had been picked up by police.

"What do you need me to do?"

"I work as a checker at a grocery store in Thousand Oaks. Call and tell them Ruth is sick, but she will try to be there tomorrow. I will call them when I can. I have the bail money. In my purse I have money, but it was taken from me. If somebody can make bail I can pay them right back."

Mikovits knew the woman's bail was probably only a few hundred dollars. "What are the charges against you?"

"Driving under the influence. But I wasn't drinking. I was sick with a cold. I took NyQuil, but it did not help my coughing and it was keeping my husband awake. He was angry and cranky because he had to go to work early the next morning. He told me to go to the store and get better cough syrup. Cops picked me up on the way to the store. I shouldn't be here."

"I'll do my best," said Mikovits just before the guard ambled up. Mikovits put her hands behind her back and placed one finger over her signet ring to conceal it, and made her way to the visiting room. Ruth was all smiles as she passed, exclaiming, "thank you" as the guard cast them both a curious look.

David and Sheralyn Littleton, a dear friend from their church, Community Presbyterian (the parishoners call it CPC) in Ventura, sat on the other side of the glass.

Mikovits grabbed the phone handset hurriedly and David mirrored her on the other side. He asked if she was okay. She mentioned to him that she hadn't been able to brush her teeth since Friday night when they had given her a small packet of toothpaste. David knew there was a general store for the inmates and had gotten the paperwork to set up an account for her, but it would not open until Monday morning.

Another day stretched before her without a way to brush her teeth, but it was just another indignity. Any time an inmate complained, the guards reminded her that jail wasn't a resort.

David looked like he had not slept. The lines on his forehead were little crevasses behind his icy-looking skin. He gave the phone over to Sheralyn, who said, "Judy, everybody has been praying for you. We can't believe this!" Sheralyn and her husband Emmet were deacons at CPC. She was part of a group that visited hospitals and prayed with people before surgery. Strong in her faith, Sheralyn typically contained her emotions to soothe those patients. When she saw the surreal image of her friend behind glass, she could see Judy was about to break into tears. Sheralyn tried to steer her friend away from emotion, to protect David who seemed fragile.

Sheralyn and Emmett had met David at CPC well before they had ever seen Judy. David would rave about the fabulous woman he had met who lived back East, while Sheralyn teased him that Judy was just an imaginary girlfriend. They finally all connected at a Superbowl party in 2001 and Sheralyn took an instant liking to the scientist. The couples traveled together and Judy enthusiastically volunteered for the church's charitable activities. "I know, Sheralyn. I'm in shock about it too. Please tell everybody I'm okay."

"I will take care of David," said Sheralyn. "He's frightened about these horrific events, but he has so many caring friends. Don't you worry."

Sheralyn was having trouble envisioning what improbable set of circumstances had landed her friend in jail. Judy was a big-hearted person. David gushed about her compassion. "I wouldn't be surprised to come home and find somebody sleeping in our living room," he joked, soliciting guffaws from those who knew Judy's open heart meant an open door policy. She and David slept on the sofa, endured the bad backs in the morning, and gave up their own bed for whomever they had welcomed in. Sheralyn hated to see her friend like this.

Mikovits knew how bedraggled she must look dressed in her orange jumpsuit with her unkempt hair, her breath surely questionable. She was grateful there were a couple inches of thick glass since she wasn't so minty fresh. David looked to be on the edge of crumbling. "Can I have him for a minute?" she asked her friend. Sheralyn handed David the phone.

Sheralyn could only hear one side of the conversation, but it was vintage Judy.

"You want me to do what?" David asked.

* * *

Back in the free-time room, Judy saw Ruth again. "So? Did you talk to him?" she asked.

"Yes, he's calling your employer." Mikovits glanced suspiciously around to see if their conversation was attracting the guard's attention.

"Oh, I am so relieved! Thank you!" exclaimed Ruth. "Were you able to talk to him about my bail?"

"Yes, that too. Let me give you his cell phone number. He now knows your situation and maybe can do more."

David and Sheralyn were at a nearby restaurant decompressing from their experience when the call from Ruth came in. David thought it was his wife and accepted the collect call, but quickly figured out it was her jail-mate. He reassured Ruth that her workplace had seemed sympathetic about her health. He was a little bewildered by her request to help her make bail, but knew his wife would want him to do something. He drove Sheralyn to her house through the still-pounding rain and then headed over to the courthouse to talk to a bail bondsman.

The process took over an hour, and then he needed to wait for Ruth to be driven over from the Todd Road facility so that he could escort her home. Prisoners also had to have someone pick them up once bail was posted. Ruth's car—her family's only car—had been impounded. At about seven in the evening, while Mikovits was in the free time room, she saw Ruth enter the area with her box of possessions and her folded-up thin mattress that could have been a Pilates mat under better circumstances. Ruth nodded to Mikovits, mouthing an exaggerated "thanks" as the guard hurried her out.

Mikovits thought for a moment about the narrow line between freedom and incarceration, between wellness and a virulent disease. Virologist Robin

Weiss once wrote, "If Charles Darwin were alive today, he might be surprised to learn that humans are descended from viruses as well as from apes."[6]

On a cellular level, there really was a pretty thin separation between inside and outside, self and other. People were made up of their interrelationships—from the cellblock to the world outside, from the virus to the inner machinery of a human cell. Interrelationships were the essence of life—both good and bad—and perhaps this is why the simple of act of connecting with another inmate made the day feel lighter.

* * *

Mikovits knew what a gentleman David was. He would immediately offer to carry Ruth's box for her, with the kind of simple sweetness that could push a person to tears after an ordeal like jail. He would also be upbeat, make pleasant small talk to keep Ruth at ease. He came across as both caring and incredibly nonthreatening to strangers, which had attracted Mikovits when she met him in 1999 and he was the human resources director of a local hospital. The difference between David and some HR folks is that he really did see all people as *resources,* much like his wife did.

Mikovits had an indulgent moment of self-pity, since this time she wasn't the lucky one. She had not seen a lawyer, nor even heard from one. She had talked to the astonished bail bondsman who proclaimed that in all of his years he had not seen a bail hold for somebody in her situation. She wondered if she should call David tonight, but at ten dollars a minute, what would she say other than tell him how distraught she was. In her helplessness, she felt perturbed. Why hadn't he been able to bail her out?

As her thoughts churned, Mikovits heard a woman's voice singing a religious melody loudly in the background, with erratic pitch. She recognized the song as "Praise God."

Mikovits glanced up saw the wiry inmate with a mop and bucket and spray bottles working on the shower. Her tune seemed so happy, defying the surroundings. She didn't mind at all if she skipped half the words, humming through verses. She kept time by slopping the mop on the tile, circling it around, wringing it out, singing her mop-microphone karaoke. Mikovits felt spurred to do something.

She walked over and asked, "Do you want some help?" Her jailmate was in her early thirties, thin with cropped dark hair, looking like somebody who might have thrilled down highways on a Harley or two. Mikovits later learned the woman was in jail for stealing a truck, a story that like many of the stories she heard in jail contained a few murky particulars.

"Sure," said the woman jovially, and handed Mikovits some of her cleaning items. "Just follow me." The woman returned to scrubbing out the shower and resumed her singing while Mikovits began wiping down the sink.

"Where did you get these supplies anyway?" Mikovits asked.

"Oh, you just ask the guards and they'll give them to you. You don't think the guards clean up, do you? They leave that to us." Mikovits hadn't even considered this.

They scrubbed for a few more minutes and Mikovits marveled at her new acquaintance's beautiful work. "You're a pro at this," she said.

The woman let out a hearty laugh and said, "Well, I should be! I do it for a living!" "Ah, that's great. My place is a mess." Mikovits flashed to imagine her ransacked home. "Maybe when you get out of here I can hire you." The woman gave her a wily little smile, as if aware of the absurdity of their situation and said, "Sure, I'm always looking for work."

Again, Mikovits hoped the movement would help her sleep. Earlier in the day she thought walking quickly up and down the stairs between the two levels would tire her out. But actual work was something different. It got into the bones in a deeper way.

They were nearly finished with the second level when Mikovits on a whim decided to take a minute to watch the football game on television. She loved watching sports, from basketball at the UVa to the minor league Reno Aces baseball team.

Mikovits had even been Annette's special guest at game five of the 2010 World Series when the San Francisco Giants won the series for the first time in history. In the picture they emailed to Frank Ruscetti, they looked as the happiest of sisters; who would have guessed that one year later Judy would be jailed by her sister. Harvey, a lifelong Giants fan, often ribbed Mikovits that he had given up his seats to her and had never gotten to see *his* beloved team at the World Series because he was too busy fighting foes for Annette and Judy.

But she wasn't in the bleachers at PacBell Park now, just straining by the second-level railing at the Todd Road Facility to see the game on the small TV. She leaned out over the atrium to make out the scoreboard.

"*Suicide!*" A shrieking call came over the loudspeakers. Mikovits peered around at the sudden, sped-up motion, heard the *thud thud thud* of the guards' hurried footsteps, the crisp steel door from the tower swinging open. Her initial thought was that someone had been found blue in a cell, and guards were rushing in to perform CPR. In another split second, she realized they were looking squarely at her.

"Stand back from the rail!" warned the guard who was now just about ten feet away from her, warily holding hands out in a protective stance and angling in a slight squat toward her.

Her cleaning partner intervened, "Oh honey, don't you know you can't stand there?" Just beyond the rail it was open air and a ten-foot drop to the first level of the free room. The guard looked at the other inmate, searching for validation that Mikovits wasn't going to make a suicide attempt.

"I'm sorry," mumbled Mikovits. "I just wanted to see the score and my eyesight isn't so great so I was just leaning—" She quickly stopped blathering and backed away from the edge.

Then the guard immediately relaxed her posture. The free time area, focused on the drama for a minute, went back to buzzing with conversation.

Mikovits went over to her jail-mate and picked up the cleaning supplies, shaken. What a crazy day for a scientist, from the early rooster call to being viewed as an imminent suicide risk.

Like jails, viruses themselves were also freedom-snatchers, basically inert until they invaded a host cell and then doled out rules, parameters, and behaviors for the host to follow. As much as free will actually mattered, viruses seemed to hijack it on a cellular level. As Michael Specter wrote in the *New Yorker*, "Viruses reproduce rapidly and often with violent results, yet they are so rudimentary that many scientists don't even consider them to be alive. A virus is nothing more than a few strands of genetic material wrapped in a package of protein—a parasite, unable to function on its own."[7]

It was easy to imagine how, given this zombie-like depiction, those with ME/CFS could feel like the living dead. Viruses toppled the inner workings of life and changed the very structure of identity, much like the jail cell was already giving Mikovits a different and harrowing vision of what was possible in her life.

CHAPTER SEVEN

The Submission to Science

At the heart of science is an essential balance between two seemingly contradictory attitudes—an openness to new ideas, no matter how bizarre or counterintuitive they may be, and the most skeptical scrutiny of all ideas, old and new. This is how deep truths are winnowed from deep nonsense.

—Carl Sagan

After the Cleveland Clinic, the National Cancer Institute, and the Whittemore Peterson Institute signed the confidentiality and collaborative research agreement on Saturday, January 24, 2009—to isolate the XMRV retrovirus from humans for the very first time and demonstrate its association with ME/CFS Mikovits quickly jotted off an email to her lab technician, Katy Hagen.[1]

When the word *contamination* later became such a hot button issue—the e-mail was an important record of Mikovits's thought process, as she briefed Katy on how to properly *de*contaminate the lab for upcoming experiments. All of the signatories understood the momentousness of what they were embarking on. They were on the hunt to isolate a new human retrovirus of unknown pathogenic potential. The implications could be even further-reaching than the AIDS epidemic. If they succeeded, their findings might create a revolution in thinking about the role retroviruses play in human health.

Most virology research dollars were spent hunting visible "killer viruses," such as Ebola, West Nile Virus, and Bird Flu. But what if such conspicuous targets were only part of the story? Scientists had been focused for over a century on viruses that brought obvious mayhem and death to society, but what about viruses that might be incapacitants, silently stealing the health of victims? As Michael Specter wrote in *The New Yorker*:

> Nothing—not even the Plague—has posed a more persistent threat to humanity than viral diseases: yellow fever, measles, and smallpox have been causing epidemics for thousands of years . . . Those viruses were highly infectious, yet their impact was limited by their ferocity: a virus may destroy an entire culture, but if we die it dies, too . . . Only retroviruses, which reverse the usual flow of genetic code from DNA to RNA, are capable of that.[2]

If a retrovirus could lower the defenses of the immune system, ushering in a host of pathogens that in turn caused complex damage, the *Invasion of the Body Snatchers* scenario became overwhelming.

ME/CFS patients sometimes called themselves the "living dead." Though other diseases had caused freakish disruptions of sleep (i.e. African Sleeping Sickness, Fatal Familial Insomnia), there were few conditions that matched the twilight reality of ME/CFS that could linger for decades. Yes, patients remained alive, but barely. Mikovits kept thinking of Dr. Mark Loveless's statement that an ME/CFS patient feels every day about the same as an AIDS patient feels two months before death, or Dr. Nancy Klimas's declaration that her AIDS patients were "hale and hearty" compared to her ME/CFS patients. "Many of my CFS patients, on the other hand, are terribly ill and unable to work or participate in the care of their families," Klimas wrote in a *New York Times* blog.[3]

Mikovits and her team now had a propulsive plan. Once they had meticulously done their research, they would submit their article to the most prestigious scientific journal in the world, *Science*. Then the evidence would cross into the global sphere for other scientists to grab the baton. With a clearer grasp of pathophysiology, they could finally work on therapeutics for the disease. But they had to be exacting. From the beginning they must eliminate *every* possible source of contamination. She had learned from Frank Ruscetti and her work with HTLV-I and HIV-I, when isolating a new human retrovirus and attempting to associate it with a disease, contamination was the foremost threat. She kept this as her highest priority. In her email to Katy Hagen, Mikovits detailed what she had learned to be the possible sources of contamination, after more than two decades of work with human retroviruses.[4]

She told Katy to wipe down everything in the lab with bleach, change HEPA filters, empty all incubators of cultured cells, also no cultures in the laboratory but primary patient culture and autoclave the glassware, instruments, and where possible, reagents. An autoclave sterilizes lab equipment in high-pressure saturated steam at temperatures above 400 degrees Fahrenheit for about 15–20 minutes. Katy methodically went through the decontamination procedures, and then carefully taped the email from Mikovits into her notebook on January 25, 2009. These were standard procedures when working with retroviruses, but of course HIV had spawned a whole generation of ever-increasing caution. When Mikovits returned a few days later, she pulled out official HIV decontamination protocols and made sure those were carefully followed in the lab as well.

Since the WPI building was still under construction, they had been working in temporary lab space, room 401 of the Applied Research Facility at the University of Nevada, Reno. Keeping the research confined to a single laboratory lessened the possibility of contamination. Mikovits flashed back to her first science job, when she was a protein chemist at the National Cancer Institute's fermentation program, and how in her opinion there had been lax safety procedures during the first growth of HTLV retrovirus in large-scale fermenters that were something like the steel tanks in today's microbreweries.

The questions about XMRV were the same as they had been about HTLV in the early fermentation lab, and much more important two years later as they isolated HIV in Ruscetti's lab. How infectious was the virus? How stable was it? Nobody knew the answers with XMRV, so it was crucial to proceed as cautiously as possible.

They also had to determine what other parties should sign the confidentiality agreement[5] and participate so the research could go forward, but keep the team limited to a minimal number due to the project's riskiness. Mikovits added Katy Hagen and Max Pfost, as signatories. Frank Ruscetti added two of his research associates. At Bob Silverman's lab, Jay dip Das Gupta was added, because he would do the genetic sequencing of the virus.

Mikovits then contacted Debbie, a nurse who had worked at Peterson's office. Besides being a nurse, Debbie was a skilled phlebotomist. Mikovits assigned Debbie to draw blood, consent people into studies, coordinate patients, and keep clinical data in order. Because the WPI had yet to be awarded its first NIH grant, Debbie could only be salaried as a part-time employee and Mikovits was limited in assistants to students and a single postdoc, Lombardi.

One of Debbie's first duties was to draw the blood of everybody at the lab to see if they had been exposed to, or infected by, XMRV. This blood work would establish a baseline for the lab workers and allow protection of the staff from unknown risks. If any of them already carried the retrovirus it could compromise the integrity of the research, and they would have to be dismissed since the workers themselves could bring contamination. A consequence of this confidential study was that relatively few people could be even aware of its existence. The graduate students running experiments in Room 401 were also a concern. If they were not directly involved in Project X (as it came to be known), they still had to be protected from exposure because of the shared facilities. Mikovits had to encourage them to finish up their projects, move them to another lab, or simply shut them down and shuffle them out without offering an explanation. She knew that being brusque and secretive about Project X was likely to generate hard feelings, especially as it would be out of character for her.[6] But those were the options.

For many reasons she decided that the lab in Applied Research Facility (ARF) Room 401 would only be used for work with patient blood samples and that no virus would be grown there. Lombardi would grow the virus in a separate lab at Viral Immune Pathology (VIP Dx), a distinct company purchased as RNASEL Enzyme Deficiency Lab (RED) LABS from Dr. Kenny de Meirleir in 2004 by the Whittemores in order to bring to market any diagnostic tests or therapeutic treatments that might be developed for ME/CFS.

Even though the WPI still didn't have its own laboratories, they would pour copious resources into the XMRV project. The project was not even on Mikovits's original list presented to her scientific advisory board just a few months earlier. The risks were incredibly high, the atmosphere necessarily covert. But Silverman and Ruscetti thought the preliminary data was so compelling and the benefits for the patients so potentially dramatic that they just had to go all in.

In a town of cheap slot machines and occasional windfalls, Mikovits and her team were ready to roll the dice.

* * *

Mikovits was routinely up at five a.m., clutching a cup of coffee, and on the phone with Frank Ruscetti who had returned to the NCI in Frederick, Maryland. They laid out plans for the critical experiments they needed to perform, determined who would do what, figured out what samples from the repository were important, and strategized how to move the daunting tasks forward. Mikovits spent substantial time writing up the plans with Katy Hagen, who had risen toward the top of the pack of research associates/students. Katy was focused and thorough, and Mikovits saw a vision of herself as a young researcher when Ruscetti had noticed her potential.

Although it was not one of the four methods they would eventually use to test for the presence of the virus, Mikovits knew it was important to show that XMRV had a natural history of infection in humans by isolating virus from patients and passing it into other cells.

Previously all of the work on XMRV biology had been done using an infectious molecular clone, which was constructed from partial XMRV sequences from RNA isolated from tissues from prostate cancer patients. One way to obtain this evidence was an electron micrograph picture of XMRV, preferably an image of the particle budding from a cell. This could be compelling evidence the virus was actually replicating and infecting other cells.

To isolate XMRV from a single human then use it to infect white blood cells repeatedly would prove that XMRV was the first member of the third known *infectious* human retrovirus family. Electron micrograph pictures had proved solid evidence with other retroviruses: Poiesz and Ruscetti presented one with the isolation of HTLV-I, and Montagnier obtained an electron micrograph image of the HIV retrovirus that became a central piece of evidence in convincing the scientific community that HIV was a human infection. Montagnier brandished the electron micrograph picture to doubters and decriers. The image was pretty indisputable, a vivid depiction of something previously unknown to science. However, there was no electron microscope available for use in the vast but sparsely-populated state of Nevada.

Most science students across the world have peered into a regular light microscope and seen cells, so they have an appreciation for the small size of a typical cell. They can learn to identify the cell wall, the protoplasm, maybe even the nucleus. But a virus is typically about a million times smaller than the cells they infect, so without much more magnified imaging the virus is guesswork, almost phantasmal, to students trying to comprehend it. The virus contains a relatively small amount of genetic information, leading to the question posted in the last chapter: is a virus is a living thing or more like a parasitic entity? Some who argue a virus *lives* say that the ability to reproduce is the hallmark of a living thing. Others claim viruses are not alive since they lack reproductive machinery and must hijack cells to propagate.

Modern virus hunting rarely includes electron microscopy, but relies primarily on molecular genetic sequence data. But to Mikovits, the appeal of an electron micrograph picture of the suspected virus was undeniable. A traditional expression in science states "extraordinary claims require extraordinary evidence." Having a vivid picture for people to see, rather than just a dry sequence of numbers on a page, would go a long way.

Mikovits was fortunate enough to have a friend back at the NCI who ran the NCI's electron microscopy facility. He was the scientist who took the electron micrographs that proved the retrovirus Gallo claimed as his HIV retrovirus had actually been taken from *Luc Montagnier's* samples. The WPI would have to pay for the work to be done, but the Whittemores could make it happen. It would take a great deal of work on the part of Mikovits and Hagen to get samples ready for electron microscopy, but Mikovits felt it was of critical importance.

The work was done in a biosafety level three lab at UNR, and Mikovits prepared to teach Hagen the procedure to get the samples ready. Cells needed to be cultured and then pelleted by high speed centrifugation, first separating virus in the supernatant from cells and cellular debris. Finally, the viral particle containing supernatant had to be concentrated at very high speed ultracentrifugation and the virus pellet stabilized.

A virus is actually composed of two parts, an outer protein shell called a *capsid* and an *inner core*, which is composed of either DNA or RNA. A virus looks like an M&M candy, with a hard outer candy shell and a chocolate or peanut interior. Enveloped virus had a lipid containing membrane called a viral envelope. The outer layer of the virus can have a different arrangement of viral and cellular proteins (which it gets while budding from the cell), which bind to cell surface receptors that allow it to enter into various cells.

These viral protein molecules have also been found to provoke many of the body's immune responses. Some researchers have speculated that many autoimmune diseases are the result of the body reacting to the *envelope proteins* (the "hard candy shell") of certain viruses, rather than an infectious process (much of the later debate about XMRV would revolve around this point). But with the M&M analogy, the chocolate, peanut, (or even more recently, pretzel) interior of a virus has caused the greatest amount of debate.

In what scientists consider a "replication-competent" virus, there would be a shell and an inner core with a complete genetic sequence.

But defective viruses have only a protein shell and a *partial* genetic sequence inside and need helper viruses to spread. The scientific community has typically responded that if a virus does not contain a complete interior genetic sequence, it should not be considered a dangerous microorganism: it's like a Tin Man without a heart or a Scarecrow without a brain.

Others, including Dr. Sandra Ruscetti (the head of the Retroviral Pathogenesis Section of the NCI),[7] have advocated for a long time that science needs to take a closer look at the disease-causing potential of the protein shells of these viruses. Sandra Ruscetti had published extensively on the spleen focus-forming virus (SFFV), a replication-defective retrovirus that nonetheless becomes very pathogenic in susceptible strains of mice.[8] What Mikovits learned from Sandra Ruscetti was later invaluable in the investigation of XMRV.

For the time being, when Mikovits and Hagen were finally able to obtain a pellet from the ultracentrifugation, it was composed of almost pure virions. Once the work was completed, the samples traveled to her friend at the NCI, who produced six electron micrograph pictures. Even looking like colorful but unearthly creatures, the images were as clear as anything that Montagnier had shown more than a quarter of a century earlier in the HIV investigation.

The evidence was becoming even stronger that they were on the trail of a new human retrovirus.

* * *

This was a critical first step. Now they had to move forward to show the virus was infectious and transmissible.

This could be the first member of the third family of human infectious retroviruses. Mikovits was simply looking for *some* evidence of the XMRV retrovirus in patient blood samples. *Gag* is the name given to one of the three major proteins that are encoded within the genome of a retrovirus. Each retrovirus has a specific *gag* sequence, almost like a fingerprint, so once a researcher has identified the *gag* sequence, she should have a pretty good idea of the identity of the retrovirus.

Since all retroviruses require the machinery of a cell to replicate, they lie dormant in cells until a signal is received that allow them to replicate to a detectable level. Transmitting the virus from the patient's white blood cells or plasma into the LNCaP cell line (an immortalized line derived from a lymph node metastasis from a sixty-two-year-old prostate cancer patient) appeared to make the virus replicate like crazy. Mikovits knew that LNCaP was responsive to hormones and cytokines (inflammatory molecules), and in fact, turned out to be the single best cell line in which to grow gamma-retroviruses. Importantly, it had never been cultured with mouse cells lines or passed through mouse tissues. It was a stroke of good luck, allowing them to biologically amplify low levels of virus from a small number of white blood cells or a minimal amount of plasma so that it could be detected. She could use this indicator cell line with those samples they had painstakingly prepared for more than two years from Peterson's patients and those with abnormal immune responses.

Another necessary step was to show *which* immune cells harbored the new gamma-retrovirus. Were they T cells and monocytes—as was the case in HTLV or HIV—or were there different reservoirs of infection? Could the retrovirus be transmitted "cell free" (by virus floating free in plasma) like HIV, or did transmission require cell to cell contact (with the virus contained within cells, like HTLV-1)? They were all key questions that a paper on the first isolation of a new human retrovirus should answer to be fully credible.

A key proof that the XMRV isolated from the patient samples was not due to contamination involved looking for the presence of antibodies to XMRV in the blood of patients. In most viral infections this was a simple and straightforward way of looking for evidence of past or current viral activity. But the question remained whether this was a reasonable way to search for evidence of a virus that had not *previously* been isolated from humans. In the first publication for the isolation of HTLV and HIV, this had been considered beyond the scope of the first paper.

Mikovits and her team focused instead on proving XMRV isolates from humans with ME/CFS were infectious and transmissible. They also attempted to confirm Silverman's finding showing that XMRV infections were limited to individuals with a specific genetic polymorphism in the RNASEL gene—that of R462Q—a single nucleotide change that essentially rendered the antiviral enzyme inactive.

Many of the most deadly viruses, such as Ebola, seemed to cause the greatest damage in those individuals whose systems did not mount an antibody response.[9] The immune system was like a sickly guard dog, sleeping through the invasion. It was reasonable to think the most severely infected patients with the virus would show no antibodies *at all* to the virus. That was why they were sick, as they could not fight it off.

By contrast, the healthiest of the ME/CFS patients or those with some degree of recovery would show the highest antibody response. Focusing on whether the isolates from XMRV-infected individuals were infectious and transmissible were "old school," time-consuming methods of finding a virus, but Mikovits preferred them. Some of the more technologically advanced methods, like the viral array chip technologies, suffered limitations, such as the failure to load the proper sequence into the chip.

If the technology wasn't right, the answers wouldn't be either. Garbage in, garbage out, as the computer programmers liked to say.

An additional safeguard was to keep the various parts of the study separate so that nobody could later question whether there had been inadvertent contamination. Pfost was doing his part of the work at a lab in the agriculture building, a place that had never grown human or mouse cultures. Hagen was working at the biosafety 3 lab. Lombardi was working at the private VIP Dx lab, which was about ten miles away from the University of Nevada, Reno campus, and patient material was all confined to the Mikovits research lab, located at the Applied Research Facility, Room 401.

Among their small group of junior researchers, Lombardi, the former stockbroker who had only a year earlier received his doctorate in protein chemistry, would take a leading position. Mikovits assigned Lombardi to work on the immunological aspects of the paper, mainly whether the RNase L enzyme deficiency, the natural killer cell markers, or any other immune indicators were correlated with XMRV infection.

* * *

While Mikovits and her team of young researchers were working on their secret project, Mikovits also had other important tasks at hand.

The International Association of Chronic Fatigue Syndrome (IACFS) was meeting from March 12–15, 2009 in Reno and the WPI was gearing up to host it at the Peppermill Resort, which had recently finished its four hundred million dollar renovation.[10] Mikovits and her team submitted four papers for presentation at the meeting.

The Seeno family, part-owners of the Peppermill, rolled out the red carpet for scientists not accustomed to the casino lifestyle. Those attending the conference got reduced room rates (some with jaw-dropping views of the snow-capped Sierra Nevada mountains and the city lights of Reno), access to ten of Reno's finest restaurants right there at the Peppermill, and a space to relax at the Spa Toscano, a full-service salon where one could relax with a Swedish or deep tissue massage.

The program promised the "Latest Advances in Virology, Genetics, Pediatrics, Brain Functioning, Epidemiology, Treatment, and Assessment" and Mikovits focused on welcoming the speakers and hosting the conference. For their keynote speaker, they had Dr. John Kitzhaber, the former two-term governor of Oregon, whose wife suffered from ME/CFS.[11] There were presentations on behavioral assessment of ME/CFS patients, alternative medical treatments, use of interferon, active HHV-6, HHV-7, Epstein-Barr virus, cytomegalovirus, and even an association between ME/CFS and mitochondrial dysfunction.

Of the four talks being presented by Mikovits and her team, two gave her pause: one that she was scheduled to give, "Identification of Differentially Expressed Viruses in American CFS Patients with a Custom Mammalian Virus Microarray," and one by her postdoctoral fellow, Lombardi, entitled, "Serum Cytokine and Chemokine Profiles of Individuals with ME/CFS Distinguish Unique Subgroups Among Patient Populations." There were many sharp minds at the conference.

One attendee was Dr. Maureen Hanson of Cornell University, who had been studying ME/CFS ever since her close relative had come down with it.[12] As Hanson later recalled, "At the meeting she [Mikovits] talked about using a microarray for viruses in chronic fatigue syndrome and I had been thinking of doing the very same experiment myself. So I was glad to see that somebody else had already done it and people were trying to find out if viruses were present."[13]

Mikovits was afraid that if anybody was really paying attention to the two talks from her group, they might sleuth out that someone was on the trail of a retrovirus. The talks could blow their cover, but they were only ten minutes long. So, Mikovits was careful with the data shown in the slides so they wouldn't scream out "retrovirus!" to somebody who really knew the science.

But the conference did not expose them. Mikovits was pleased the talk she presented on behalf of her NCI colleague and the talk by colleague and UNR epidemiologist—Julie Smith-Gagen—were noted in Harvard professor Komaroff's summary as among the most intriguing at the conference. There were some rumors buzzing around the conference of a new *virus* isolation, but the research team was tight-lipped, and their data was preliminary. If attendees later felt they had been misled by the attenuated talks, well, the information would break with a bang in a couple months.

The data was just too sensitive to be prematurely released.

* * *

As the team moved ahead, however, there were a few tiny fissures in the unity of the project.

When Mikovits had returned from the earlier San Diego conference and discussed the suspicions of a retrovirus with the Whittemores, they all agreed to bring Peterson in on the research. He would be integral to choosing the patients most likely to harbor the virus, and Mikovits was concerned that they might be constrained if they were unable to obtain fresh patient blood.

The Whittemores asked Peterson to sign the confidentiality agreement.[14] He asked what it was about, but the Whittemores said they would tell him after he signed it. He wanted his lawyer to take a look. This reasonable volley back and forth somehow began the deterioration of the relationship with Peterson, but the reasons were never clear to Mikovits.[15]

Mostly, the process was efficient and congenial, as well as exciting. Every Wednesday there was a meeting at the luxurious offices at the Red Hawk resort with Annette, Harvey, Mikovits, Peterson, and Mike Hillerby, who was one of the bosses at the Wingfield Nevada Holding Company. They discussed what was not covered by the confidentiality agreement, and only that, carefully avoiding any mention of the details of their secret project, the one named "Project X" in their computer folders now.

Harvey would often take Peterson aside and tell him that the confidentiality agreement needed to be signed, because he had an important matter to discuss. But even the "king of the legislature" could not get through to the doctor who had landed at ground zero of the modern epidemic of ME/CFS.

Nobody knew why.

* * *

Silverman also threw a wrench into the works.

In early April of 2009, Mikovits contacted Silverman and asked, "Okay, you said you've done this in hundreds of samples. Can you provide the peripheral blood mononuclear cells, the blood data, for those where you found XMRV negative for everybody except those with the R462Q variation?"

The R462Q variation was the genetic aberration that Silverman credited with causing the crippling defect in the RNase L antiviral defense enzyme. Mikovits wanted those data from blood, rather than tissue, since they were using blood samples from the ME/CFS patients and XMRV had never been isolated from the prostate cancer patients tissue or blood. They were trying to determine if the R462Q variation correlated to evidence of XMRV in blood, and thus to the RNase L defect, and needed to make this comparison using blood instead of tissue.

Silverman replied, "Oh, we never looked in blood."[16]

The control samples he had done were of prostate tissue for people who either didn't have prostate cancer but had resections for things like benign hyperplasia, or from other tissue surrounding the diseased area, and thus tissue samples were what was available.

This revelation injected an unanticipated degree of uncertainty into the XMRV investigation. It was clear that Silverman could identify the XMRV virus in *tissue*.

But it might be another thing entirely to locate the virus in *blood*.

This was in part due to the problem of viral latency, the ability of a virus to hide in certain organs and not be detectable in the bloodstream. This was a dilemma that kept virologists awake at night. Does a negative viral test based on a blood sample really mean the patient is negative?

Could the virus be latent and inactivated in organs and tissue, in the DNA, as theorized by Cheney? Could a relatively low amount of viral activity cause symptoms in the patient but not otherwise leave traces in the blood? If the virus was in some way modulating the immune response to insure its survival, it became much more complex, stealthy, and puzzling.

So Mikovits, just about a month out from submitting the paper, suddenly had her back against the wall. While Mikovits and Ruscetti had looked at the blood of healthy people, the non-contact controls for viral expression, they had never looked at the much higher number of samples needed to be examined for a genetic mutation like RNASEL R462Q. She had to find more healthy blood donors since Silverman could not offer blood samples.

Fortunately, the cytokine paper the Mikovits team had started in 2007 yielded an unexpected bonus. In preparation for the study, they had acquired a number of normal blood samples. The samples had been collected by a UNR researcher and a physician in Reno in the course of a study to examine normal immune responses in women. It was Mikovits's good fortune that the samples had been perfectly prepared for her intended use. The samples contained both plasma and DNA and a lot of flow cytometry data, which would allow them to look at populations of natural killer cells and differences in lymphocyte populations. All they had to do was take the DNA and plasma for XMRV PCR and serology and do cocultures with LNCaP and RNASEL genetic studies.

Other controls were also tested blindly in Frank Ruscetti's lab at the National Cancer Institute. He was also able to get fresh controls through the National Institutes of Health, who were always collecting samples as part of their ongoing mission of protecting public health. In addition to the samples from the cytokine study, the WPI was able to purchase a hundred frozen samples from a paternity testing lab in Nevada. The paternity samples offered an advantage in that they were age and sex matched according to geographical locations, which could be identified by their zip codes.

One of the concerns in terms of backlash was that people might make the mistaken assumption that the control samples were from those who might have been linked to the Lake Tahoe outbreak of 1984–1985. The mix of samples from around the United States with a concentration in Nevada, as well as the controls Frank Ruscetti was receiving through the NIH, should make it clear the team was looking at the prevalence of the virus in the *general* population. They combined the results from controls for the Nevada group, which was about four percent positive for XMRV, and the results from Frank Ruscetti's group at the NCI, which were close to 3.5 percent. The control group eventually included three hundred and twenty subjects who had no history of ME/CFS.

Twelve of the controls were positive for the XMRV retrovirus, an infection rate of 3.75 percent. If those numbers could be extrapolated to the US population it meant that somewhere between *ten to twelve million Americans* were infected with the virus. Twelve million people would equal close to the total 2012 populations of New York City (8.337 million) and Los Angeles (3.858 million) combined, obviously a devastating number.

The comparison of ME/CFS fatigue to an atom bomb suddenly seemed eerily apt, and not just because of how the patients felt. It was as if a sneak attack had been launched against the American public, and even now they didn't see the smoking wreckage.

Before they could finish up their manuscript and send it off, one event took precedence over all other things: a soiree that every member of the WPI was expected to attend, the Andrea Whittemore wedding.

* * *

Andrea Whittemore and Brian Goad said their vows on April 25, 2009 in Reno.[17] It was a quintessentially lavish Whittemore affair, displaying the wealth, political influence, and familial closeness that led Mikovits to later dub them the "Kennedys of Nevada." The wedding was slated for eleven o'clock at the historic Trinity Episcopal Church on the banks of the Truckee River that twisted through downtown Reno.

Frank and Sandy Ruscetti flew out for the wedding and the Whittemores put them up at their luxury twelfth floor condominium at the Palladio, a swanky downtown development. The unit had Western views, which took in not only the sunset over the Carson Range whose peaks reached more than five thousand feet into the sky, but a section of the Truckee River and the city of Reno. Mikovits's condominium was just a block away. The Ruscettis were also there to work on the final sections of the paper with Mikovits. Sandy Ruscetti's expertise in mouse retroviruses and the protein envelope were of critical importance.

* * *

The wedding was on a dazzling, sun-dappled April day typical for Reno: pleasant and bright but not too hot. The wedding party included many members of the large Whittemore family. The church's old-fashioned pipe organ added to the grandness of the space, and once Mikovits showed up with her group, there were at least two hundred people seated. She and David slid into a pew with Frank and Sandy Ruscetti, as well as Katy Hagen and her parents. Katy's father was Judge David Hagen and her mother Peggy was one of Annette Whittemore's best friends. The service, traditionally Episcopalian, reminded Mikovits of her girlhood Catholicism, and she was delighted to see Andrea—the woman who lost much of her girlhood to a ravaging disease—have this opportunity.

The wedding ended at just past noon, but the reception wasn't scheduled until six, giving Andrea time to recuperate a little before her next public showing. Along with David, Frank and Sandy, and Katy and her parents, their group headed over to the Silver Peak for some lunch and drinks. When the meal was over, Judy and David went with Frank and Sandy to the Whittemore condominium at the Palladio and spent a few hours chatting about the remaining issues in the paper.

The reception at the Peppermill Casino gathered together all the usual political dignitaries. But the highlight of Andrea's wedding reception was without a doubt the father-daughter dance. Pictures of Andrea's porcelain skin and slightly forced smile with a nasal oxygen cannula were later featured in the *New York Times*, but for now, the dance floor cleared to a lighted silence, and then big Harvey took the floor with his pale, delicate daughter, and they floated around to the music.

* * *

Unbeknownst to Mikovits, and probably most of the guests, beneath the surface of the celebration were broiling issues that threatened to destroy the Whittemore empire and the accomplishments of the WPI.

In a lawsuit filed on January 27, 2012, by their business partners in the Wingfield Nevada Group Holding Company, the Seenos alleged that prior to Andrea's wedding, Harvey had illegally obtained a forty-four million dollar loan and spent it all in fifteen months time.[18]

According to the lawsuit, Harvey had also made a confession.

> "As admitted in his confession, Whittemore lied to the Seenos when he represented to them that he would use his personal funds toward capital contributions to Wingfield. To the contrary, Whittemore admitted in his confession that he used bank proceeds for lifestyle choices and to make other investments . . ."[19]

It also alleged that Harvey did not even have enough money to pay for his daughter's wedding. He put the wedding expenses on the Seeno and Wingfield "tab,"

> "One such event was his daughter's wedding, which included five-star food and wine tastings, bridal showers and a rehearsal dinner leading up to the wedding. These events included the purchase of tuxedo uniforms for Red Hawk wait-staff, special china dinnerware named after his daughter, special custom lighting, exterior and interior alterations to Red Hawk premises and many other purchased items and extraordinary costs and expenses which, except for a meager $10,000 paid by Whittemore, were all paid for by Wingfield at Whittemore's direction.[20]
>
> Wingfield's expense for the affair totaled approximately $200,000 which was charged to Wingfield without the Seenos' knowledge, approval or consent."

Harvey was already legendary in Nevada as "king of the legislature," a "Power Ranger," and a close confidante of the Senate Majority Leader of the United States Senate. But according to the Seeno lawsuit, in 2009, he barely had ten thousand dollars to contribute to his daughter's wedding.

Mikovits, her team at the WPI, and her collaborators at the NCI and the Cleveland Clinic, had no idea that the vaunted Whittemore real estate empire was looking more and more like a house of cards.

*   *   *

On the afternoon of May 6, 2009, Mikovits emailed the *Science* article showing an association between the XMRV retrovirus and patients with CFS, along with the accompanying figures.

April and early May of 2009 were consumed with the furious writing and revision of the paper. The article went through a dozen revisions, with Silverman, Mikovits, Ruscetti, and the coauthors all rewriting various parts until it was finally in impeccable shape. To celebrate the victorious completion of the edits, Mikovits took her small team out for dinner.

Mikovits couldn't help but notice how weary everybody looked. They had worked feverishly over the past four months, and though their faces might not show it, they had reason to be deeply proud. *Science* was supposed to respond within thirty days of the submission, so for the first few days Mikovits tried to block thoughts of the paper out of her mind.

In the review process, the journal would likely request additional items. Even if they didn't, there were still aspects of the project to complete. She would take a breather, relax a bit, and then plunge back into the work that had been put on hold after the meeting of the International Association for Chronic Fatigue Syndrome the past March.

The *Invest in ME* meeting was at the end of May. It was one of Mikovits's favorites and she had to prepare her talk on immunology and the viral microarray. She was so hopeful that she might have an answer from *Science* by the meeting that her last slide was an electron micrograph of XMRV.

At the conference, Hillary Johnson delivered an electrifying speech called "The Why," about the systemic ostracizing of ME/CFS patients, stating,

> "Hate speech incites acts of discrimination against the victims of such speech.
> Think of all that is denied M.E. patients as a result of being characterized as
> malingerers, attention-seekers, neurotic and emotionally weak, or as David

Bell says, 'Nutballs and fruitcakes'? How did the Soviet Union discredit its dissidents? It called them mentally ill. Labeling M.E. a psychiatric disorder is a political act, a form of social violence."[21]

It seemed to underscore the timeliness of Mikovits's research that would hopefully end tactics to discredit the sufferers. She was so moved by Johnson's speech that she called Frank Ruscetti at a late hour and begged him to let her reveal the secret of the retrovirus. He reminded her how important it was to have the paper published first.

Mikovits remembered the January day six months earlier when she and Lombardi had showed the preliminary data to Silverman and Ruscetti, and Silverman had exclaimed, "This will change everything for the patients."

On the way back to Reno, she stopped in Frederick, Maryland, to visit family and work on the ongoing collaborations with Frank Ruscetti and her former National Cancer Institute colleagues. They anticipated one of the questions a reviewer might ask was: "If your original hypothesis was based on the RNASEL, R462Q SNP correlation in the prostate cancer patients and the RNASEL deficiency in ME/CFS patients, did you find the R462Q SNP correlation of the RNASEL gene in ME/CFS patients?"

They wanted to prepare for this essential question. Mikovits was concerned because the work had seemed beyond the reach of the WPI's limited resources. She knew the experts in these types of genetic analyses were Drs. Michael Dean and Bert Gold at the NCI, so she invited them into the project.

Curiously, their team did not find an association between the RNASEL genetic variation and patients with ME/CFS.

It was still true per earlier research that the RNase L anti-viral defense enzyme was working poorly in patients with ME/CFS. But they didn't know which genetic variation was causing it or whether the faulty enzyme might be caused by something else altogether.

As frustrating as this development was, these little hiccups commonly happened in research. The finding was unexpected, and something of a black eye for Bob Silverman whose original hypothesis now appeared in jeopardy. What appeared to be true one day might be disproven the next.

But that was the nature of science.

* * *

On Thursday, June 4, 2009, the editor of *Science* sent out the journal's review of the paper. Mikovits was in Frank Ruscetti's office when the email arrived and the two of them quickly raced through it. The journal was "intrigued" by the findings but had a number of reservations about the work. There were concerns about the possibility of contamination, the genetic background of the patients, whether the virus might be a cofactor, and whether the finding might be a PCR false-positive.[22]

Mikovits and Ruscetti read through the comments of the referees, and perhaps because they felt too close to the original work, they sent a copy of the comments to a colleague and WPI advisor.

The colleague was much more optimistic than Mikovits and Ruscetti had been about the comments. "You're in!" he wrote back enthusiastically, noting the objections were easily answerable and the additional research could be completed in a fairly brief period of time. The key was showing an immune response, which would eliminate the possibility of contamination.

Mikovits dove back into the research, adding a former student and brilliant young scientist who could develop the phylogenetic tree of XMRV. The phylogenetic tree would show the evolutionary relationship between XMRV and its closest viral cousins. *Science* asked for the immune response and Silverman would say, "I hope you can find one [an immune response], because Abbott [Labs] cannot." Ruscetti would confirm this statement when he attended the retrovirology conference in Brazil over the Fourth of July holiday and said Abbott had not been able to find an antibody in more than 1,000 blood samples.

By July 12, 2009, Mikovits felt the manuscript was ready to be resubmitted. In the letter she wrote:[23]

> All of the referee's concerns have been addressed with new data. As a result, we feel our new data strengthens the conclusions that the novel human retrovirus, XMRV is present in most patients with chronic fatigue syndrome (CFS).
>
> We show that XMRV detected in humans is not a lab contaminant by: (1) providing a phylogenetic tree relating all human and mouse XMRV and XMRV sequences as requested by reviewer #1 (supplemental fig 2) showing that XMRV strains identified in this study and in the original report on XMRV cluster as a group separate from XMRV; (2) demonstrating the presence of circulating antibody to XMRV Env protein [the envelope or protein shell of the virus] plasma as requested by Advisor A (figure 4C and S5) . . .
>
> Also, as requested by all the referees, we have now provided data on the RNaseL genotype for the R462Q variant that had originally been associated with susceptibility to XMRV in prostate cancer. In CFS we found no cor-

relation between XMRV infection and the R462Q variant. As reviewer #1 pointed out, this result makes it more likely the general population is at risk for infection . . .

What would they say? Would they accept it or ask for further revisions? Imagining they were in a fairly private dialogue with *Science*, nobody in the team could have predicted the bombshell that came next.

CHAPTER EIGHT

The Invitation-Only July 22 Meeting

Oh my God! You mean all those sequences we saw in the 1980s were real?

—Dr. John Coffin[1]

On June 12, 2009, Dr. Stuart Le Grice, Head of the Center for Excellence in HIV/AIDS & Cancer Virology at the National Cancer Institute, and Dr. John Coffin, Special Advisor to the Office of the Director, also at the National Cancer Institute sent a letter to a very select group of researchers. The letter read in part:

> In 2006, XMRV, a human retrovirus closely related to murine leukemia virus, was identified and associated with enhanced susceptibility to prostate cancer. Although the public health impact of this was not immediately clear, a series of presentations at the most recent Cold Spring Harbor meeting on Retroviruses provided additional support for this linkage, and also suggested that the number of individuals infected with XMRV is significant enough to be a cause for public concern.
>
> The CCR is convening a small group of intramural and extramural scientists and clinicians with interests in this area to draft a current status report and suggest next steps for NCI leadership. This meeting will take stock of how the CCR, as a leading retrovirology center, might coordinate efforts to address this potential emerging health issue.[2]

This letter from two of the leading scientists at the Center for Cancer Research division of the NCI, nearly four months before the publication of their article in the October 8, 2009 issue of *Science* clearly shows that many researchers besides Mikovits were extremely concerned about the public health threat posed by this retrovirus.

Additional XMRV research had been presented at the Cold Spring Harbor meeting on retroviruses, but many other researchers had presented as well. A team led by Dr. Ila Singh of the University of Utah presented an abstract on the prevalence and distribution on XMRV in human prostate cancers.[3] They noted "We found 23 percent of the 233 prostate cancers to stain positive for XMRV." In conclusion they reported, "The presence of virus in malignant cells invokes classical pathways for retroviral pathogenesis, i.e. inactivation of a tumor suppressor or activation of an oncogene by retroviral integration, as possible mechanisms of tumorigenesis."[4]

In plain language, the retrovirus was likely leading to cancer by either inactivating a gene which prevents cancer, or activating a gene which promotes cancer. These methods of inducing cancer were well known to the retrovirologists assembled at the Cold Spring Harbor Meeting.

Researchers from Japan looked at XMRV infection in both healthy patients and those with prostate cancer.[5] They wrote, "Although our study has a limited sample size, the prevalence among blood donors as determined by identifying XMRV specific antibodies was found to be 1.7 percent, while that among prostate cancer patients was found to be 6.3 percent . . . The results of genomic PCR performing on the PBMCs indicate that XMRV is situated in a few fractions of blood cells and can spread through blood even though the virus replication appears to be very low."[6]

Any later assertions that the concerns over XMRV were generated solely as a result of Mikovits's research must be rejected as a false narrative.

The question, which would trouble Mikovits in the years that followed is why so many researchers, reporters, and journals were not interested in telling this very important truth.

* * *

The special meeting to discuss the threat posed by XMRV was held on July 22, 2009 at the National Institutes of Health in Bethesda, MD. As described in the introduction to the confidential summary of the meeting:

> In 2006, the human retrovirus XMRV (xenotropic murine leukemia virus-related virus) was identified and reported to be associated with certain cases

of prostate cancer. Although the public health implications of this finding were not immediately clear, a series of presentations at the most recent Cold Spring Harbor Laboratory meeting on Retroviruses provided additional support for this linkage and suggested that the number of individuals infected with XMRV is significant enough to be a cause for public concern. In view of these developments, it was deemed appropriate for NCI to convene a small group of intramural and extramural scientists and clinicians with expertise in this area to provide the NCI leadership with recommendations on future directions.[7]

Even before Mikovits's data had completed the peer review process, these well-connected scientists were discussing and working on XMRV. That the organizers even included Mikovits and her research in the proposed gathering was an ironic twist of the normal investigative protocol of science.

The general public, less well connected scientists, as well as long-suffering patients with ME/CFS were blinded from results until publication, but the scientists could apparently talk about them to their hearts' content. Mikovits suspected the "anonymous" referees hadn't refrained from the chatter either. It is well known amongst scientists that peer review is anything but confidential and that experts discuss key papers prior to publication.

From the style of the comments, Mikovits and the authors had carefully considered whether any of their three anonymous referees were in the room. She suspected at least one of them was Coffin but later she learned this hunch was incorrect, though any number of his colleagues could have alerted him to the paper's existence.

Coffin also co-organized this hastily convened workshop, a fact that stood out to Mikovits as she reviewed the roster. She'd never met Coffin before but knew him by reputation. In addition to his position at Tufts University, Coffin held a curious title, "Special Advisor to Director," for the HIV drug resistance program at the NCI.[8] In 1995 Coffin had been hired by an inter-personal agreement (IPA) from Tufts University to run the HIV drug resistance program. An IPA is only supposed to last for three years, but Coffin stayed on for ten years. This was against regulation, so they created a new title for him.

Mikovits had a mental image of Coffin and most of his contemporaries from Frank Ruscetti's stories of attending meetings with him in the late 1970s, when Coffin and many other experts had advised Ruscetti not to pursue research on human retroviruses, stating human retroviruses don't exist.

Ruscetti recalled, "There was a prejudice, and still is a prejudice today against a retroviral etiology in human disease, even though a couple have been found. They think those are the outliers. They think animals are different than man. The basis is that they think the human immune system is much more effective in taking care of viruses. In the last 30 years, including HIV and HTLV-1, there have been 109 new human disease-causing pathogens identified. And 65 of them are animal viruses. What does that tell you?"[9]

Mikovits hoped the decades-long passage of time and the ravages of the AIDS epidemic, which provided him with tens of millions of dollars of research funding had since mellowed the man.

* * *

Twenty-two scientists attended the workshop, according to the summary, including many luminaries in the field.

The meeting included representatives from the HIV Drug Resistance Program, Columbia University, Tufts University, the Cleveland Clinic, the Fred Hutchinson Cancer Research Center, the University of Utah, the Laboratory of Cellular Oncology (NCI), the Medical Oncology Branch (NCI), the Laboratory of Tumor Immunology and Biology (NCI), the Urologic Oncology Branch (NCI), the Division of Cancer Epidemiology & Genetics (NCI), the Laboratory of Experimental Immunology (NCI), the Laboratory of Cancer Prevention (NCI), the AIDS and Cancer Virus Program (Science Applications International Corporation—a Fortune 500 company with approximately 40,000 employees worldwide), the Centers for Disease Control and Prevention, and the Food and Drug Administration (FDA).[10]

Mikovits later recalled feeling she had made her presentation before representatives of most public health hierarchy of the United States government.

* * *

The confidential account of the meeting listed Dr. Coffin's presentation first. It was summarized in science-speak that belied the explosive nature of his words:

> Dr. Coffin discussed the properties of XMRV and its relationship to xenotro-
> pic murine leukemia virus (XMLV). He pointed out that different XMRV

isolates are very closely related to, yet distinct from, endogenous proviruses found in the sequenced genome of an inbred mouse.

These observations alleviate concerns of laboratory contamination with virus or DNA from laboratory mice but are consistent with very recent, perhaps ongoing, transmission from a wild-mouse reservoir into the human population. However, it is also possible that another animal species has been the vector for zoonotic infection.[11]

In lay terms, a couple of powerful and prescient points stand out. First, XMRV has a very close association to a known murine leukemia virus. The stretches of genetic code that were sequenced from XMRV were closely related to the previously characterized murine leukemia virus, but different enough to support its identification as a newly-discovered virus that could not arise artificially from laboratory contamination, but likely had a zoonotic transmission.

In other words, Coffin asserted that the virus probably leapt the species barrier from animal to human, or in this case, "from mice to men" in the kind of virulent largesse that presents a substantial public health threat. Coffin stated the data was "consistent with very recent, perhaps ongoing transmission from a wild-mouse reservoir into the human population." Each word underscored the possibility of a widespread, devastating outbreak.

The differences between this new virus and the previously known mouse leukemia virus, was enough in Coffin's view to, "alleviate concerns of laboratory contamination with virus or DNA from laboratory mice." When Ruscetti heard Coffin say this, he leaned over to Mikovits and whispered, "It does not have to go directly from mouse to man, there could be an intermediate animal such as a cow or a goat."

At least at this initial meeting, some of the highest officials in the United States public health hierarchy had cogently concluded that *contamination could not explain the XMRV results.*

What set of events had transpired which allowed this apparently recent cross-species jump from a mouse reservoir into the human population? It was possible that the mouse retrovirus had jumped first into another species, prior to entering into humans, the way a Lyme disease-causing spirochete jumps from field mice nymph deer ticks and ultimately to a mammalian recipient.

Alternatively, the transmission could have happened in the course of laboratory investigation, in other words from a scientist directly acquiring the virus from a laboratory animal in a quotidian moment that went unnoticed. It had not occurred to any of the scientists at the meeting that this virus could have occurred by direct laboratory transmission without any *natural history* in external animals. This would have been the first in the history of retrovirology. No scientist is ever held responsible for exotic biology that goes beyond anything that has ever been observed. As Mikovits said in a later *Discover* article, "How could anyone have known?"

So much was unknown in this area. Only further research could prevent untold damage if the virus was indeed wreaking havoc on the human population. Another area of concern was that some species of wild mice "have very large numbers of related endogenous proviruses and that recombinant viruses with long terminal repeat sequences (LTR) very similar to XMRV are capable of insertional activation of proto-oncogenes in animal models of lymphoma, consistent with a potential for oncogenesis by this mechanism."

In plain English this meant that the part of these viruses, which promote the production of viral particles when integrated in the host could promote the abnormal production of host genes leading to cancer. As Mikovits saw it, cancer was the end of the line for the damage that this retrovirus caused in people.

But for the millions who suffered from ME/CFS it was not the only damage that this retrovirus could cause.

* * *

Ila Singh, a slender, soft-spoken Indian woman from the University of Utah, dropped a few more bombshells. Her findings were almost as shocking at those of Mikovits.

Singh had spent several years working with Ian Lipkin and had done a postdoctoral fellowship with Dr. Stephen Goff, both at Columbia University.

As recounted in the confidential summary of the meeting:

> Dr. Singh presented immunohistochemical and PCR data on XMRV infection. She observed XMRV expression in malignant prostate epithelial cells . . . Dr. Singh examined a total of 334 human prostate samples and determined that 27 percent of prostate cancers were XMRV positive. Her finding that 6 percent of the nonmalignant controls were also positive has implications for the general population . . .
>
>   . . . Dr. Singh also found that the presence of XMRV in patients was independent of the RNASEL mutation. Other issues addressed by Dr. Singh included (i) whether women are susceptible to infection with XMRV, (ii)

whether XMRV is present in tissues other than prostate, and (iii) whether
XMRV itself would be an early predictor of prostate cancer.[12]

Singh's work confirmed many of the things Mikovits had found in her own
research as well as some things that she had suspected but had been unable
to prove. Singh also casually released some real surprises that should have
struck everybody in the room with the power of a thunderbolt.

The discovery that 27 percent of the prostate cancer samples Singh had
examined were positive for XMRV was simply devastating. Her sample size
of 334 human prostate samples was large enough to give some confidence
that it was a valid finding. One in six men eventually develop prostate cancer
over the course of a lifetime and better than one in four of those men with
prostate cancer showed evidence of XMRV infection.

Singh's work suggested XMRV appeared to be a factor in the development
of prostate cancer, and in the severity of the tumor. In other words, even if
XMRV proved not to be a causal factor in prostate cancer, Singh's research
suggested it could accelerate the disease, creating a grimmer prognosis.

Another startling finding was that 6 percent of the non-malignant
prostate tissue controls were positive for XMRV. If 6 percent of the men
in the United States were infected with this retrovirus it was a staggering
number.

Mikovits thought again of the similarity to HTLV-1 in Asia and how
it infected many individuals but only 5 percent of those infected would go
onto develop leukemia or the associated neurological disease, HAM/TSP.
The problems this retrovirus caused wouldn't show up in a pattern like HIV,
striking down entire swaths of the population with visible symptomatology
if left untreated, but it would be more diffuse, harder to recognize, and thus
harder to attack.

Given a healthy immune response and no other complicating factors, it
seemed this retrovirus could live in many men for years and even decades
without causing any noticeable problems. But as the men aged, and the
functions of their immune system started to break down, the retrovirus
might begin the chain of events that led to prostate cancer. This virus could
be a sleeping monster and the question that would probably need to be
asked at some point was: what caused it to wake up and go on a rampage.

As *Science Now* reported on Silverman's work in 2012, "Silverman's work
on XMRV had started much earlier, around 2004. His group suspected that
a virus might be involved in some cases of prostate cancer because men with
mutations in the gene for RNASEL, which is involved in innate immunity,

have a higher risk for the disease."[13] If early treatment or lifestyle changes could prevent progression, wouldn't some men like to know?

As Singh and Mikovits had both found "the presence of XMRV in patients was independent of the RNASEL, R462Q mutation." Silverman's finding of XMRV infection *only* in men with the R462Q RNASEL mutation seemed to be a mistake. It would not be his only false step in the investigation. As Dr. Singh discussed later in the meeting, one could *not* use a microtome (a tool used to cut extremely thin slices of material called sections) to process tissues that had ever been used with mice or contamination might result.

It was not until years later that Mikovits learned of the possibility that the microtome used in the original study may have been contaminated. However, the finding of defects in the RNase L anti-viral enzyme pathway was still a viable hypothesis. It's just that they didn't know why the antiviral pathway was not functioning to destroy the invading XMRV.

When Singh was finishing her presentation she noted that one of the questions that remained was whether women were susceptible to this infection.

Frank Ruscetti kicked Mikovits under the table. "Why don't you raise your hand and tell her?" he whispered.

Mikovits leaned over to Frank. "I'm not giving away my punch line," she replied, with a grin.

Singh ended her talk by noting that the other two questions which remained were whether or not XMRV was present in other tissues besides prostate, and whether or not XMRV on its own could be used as an early predictor of prostate cancer.

After Singh's talk, Mikovits felt the excitement in the room increase dramatically. Scientists were talking spiritedly with each other, praising Singh, and it was difficult not to feel that science was on the verge of an important discovery.

* * *

Then Dr. A. Dusty Miller, a scientist from the Fred Hutchinson Cancer Research Center in Seattle, Washington, threw a bait and switch to a still-rapt, but momentarily quieted audience. "Dusty" as he preferred to be called by his colleagues, was a slender man with an easy-going smile, but strong determination when he believed in something.

He presented two critical XMRV findings.

The first substantiated a theory that the XMRV retrovirus could infect all types of cells, and the second, more controversially in the eyes of Mikovits, supported the idea that XMRV *might not be something to worry about at all.* These ideas seemed almost diametrically opposed enough to cancel each other out.

The confidential summary of the meeting detailed his findings this way:

> Dr. Miller discussed host range and receptor use of an XMRV strain isolated from 22RvI cell line, which was established many years ago. 22RvI cells were derived from a patient with Gleason grade 9 prostate cancer in 1991, following multiple passages through nude mice, suggesting that XMRV has been present in humans since at least 1991 . . .
>
> . . . Preliminary data from the Miller lab suggest that XMRV has transforming potential in cell culture. However, unlike the oncogenic mink cell focus-forming retroviruses, his data suggest that XMRV cannot multiply reinfect cells in culture, causing cytopathicity [cell death].[14]

One of the most critical findings was the discovery of an XMRV strain derived from prostate cancer tissue which had undergone "multiple passages through nude mice," a common type of laboratory animal.

This statement of the natural history of XMRV could certainly account for ME/CFS being seen in the 1980s. But then Miller would later illogically claim that since the modern ME/CFS was written into the medical literature before 1991, the retrovirus could have nothing to do with the disease. This was as odd as an archeologist stating that his carbon data from a single site marked the genesis of an entire civilization, without digging any further. The genesis of a disease could not be pinpointed from such a small amount of data.

Miller also later claimed that the findings of XMRV by Mikovits and others was the result of contamination by the 22RvI cell line, disputing Coffin's earlier implication that contamination was not possible.

Mikovits would later show that the 22RvI cell line had never been in her lab and that there was no evidence that the cell line had ever even been in the whole state of Nevada. But Miller would increasingly dig in his heels over time and claim that, for a number of reasons, the 1991 genesis of XMRV was the only reasonable possibility.

Miller felt the discovery of an XMRV strain in the 22RvI cell line was a unique event, something that could not have happened prior to that date, and he was insistent on his theory. Retrovirologists later came down on his proposed "Immaculate Recombination," defined as an event that could happen at only one place, at one time, in all of evolutionary history, but it was surely a bold statement on XMRV's origins.

His evidence led Mikovits to another, perhaps more alarming conclusion.

If XMRV had inadvertently been created once in a lab, *it could happen again and could have happened before.*

The process might be "immaculate" in the sense that it arose in a seemingly sterile environment, but that did not mean the recombination had a one-time, one-place origin. Recombination could, instead, be an event caused by synergistic factors that were common in unrelated labs, at different times.

Thus maybe this recombination or one closely related to it had taken place a number of times in the past, whenever mouse and human biological materials had shared the same test tube, petri dish, or even when they had simply shared the same lab.

* * *

In July 2011, scientists published evidence that XMRV could spread through a laboratory by aerosolizing it, an even more alarming event. Drs. Yu-An Zhang, Maitra Anirban, and Adi Gazdar, et al of the University of Texas, Southwestern, published a paper in the journal *Cancer Biology and Therapy* in which they investigated the presence of XMRV-related retroviruses in human cell lines which had been established from mouse xenografts.[15] (Xenografts are the transplantation of tissue from one species into another for purposes of research. Human cancer cells are often xenografted into mice because the cancer cells grow so abundantly in the mouse tissue.)

Gazdar was well-known for his investigations into Simian Virus #40 (SV40), a monkey virus which had contaminated both the Salk and Sabin polio vaccine and had produced evidence of carcinogenity. Prior to working at the University of Texas, Gazdar headed the Tumor Cell Biology Section of the NCI. The medical establishment had in fact ultimately concluded that the Salk and Sabin polio vaccines were contaminated with SV#40 (a result of the growing of the virus in monkey kidney cells), but they also concluded (erroneously, according to Gazdar) the virus did not pose a threat to human health.

Gazdar simply did not share in that view of SV#40's harmlessness. (An excellent account of the entire SV#40 controversy can be found in the book *The Virus and the Vaccine* by Debbie Bookchin and Jim Schumacher.[16]) He and his team examined tissues from "xenograft tumor cell lines from seven independent laboratories and 128 non-xenografted tumor cell lines."[17] The cell lines were tested for mouse DNA and came up negative.

What they found would confirm the fear that the passage of human tissue through mouse biological materials presented the danger of creating a recombinant retrovirus, but would also raise the specter of a completely unexpected threat.

Gazadar and Zhang reported:

> Six of 23 (26 percent) mouse DNA free xenograft cultures were strongly positive for MLV [murine leukemia virus, the same family as XMRV] and their sequences had greater than 99 percent homology to known MLV strains. Four of five supernatant fluids from these viral positive cultures were strongly positive for RT (reverse transcriptase) activity. Three of these supernatant fluids were studied to confirm the infectivity of the released virions for other human culture cells.[18]

This meant that a little more than a quarter of the xenograft cultures showed evidence that they had picked up a murine leukemia virus during its passage through the mice. This was confirmed by both reverse transcriptase activity (the enzyme an RNA or retrovirus releases in order to transform its RNA into DNA so it can take over the machinery of a cell) as well as the ability to infect other human cells.

This was truly a Frankenstein scenario.

The ability to infect other human cells was an important demonstration of the danger of these viruses to the human population. A certain portion of health advocates claim that natural and internally-occurring, or endogenous, human retroviruses, if stimulated by chemicals in the environment, could be the true cause of many contemporary illnesses. This argument certainly raises an important discussion—that is, do ubiquitous chemicals in daily life "switch on" internal "passenger" retroviruses to make them suddenly steer the body toward disease?

But if a pathogen shows the ability to infect other human cells it demonstrates that the virus has not become "endogenized," because it is now "acting out" instead of merely "acting in" on its host. A virus that has become endogenous will instead integrate itself into the human genome and then it normally loses the ability to infect, so its damage is not endlessly disseminating. Some virologists view loss of the ability to infect other cells as the price the virus pays for living in symbiosis with its host. It assimilates into the structure and function of the body, becoming a visceral occupant rather than an enemy faction. Some researchers claim that these endogenous viral genes may make up as much as 12 percent of the human genetic code, so they are common.[19]

But if *this* virus had not become endogenized, it meant it had probably made the jump to humans within recent history. For those who understood the potential dangers of viruses and the promiscuous manner in which they could evolve and recombine with other pathogens to create even more virulent strains, the next finding by Gazdar and Zhang was the stuff of apocalyptic nightmares. *"Of the 78 non-xenograft derived cell lines maintained in the xenograft culture-containing facilities, 13 (17 percent) were positive for MLV, including XMRV, a virus strain first identified in human tissues."*[20]

These were not cell lines that had been passaged through mouse tissue.

These were cell lines that had simply been in the same laboratory as cell lines which had been passaged through mouse tissues.

Among the plausible explanations for this alarming occurrence: *these viruses could become airborne.* In other words, the virus created in the lab could escape by simply floating away.

* * *

Those who lived through the terror of the early days of the AIDS epidemic well remembered the fears of many people who watched dramatic symptoms leading to wasting and death ravage whole communities. Many assumed the HIV retrovirus could be spread by *casual* contact, such as a kiss, the holding of hands, a hug, or even a sneeze. Political and cultural hysteria soon followed as people divided themselves into opposing groups with a panicky scramble for safety, often blaming and condemning those who were ill. But the fears had been unwarranted. AIDS had fallen prey to a long tradition of mystifying modes of transmission before pathogen theory showed that cultural hand-washing of human responsibility should not supplant actual sterile practice. Rubbery latex gloves and condoms ironically made flexible thinking possible again.

However, now scientists worried that XMRV might be the realization of that nightmare scenario that many feared with AIDS, where just sharing a closed air space with an infected person could put one in serious danger. This horrifying image so worried the Department of Health and Human Services, that its director of health of blood supply, made an urgent phone call to Frank Ruscetti and said, "Our number one priority is to determine if XMRV is a health risk."

The last finding of the Gazdar and Zhang paper provided strong evidence that XMRV might have the capability to become airborne, rather than contaminate cell cultures in some other manner. They wrote, "By contrast, all 50 cultures maintained in a xenograft culture-free facility were negative for viral sequence."

If the cultures were kept in facilities without mouse xenograft cell cultures, the risk of contamination appeared to be essentially zero. In their concluding remarks, the Gazdar and Zhang team wrote,

> XMRV, originally identified in human tissues, is a retrovirus closely related or derived from XMLV [xenotropic mouse leukemia virus] and believed to derive from the recombination of PreXMRV-1 and PreXMRV-2 proviruses in mice. These findings clearly demonstrated that xenograft cell lines, currently widely used as experimental models, can potentially acquire XMLV including XMRV after mouse xenografting.[21]

And how did XMRV infect these cultures? Was the virus simply wafting out of its mouse xenograft cultures to infect other cell lines? Or were the lab workers themselves being infected with the retrovirus and then spreading it? No one could say.

When Mikovits read the Gazdar and Zhang paper in July of 2011 she couldn't help but remember the fears she'd had as a young technician in the fermentation lab about the potential for infectious spread amongst her fellow lab workers as they performed the job of purifying HTLV-1, the first identified human retrovirus from the seminal study of Poiesz and Ruscetti, from hundreds of liters of tissue culture. It was what made the unsuspenseful meticulousness of a lab palpably risky, especially when lab workers handled an unknown pathogen.

The data Miller presented at the July 22, 2009 meeting had some bearing on the question of how infectious this retrovirus was likely to be. The confidential summary recounted that "The wide XMRV host range was demonstrated by its ability to infect human fibrosarcoma cells, rat fibroblasts, canine kidney epithelial cells, and mink lung epithelial cells."[22]

In other words, XMRV favored many species, making it difficult to reign in.

XMRV had in fact demonstrated the ability to become *airborne* and infect tissues in four different species, especially endothelial cells, which are usually located on the inside lining of blood vessels where the virus could be released into the blood.

XMRV was shaping up to be a very dangerous virus.

* * *

Mikovits was slated as the last speaker before the roundtable discussion. The confidential summary's account of her talk eclipsed the others in length as the most scientifically sound presentation, giving an unspoken clue to its centrist role in the hastily-convened meeting. It warranted the longest summary to the officials who had realized it contained complete data on XMRV in humans:

> Dr. Mikovits presented data from a Nevada cohort linking XMRV infection with chronic fatigue syndrome (CFS) and, potentially, lymphoma in 300 CFS cases identified from 1984–1987. RNA, DNA, plasma, and frozen PBMC were collected from (about) 100 of these patients between 9/06 and 7/07. Of these, 68 CFS patients were positive for XMRV. Equally important was that 3.75% of local controls were XMRV positive.
>
> As with Dr. Singh's work, these results have implications for the general public. If these numbers are reproduced in a larger survey, they would suggest that as many as 10,000,000 Americans are infected with XMRV. Similar results were generated from a Florida/Carolina cohort of CFS patients, i.e., 60% positive for XMRV *gag* from fresh PBMC. XMRV was also detected in saliva samples and plasma transmission was demonstrated from >80% of CFS patients' plasma.

In these studies, there was no correlation with XMRV infection and RNSEL mutations.

> . . . Finally, Dr. Mikovits, with NCI coauthors Francis and Sandra Ruscetti and Cleveland Clinic coauthor Robert Silverman, indicated that a manuscript showing XMRV enrichment in individuals with CFS was under review; however, experiments investigating a link to lymphoma were still under investigation and were provided in preliminary form to alert NCI leadership, given the potential public health implications.[23]

The confidential summary gave a good account of her presentation. Mikovits expected the research to generate controversy and understood there might be a pitched battle that would last for years, as it had with HIV and AIDS, but this?

The research was supposed to neatly transit to *Science*, receive anonymous reviews from referees whose job was to parse out scientific concerns, and then if she carefully and methodically assuaged their objections, the article would be published. After that point, the research would enter the scientific marketplace of ideas in which others would either confirm or disprove the findings.

This was something else entirely.

This was a glass-walled confessional where not just the scientific content, but her very ideas were seen as something to scrutinize or perhaps tame, before dispersal to the general public. The scene felt like one of those science fiction movies where harried government authorities in a hollowed-mountain bunker try to come up with a plan of counterattack for some expected alien invasion before alerting the public to the threat.

Mikovits couldn't decide whether the government effort made her feel better or worse. Were they more worried about the danger of XMRV or the potential public stampede (or others in science stampeding onward) if the danger became known?

She felt ethically mandated to defend her research, and if possible, shorten the time the patients afflicted with ME/CFS and other diseases had to wait for effective treatments.

Too many patients slogged through stalled lives already.

* * *

In the brief time between the July 13 resubmission to *Science* and the July 22 meeting, Mikovits worked extensively on the slides for her presentation.

She began by defining Me/CFS, using the CDC's own criteria, (Fukuda, 1994) which included:

> "persistent or relapsing fatigue of 6 months or longer in duration," other known medical conditions excluded by clinical diagnosis, and at least four symptoms from a list of eight, including impaired memory or concentration, sore throat, tender cervical or axillary lymph nodes, muscle pain, multi-joint pain, new headaches, unrefreshing sleep, and post-extertional malaise lasting more than 24 hours."[24]

Mikovits defined the cohort as being from a cluster of 300 cases of ME/CFS, which had been identified from Incline Village, Nevada from 1984 to 1987.

The summer-long effort of her young team to collect RNA, DNA, plasma, and frozen peripheral blood mononuclear cells from about a hundred individuals in both September of 2006 and July of 2007 was completed. RNA, DNA, and plasma had been collected from 320 normal regional controls between 2004 to 2008 by her colleagues at UNR. Mikovits was especially proud of one of their most telling innovations, the inflammatory cytokine/chemokine signature.

Using this information, Mikovits was able to augment her presentation, and suggest that CFS was "a multi-system disorder manifested by inflammatory sequalae including: antiviral enzyme RNase L dysfunction, low natural killer (NK) cell numbers and function, and innate immune activation."[25] On an overhead projector she then displayed her PCR gels of XMRV gag and envelope proteins. She flipped to another compelling slide of the phylogenetic analysis of XMRV that placed the retrovirus in its evolutionary family.[26] Then she clicked to another PCR gel that demonstrated XMRV protein expression in activated peripheral blood mononuclear cells, compared with an absence of such expression in normal donors. Summarizing the data, she was also able to show XMRV protein expression in B and T cell lymphocytes, which had been activated and purified.

The next few slides showed the protein gels, known as Westerns. She demonstrated how XMRV had been transferred from both activated peripheral blood mononuclear cells and patient plasma to LNCaP cells. In order to study a virus, one has to be able to grow the virus in very large quantities in the laboratory.

To provide an even more dramatic visual, she included two electron micrograph pictures showing the retrovirus budding from a ME/CFS patient's white blood cell. All that was known from previous studies was the demonstrated presence of the XMRV retroviral sequences and Singh's new data showing XMRV protein in prostate cancer biopsies.

The next group of slides evidenced antibodies to the envelope of the XMRV retrovirus in several of the CFS patients and showed how the suspected RNASEL variant R462Q (Silverman's claim) did *not* correlate with XMRV infection. In other words, despite an overlapping finding of RNase L defects between prostate cancer and ME/CFS, Silverman's claim that XMRV was directly linked to a specific genetic mutation did not prove true in their research on ME/CFS.

She then displayed her next slides: preliminary results from a ME/CFS cohort of Cheney's in Florida and Carolina. Interestingly, the Florida and Carolina cohort showed differing results for XMRV infection based upon which assay was used.[27]

When looking for just XMRV gag proteins in fresh peripheral blood mononuclear cells, 9 of 15 samples were positive for an infection rate of 60 percent. But when combining western blot for XMRV envelope and gag proteins upon co-culture of plasma or peripheral blood mononuclear cells with LNCaP, 13 of the 15 samples were positive for an infection rate of 86.7 percent. If one simply looked at whether the plasma samples contained antibody to the XMRV envelope, the rate was 8 out of 15 for an infection rate of 53 percent.

The important takeaway seemed to be that depending on the assay used the infection rate among the same 15 samples could vary from a low of 53 percent to a high of 86.7 percent. None of these were miniscule or chance-driven percentages, but to assure the highest accuracy, careful processing and using multiple tests was key.

Toward the end of the presentation, Mikovits showed a slide listing twenty Nevada CFS patients with cancer or suspicious biopsy findings. As she spoke, the somberness of these slides seemed to sink in; this audience understood the language of cancer more concretely than the vague misnomer of "chronic fatigue syndrome." Mikovits noted how quiet and attentive the entire group was as they riveted their gaze on the slides and took in all the cases of mantle cell lymphoma, thymoma, and myelodysplasia. These words seemed to humanize the presentation for them.

Of her many former colleagues from the NCI all understood the gravity and lack of knowledge of pathophysiology of cancer, despite more than forty years of study. Mikovits could almost see in their eyes a shift in their thinking as they considered reawakening an idea which had long ago been proposed but for which over the years there had been precious little evidence.

Could many cases of cancer be the consequence of a decades-long retroviral infection? The implications of this would be mind-boggling, shifting at least some cancer research into the field of infectious disease.

* * *

In the final slides Mikovits put forth her conclusions and the formidable challenges that lay ahead. First, she conveyed that infectious XMRV had been found in lymphocytes and plasma from slightly less than 75 percent of ME/CFS patients. She noted that XMRV in ME/CFS and prostate cancer were closely related and formed a distinct phylogenetic (evolutionary and genetic) branch.

She also reiterated that her team discovered an immune response had been detected in some ME/CFS patients and their evidence indicated the retrovirus caused a neuro-immune deficiency predisposing the patients to cancer. Alarmingly, she also put forth that the rate of positive infection in the general population suggested that the public was at significant risk from XMRV and that millions in the United States alone could be impacted.

While it was unclear how many detours this pathogen could take as it spawned illness, it seemed that it could be a factor in both extremely debilitating and deadly disease.

She underscored the challenges ahead: development of an accurate diagnostic test, therapy for those already infected, and perhaps a vaccine to protect the public, if one could be developed. When she finished talking, she scanned the room and saw barely a fidget or the flick of a pen cap. The energy was taut as the scientists absorbed the implications of what Mikovits had just said. They all seemed dumbfounded, the slides now after-images of a horrific, uncontained crisis.

A large-scale scientific calamity had been brought to light and most of them looked wary of blinking. Finally, a scientist broke the dead air, standing up to ask a question that prompted a barrage of questions directed at Mikovits, as well as animated whispering among the other scientists.[28]

Frank Ruscetti later told Mikovits that at that moment, Coffin leaned over to Dr. Vineet Kewal Ramani, also of the HIV Drug Resistance Program, and said, "Oh, my God, you mean all those sequences we saw in the 1980s were real?"

When Ruscetti told Mikovits about overhearing Coffin's remark, she knew immediately what it meant. It was if she and Coffin had stumbled into the same walled city of the walking wounded, and were now privy to a macro-lensed view of suffering as well as its microscopic guts and gore, they now saw the same epidemic clearly[29]

In the months and years to come, Coffin would be on both sides of that wall, first championing the work, and then when it appeared that XMRV had no natural history of human infection, seemingly doing everything in his power to bury Mikovits, though the mistake had been made years before in the original prostate cancer research.

Curiously nothing was done to the investigators studying XMRV in prostate cancer.

* * *

All of this had, remarkably, happened before lunchtime.

The gathering broke for a midday meal and Mikovits was so wired from nerves and excitement she wondered whether she would even be able to get her wobbly legs out of the room. Adrenaline pulsed through her, but after she strode from the auditorium, she found it easy enough to scarf down a few calories because she knew the afternoon session would be mercilessly drawn out with so much ground to cover.

Some folks were abuzz with excitement at this point and others looked simply tired, but as she made her way back to the workshop she felt ready to lean in and assert her presence in what felt like a seminal moment.

* * *

The ensuing roundtable discussion covered a panoply of concerns, as detailed at length in the confidential summary. They were broken down into four areas; epidemiology, basic research, procedural issues, and action items.

Regarding epidemiology, the group agreed,

> "the most critical research goal is to rapidly assess the prevalence, distribution, mode of transmission, and diseases associated with XMRV infection in the human population."[30]

The group noted:

> "Abbott Laboratories has been licensed to develop diagnostic assays and has developed a first-generation anti-body based diagnostic, the prevailing opinion was that the Abbott antibody test is not yet suitable for these purposes."

However, the NCI was designated as a "repository for coordination of samples for distribution to research groups in order to determine the most appropriate assays." The group tasked the Division of Cancer Epidemiology and Genetics at the NCI to play a lead role in these initial efforts. The confidential summary further urged that, "Studies should address the mode of transmission (human to human, mouse to human) as well as the mechanism."[31] Some members had raised concern that sexual transmission among humans was a reasonable possibility, adding to the importance of further research. Could XMRV be spread like HIV? Were unknowing victims having unsafe sex?

Another major concern was that the scientific record showed mouse leukemia viruses already caused a number of immunological and neurological diseases in animals, creating an urgent need to determine whether such a pattern was also evident in human beings as Mikovits's presentation suggested.

"MLVs related to XMRV infect many tissues and organs and cause a
wide variety of malignant and non-malignant (e.g., immunological and
neurological) diseases in mice. Broad surveys to look for such associations
with XMRV infection in the human population are critical to understand its
full impact."[32]

As noted in the last part of the epidemiology section, participants also voiced concern about the blood supply, animal handlers, and lab workers.

"Serious issues regarding protection of the blood supply, as well as transmission
of XMRV or related viruses to risk groups such as animal handlers and
laboratory workers who handle cell lines that were derived from mice, need
to be addressed."[33]

This, of course, could put more unknowing people at risk, from wounded
hemophiliacs to veterinarians. Once epidemiological issues had been cleared
from the agenda, the workshop participants moved swiftly to the research
issues.

* * *

First, they addressed in the basic research discussion was the finding that
the RNASEL R462Q genetic variant did not appear to be associated with
XMRV infection. "In view of presentations from several individuals (Silverman, Klein, Singh, Mikovits), the role of RNASEL in XMRV infection
remains ambiguous and needs to be clarified."[34]

The question over the genetic variant—which about ten percent of the
population carries—was important whether genetics made some individuals
into resistant carriers of the pathogen (who might simply shuttle it along yet
never grow ill), or whether some people could be more likely to contract
XMRV.

Next, they discussed whether XMRV was "a causal agent or passenger"
in prostate cancer.

"The group expressed the opinion that "the clearest evidence will come from
analysis of clonality of integration sites in tumors, particularly if they are well
placed to affect expression of known proto-oncogenes."[35]

Put simply, the participants asked whether or not the retrovirus could integrate itself into places in the DNA that were known to be involved in the

development of cancer? If the virus *did* integrate into such sites, the researchers would have strong evidence that the infection was in fact *causing* cancer.

The summary prodded the medical community to develop "reliable and inexpensive animal models mimicking XMRV infection and pathogenesis." Even within the dry language, exciting momentum and collaboration seemed to shimmer on the horizon. The summary's tone conveyed both an appropriate urgency and a respectful gravity.

The text advised it was of particular importance to develop these expanded animal models and determine the mode of transmission "especially in light of the report from Mikovits that XMRV can be detected in saliva." It also noted that "additional studies are necessary to determine the role of XMRV infection in neurodegenerative disease associated with many CFS patients,"[36] suggesting a longitudinal commitment to these patients that had not been previously seen.

The group also seemed to take the ME/CFS-cancer link seriously, noting that,

> "the suggestion of enhanced lymphoma incidence associated with CFS should be vigorously investigated, particularly the possibility of direct involvement of the virus in these malignancies."[37]

Jumping to the question of what treatments might be currently available using drugs already approved, the summary boldly asserted that, "Initiating a clinical trial of CFS patients treated with available antiretrovirals (such as AZT) was proposed." In other words, the group recognized that CFS was life-robbing enough to warrant an immediate trial of drugs typically used for AIDS. However, Coffin compellingly argued that the high degree of sequence similarity between different XMRV isolates suggested low-level replication, "which poses a problem for retrovirals," since existing drugs tend to be the most effective on a "fast-moving" virus like HIV.

Another issue was that even though,

> "XMRV is closely related to proviruses found in inbred lines, it is not identical to any of them, and even very small sequence differences are known to have large effects on species and tissue tropism and pathogenicity. At the moment, there is no reason to believe that XMLVs are human pathogens; however, research to investigate this issue is urgently needed."[38]

The document illuminated one point: a great deal of basic research needed to be done to answer some very important questions about XMRV.

* * *

Regarding procedure, the roundtable meeting tackled some riveting concerns around biosafety, implying the seriousness of the risk to lab workers, and an escape of the virus beyond lab walls.

One participant asked:

> "considering the oncogenic [cancer-causing] potential of XMRV, what are the appropriate biosafety guidelines? Guidance from work with HIV should prove useful, given the basic similarity among retroviruses. Traditionally, oncogenic retroviruses are handled in the research laboratory under BL-2 [bio-safety 2] conditions. Dr. Hughes should take a lead role in discussing such issues at NCI-Frederick."[39]

Second, participants addressed the significance of cell lines that had already been passaged through mouse tissue.

> "How does one deal with human cell lines that have been passaged through mice and have a high likelihood of carrying infectious XMLV? This is a major concern for researchers using these cell lines, for issues of biosafety as well as experimental contamination."[40]

Similar concerns were raised about anyone who worked with the mice.

Of course, the potential runaway train of public hysteria and panic were also addressed.

> "With these concerns in mind, how does one present the potential health hazard associated with XMRV infection to the general public? Given the publication of many of the results presented at this meeting is imminent, this issue is particularly urgent."[41]

Because so many of the research findings were on the verge of being published the participants generally concurred,

> "The NCI should have a procedural plan in place as an immediate follow-up to planned press releases from the participating institutions involved in the association of XMRV infection with CFS. Other NIH institutes (NIAID, NHLBI) and federal agencies (FDA, CDC, RAC) should be notified of the NCI response prior to release of this statement."[42]

If one reviews the summary from a retrospective distance, the participating agencies seem to be as calibrated as a Swiss watch, showing both appropriate public concern and the want of a timely response. Given the relative lack of discord, it is almost impossible to imagine this same group scattering or participating in infighting. There is little ominous futurity in this well-calibrated watch *group*, where the response seems both optimistic and appropriate.

At this very early point of investigation into XMRV there was evidence of it being involved in a quarter of prostate cancers, an overwhelmingly majority of ME/CFS cases, a connection to various types of lymphomas, and an infection rate among the general public which varied in studies from 3.75 percent in the Mikovits cohort to 6 percent in the Singh group, so most of the participants seemed aware that they were dealing with a significant potential threat to public health.

They were, in fact, revving up to go ahead with an action plan.

* * *

The action plan broke down into five parts.

First, Le Grice, the cochair of the meeting, with Coffin, would summarize the the meeting and circulate the report to the participants for their comments and corrections.

Second, a special team would meet two days later, on July 24, 2009, to discuss pressing action items.

> "The consensus was that the most immediate actions should be the evaluation/ improvement of currently available diagnostic reagents, which would enable epidemiological studies to better understand the extent of XMRV infection in cancer patients and the general public."[43]

The third action item concerned identifying the best way to develop a reliable diagnostic test.

> "Also, the general opinion was that a diagnostic test developed by Abbott is not yet sufficiently reliable, although details were not available. Thus, it was suggested that NCI should play a leading role in providing a panel of positive and negative samples for extramural researchers to test individual assays."[44]

Abbott Laboratories did not have a reliable diagnostic test and it seemed the participants were skeptical of Abbott's ability to develop one.

The fourth action item belied a curious paternalism regarding how the roundtable group wanted to release information on XMRV to the general public. In full, this part of the confidential summary reads:

> "A sensitive issue is the need for NCI senior leadership to issue a statement in parallel with a planned release linking XMRV with CFS, which is planned for simultaneous release by three institutions. The response to a report that an MLV-based retrovirus is potentially oncogenic and infects humans could range from a responsible scientific reaction to the hysteria that accompanies internet reports, and the NCI should be positioned to make an appropriate statement.
>
> Dr. Coffin will prepare a draft letter for distribution/discussion among meeting participants and ultimate submission to the NCI Director.
>
> Dr. Coffin has also proposed that Jon Cohen of *Science* be brought in to provide the appropriate perspective for the public.
>
> Sharing this information with other NIH (NIAID, NHLB) and federal (CDC, FDA, RAC) institutions and at the appropriate time is important."[45]

* * *

Is it possible that the "responsible" reaction to evidence that a retrovirus is involved in more than a quarter of cases of the most common type of cancer among men, in a debilitating disease like ME/CFS, which is estimated to affect one million people in the United States and seventeen million people across the world, as well as in many cases of lymphoma, would be a little bit of hysteria?

Would the public's interest in having a clear and unobstructed answer to these questions be such a bad thing? Weren't the scientists involved in "public" health?

As the roundtable group deliberated about how to best manage this information, they needed to appoint their designated writer. The man who later evolved into Mikovits's nemesis had an answer. "Dr. Coffin has also proposed that Jon Cohen of *Science* be brought in to provide the appropriate perspective for the public."

While some might see nothing wrong with this arrangement and might eagerly anticipate having the "appropriate perspective" offered by "Jon Cohen of *Science*," others might view this as something more akin to a well-oiled corporate communications strategy, rather than the rough and tumble of a free press. After all, scientific journals were not supposed to play dual roles of reviewing scientific papers and doing public relations for a national health crisis: this could surely be considered a conflict of interest. If scientific curiosity was legitimately based on objectivity, shouldn't data be presented without the filter of a public relations strategy?

What a person thinks about such affairs will likely be determined by their belief about whether any group of individuals working in isolation away from the eyes of the public, can be trusted to decide which information should be presented to or withheld from the citizenry. It is well-known among scientists that the most important part of a publication is not the data you show but the data you *do not* show. Unfortunately, Mikovits had been trained by Ruscetti to show all of the data.

One thing was clear. The group that met at the National Cancer Institute on July 22, 2009, to discuss XMRV determined that this information had to be carefully managed. And "Jon Cohen of *Science*," was Coffin's ideal man for the job.

* * *

In the fifth action item, the roundtable group talked about the possible threat to researchers who worked in labs containing cell lines that had been passaged through mouse tissue.

To Mikovits, the workshop, although unexpected and unusual in many respects, had been a complete success.

As soon as she was able to pry herself away from the other participants, she eagerly called Harvey and Annette Whittemore who were in Reno awaiting word from the conference. She knew the Whittemores would understand a sports analogy, so she exclaimed, "World Series, game 7, ninth inning, grand slam, walk-off!" After giving them that positive report of success, she delivered the highlights of the rest of the meeting.

Mikovits knew from experience that the NCI would start to go into scale-up mode to study all the questions raised about XMRV. She had first joined the NCI as a young lab technicin shortly after HTLV-1 had been discovered. Now she couldn't help but feel she'd been present at the birth of something similar to HIV with XMRV. Judy Mikovits and Frank and Sandy Ruscetti celebrated that night at a restaurant in Frederick, Maryland that Mikovits had frequented when she lived in the area.

Frank drove her to the Washington/Dulles airport the next morning and dropped her off. She was in an airport lounge when her cell rang. The woman quickly identified herself as an editor from *Science*. "I just sent you an email," she said hurriedly. "Did you get it?"[46]

The urgency in the woman's voice was not something one expected from the editor of the world's most prestigious journal of "original research." The editor asked that Mikovits send all of the data from the manuscript because they needed to start working on editing the article.

Mikovits explained she was in an airport lounge and didn't have the figures on her iPhone.

"Okay, then just send me the text, the Word document. We'll get working on that."

Mikovits told her she would do that immediately.

After sending the Word document to *Science*, Mikovits took another sip of wine and allowed herself a wry smile. Did she really believe that the phone call from the editor at *Science* had come because the three referees had finally made it through the paper and deemed it worthy of publication in just ten days?

Not a chance in hell.

In all likelihood, the lead of those three "anonymous" referees were among the twenty-two leading scientists gathered for the XMRV workshop at the NCI, walking incognito among the attendees while they ate their lunch.

It was also no stretch of the imagination to say her work had been vetted by researchers from the leading research institutions, including Columbia University, the Cleveland Clinic, Fred Hutchinson Cancer Research Center, and the University of Utah, multiple branches of the NCI, the CDC, the FDA, and industry, in the form of the Science Applications International Corporation.

So who did she have to thank for the expediting call from *Science*?

It could have been John Coffin, who just happened to mention that Jon Cohen from *Science* should be the one to provide the "appropriate perspective for the public."

She phoned Harvey and Annette Whittemore again, told them of the conversation and said, "We're in!"

CHAPTER NINE

Day Four in Jail

I had gone in and found we were locked out of our labs. The key card access didn't work and it appeared when I looked in that Judy's drawers had been gone through in her desk. And it was weird because her office door was open. She always locks the doors, and this door wasn't even shut. It was cracked open. Her stuff had been gone through so I let her know.

—Max Pfost, on what he observed at the WPI
on the day after Mikovits was terminated.[1]

Monday, November 21, 2011

A jarring 5:45 a.m. breakfast announcement roused Mikovits from her unyielding jail cell bed, and she thought this was the day she would finally see her lawyer and make bail.[2] She stretched her limbs, marveling how even a mattress could be made to feel punitive. She was comforted by the belief that whatever pressure was meant to be exerted on her, the facts would now begin to sort themselves out.

It had been the perfect moment to arrest her, a Friday afternoon at the end of a waning work week, and who knew if anybody had expected this mysterious "bail hold" to keep her in jail over the weekend. Mikovits rationally concluded that by late Friday afternoon, the situation had so baffled and hampered the authorities, that they decided to address her case after Monday morning coffee. Even though Mikovits did not realize it then, she had not been abandoned over the weekend by either her family or her professional colleagues.

Two days previously, David had told her that Frank Ruscetti and Ian Lipkin of Columbia were trying to reason with Harvey on Judy's behalf, an account later verified by Ian Lipkin in the pages of the journal *Nature*.

> Before she was incarcerated, I tried to negotiate some sort of agreement between her and the institute, so that this would not happen. One of the problems was that she really did need access to her laboratory notebooks so that she could answer questions from scientists and patients, and she was not given an option to do that.[3]

Frank Ruscetti would also provide an account of what happened in the wake of Mikovits's arrest in a declaration from May 17, 2012. Ruscetti explained:

> "After Dr. Judy Mikovits was arrested and incarcerated in the Ventura County Jail, I spoke with several members of the scientific community, including Ian Lipkin, MD, Mike Busch, Mary Carrington, Mike Dean, Frank Maldarelli, Steve Bartelmez, and others. We were all in a state of shock and outrage that a scientist could be arrested for possessing data notebooks. As demonstrated below, WPI's data notebooks do not contain any trade secrets or proprietary information. Moreover, research scientists are always provided with copies of their federally funded research notes."[4]

In the seemingly murky waters of intellectual property, both sides superficially seem to have merit. From the WPI's perspective, shouldn't an institution own the results of research that it sponsors to exploit for financial gain? However, the WPI is a non-profit, taxpayer-supported institution. Shouldn't everything from such a facility belong to the taxpayers and donors who support its research? Conversely, doesn't an academic researcher possess a personal responsibility to publish articles based on that work in scientific journals and add not only to the store of human knowledge, but also to his or her own academic reputation and that of her students? To Mikovits the truth was straightforward from the perspective of science.

She was the principal investigator on three government grants (R01, Lipkin, and DOD). These awards came with duties and obligations, with her paramount responsibility being the protection of public monies as well as upholding the integrity of the scientific work that such funds had purchased. She believed there was nothing foggy about it, as her moral compass pointed right outside of the institution's doors toward the greater public need.

But even if each side felt strongly about their position, there were others for whom the entire affair was astounding. Lipkin later recalled:

> To try to adjudicate some sort of agreement so she doesn't go to jail, it's a little beyond the pale. I hope I never do anything like this again.[5]

But Mikovits didn't have the luxury of walking away from the situation. Her professional reputation was on the line. She believed the real reason for her current legal difficulties was because (1) of what she had said in Ottawa in August when the results of the Blood Working Group had been revealed, and (2) she had prevented Lombardi from using the grant-funded cells and thus insured that the WPI could not claim salary support for him as he had done no work in the research program that year. For others it may have been a question of interpretation, but Mikovits felt allowing a researcher to do a day of work on a grant, then claiming part of his salary for a much greater time period, was unethical.

Mikovits believed the information, contained in her notebooks and personal e-mail record, was evidence to her claims and would likely convince the NIH to allow her to take the grants to another institution that could support them.

She did not fear anything in the notebooks. That's why she was trying to protect them. Let the world read them. She had nothing to hide.

* * *

At around four o'clock on a dragging afternoon, jail personnel received a call that Mikovits had a pastoral visit.[6] Free-time didn't come until much later, and Marie had slept most of the day. She was prepared and led to the visiting area.

Waiting for her in that space—the one without a Hollywood plastic window but with a rather bland institutional table and chairs—was no pastor but her fill-in attorney, Tom Adams, who was covering while Dennis Neil Jones was recovering from a knee replacement.

"Hi, Tom," said Judy lightly, not sure of the proper formalities.

He was friendly enough in return. "Hi, Judy."

She thought of all the retrospective things she could take him to task on now, such as "See?—wasn't I correct that people were following me before I was arrested?" and "do you remember how I called you up on Friday morning and you assured me that there was no warrant out for my arrest?" How could somebody be arrested when there was no warrant out?

She *wished* she had been irrationally paranoid instead of right.

Tom leaned forward soberly. "Judy, Max Pfost called your husband and told him you had the notebooks and *you knew* you had the notebooks."

"I don't have the notebooks. I never saw the contents of the bag of materials Max secured when he found my office open and ransacked the morning after I was fired."

Mikovits saw a look of disbelief sweep over Tom's face as he firmed up: "Judy, you need to tell your lawyers the truth." Maybe it was the lack of sleep or the petty indignities of being in jail, but Mikovits felt spitting mad.

"Tom, I did tell you the truth! You had selective hearing before my arrest. You didn't even ask me what happened on those crucial days of October 16th and 17th. You asked what happened within seventy-two hours of when I'd been fired and if I had ever been back to the WPI. Then without any warning, I'm in jail. The affidavit you tried to file, which the court mysteriously rejected, is inaccurate to boot because you didn't listen. To top it off, I am blindsided and dragged off to a cell, when you claimed there wasn't even a warrant out for my arrest! How exactly could this happen?"

"Well, just tell me what's going on."

"You tell *me* what's going on!" she snapped.

Tom put his elbows on the table, as if settling in to let the story uncoil. "Okay. Max called David and said, 'The jig is up!'"

Mikovits almost laughed at the hardboiled, archaic expression. *The jig is up!?* Only Harvey Whittemore talked like that: it had to be a borrowed phrase.

Max Pfost and Harvey Whittemore had a complicated relationship. As Max recounted in a 2014 interview "I could never figure him out. Sort of a sly guy. I kind of felt like he was a friend, a nice guy, but I think he manipulates things. Yeah, I think manipulator is a good word."[7]

Later, Max would tell Mikovits that Harvey Whittemore was listening on the phone line when he had phoned David.[8] After Mikovits had been fired and Max had kept her informed of eerie happenings at the WPI, like the shutting down of the lab for a week and the brusque removal of everything from the labs and from her office, Max grew fearful that his taking of the notebooks would be discovered.

These fears would lead Max to make some decisions he would later regret, especially the trust he would give to Harvey Whittemore. As Pfost recalled in that same 204 interview, "It was after I was meeting with Harvey's laywers and they were grilling me pretty good, and they dropped me back at home. And I was sitting there, worked up, and so at the time I felt like Harvey was looking out for my best interests, this whole thing. Because he was putting in my mind that Judy was out to set me up [with] this whole taking the notebooks so it looks like I'm the fall guy."[9]

Tom continued the story, saying that David didn't know what Max was saying. But Max had pressed on, saying he knew that Judy had the notebooks. David told him they'd unpacked all the boxes from Judy's Reno condominium after she was terminated and there was nothing there.

Mikovits sat back. What she'd just heard about Max was deeply disquieting because she had not taken the notebooks, Max had, and Mikovits did not know where they were. So why was Max being evasive? It was hard to fathom a motive even though his fear was palpable. She thought maybe Max was being truthful yet playing an unintentional shell game with the notebooks. Or maybe Max had taken the notebooks to copy them, in order to protect her when he left in the early morning hours of October 17th to meet a friend. Or maybe Max had packed them so she would not know their exact location, just in case she was stopped and questioned by someone connected to Harvey Whittemore.

"So, do you have the notebooks or don't you?" Tom probed.

"Well, I don't know," Dr. Mikovits replied. She would have to describe the night of October 16–17, after her plane landed in Reno at one o'clock in the morning, up until Max left at eight a.m. for his job at the WPI. She took a deep breath and plunged into her story.

She wanted to impress upon Tom Adams the absolute importance of "the notebook" in scientific research.

From the days of Leonardo Da Vinci, the notebook of a scientist provided an incorruptible record of one's research, thoughts, questions, and conclusions at a specific point in time. Da Vinci's own notebooks display his forward-thinking contribution to history; they show graceful sketches of a manned glider based on the wings of birds, blueprints for the first tank, even the principles that would enable helicopters to fly.

Protocol might have changed over the centuries, but the basic pattern remained consistent. First and foremost, a daily record of labwork was kept. Any attempt to alter the contents, such as ripping out the page for a less-than-successful day, was obvious. The system is low-tech but incredibly effective. The notebook had to remain intact.

The notebooks of Mikovits held the key to resolving the controversy raging around XMRV, the validity of the diagnostic test by VIP Dx, and so much more. Mikovits didn't care if the notebooks proved or debunked her theory. They simply needed to be maintained and protected so future researchers could determine whether she had been right or wrong. That was how science moved forward atop a solid cogwheel of proof and disproof, made up of simply-scribed data in a basic notebook.

Beyond the imperative of science to present the unvarnished truth, Mikovits had a deeper motivation. The research contained in those notebooks had been bought and paid for by US taxpayers who wanted answers, and they held names linking to coded identifications of more than 800 study participants whom she had vowed to protect with scientific confidentiality. From a populist and a scientific point of view, they did not belong to a singular institution or person.

* * *

She had to impress upon Tom the other critical piece: that she believed the Whittemores were in serious financial trouble, which she felt was enough to fog anyone's mind and generate impulsive behavior. This claim would be echoed in the lawsuit later brought by the former partners of the Whittemores, the Seeno family of Concord, California, which would paint a similar picture of desperation.

With that in mind, Mikovits began to recount the events for Tom Adams.

Annette Whittemore fired Mikovits over the phone at around 4:30 p.m. on Thursday, September 29, 2011.

It was an uneventful day prior to the Lombardi brouhaha. Mikovits had left the Whittemore Peterson Institute about a half hour earlier after going to the lab to tell her research associate Shanti Rawat that she would be taking the next day to move David's possessions out of the Condo they had been renting the past ten months from the Whittemores as her Riverwalk condo was being repaired from mold. Mikovits would then be moving her things back to the Riverwalk as her lease was up September 30. David would be moving back home to Oxnard and Judy would commute to Reno beginning in October of 2011. It had all be prearranged when Judy gave the Whittemores notice of her intentions by email in August of 2011. Mikovits told Rawat she was placing the office keys in the lab drawer as usual and that she would fly back on Monday morning, mentioning that the Whittemores were aware of her plans, and that Max was helping her move. Mikovits added a cordial "Have a nice weekend."

The call from Lombardi came in on her cell phone as she was walking down the street. Lombardi said, gruffly, "Tell me where the cells are. They were sent to me."

Mikovits snapped, "Didn't I tell you that you were no longer part of the research program? I will be back on Monday and we can discuss it then." She hung up quickly. Annette Whittemore swooped in immediately, calling Mikovits's cell phone about a minute later and no longer playing nice.

Annette screamed, "You give him those cells! They were shipped to him."

"I told you Lombardi is no longer in the research program. I am the Principal investigator (PI) and responsible for all resources on all projects in the program. Those cells were purchased off the R0I grant without my knowledge or permission. That is illegal. I told you given his failure to perform in the BWG, Lombardi was no longer in my program."

Annette seemed frantic. "I have had enough of your insolence."

"So it is rude to stop fraud and misappropriation of federal funds? I am the research director and I said *no*."

"You are no longer the research director," Annette proclaimed, "You you are fired!"

At first Mikovits just stared at the phone. As she stood on the street, Max arrived to take her to pick up the rental car from the airport. Max saw her face and immediately asked what was wrong.

"Annette had a hissy fit and fired me, but she'll get over it." There had been an incident the past spring where Mikovits had threatened to quit because of what she viewed as a conflict of interest involving Lombardi and VIP Dx, which she thought it had been resolved. That's one thing which was so difficult to later convey to people. For the vast majority of her time working in Nevada, Judy had trusted the Whittemores. They had worked through their problems, as do people who deeply respected each other.

"Well, what do you want to do?" asked Max.

"We have a job to do. I have to move David's stuff out of Harvey's apartment," she replied.

Little did she know that this time there would be no reconciliation.

* * *

The first inkling that things were not going to bounce back came at five a.m. the next morning when Max called. "I'm really concerned because I can't get into the WPI building. My key card access isn't working," he said.

Max wanted to run a gel for an experiment and as most students, he usually worked odd hours. He was upset by the events of the previous evening and thus arrived at the WPI around five in the morning. He had put the gel in the refrigerator the day before and wanted to get it started since it usually took several hours to run the experiment. Mikovits was an early riser as well, but the experience of being fired had made her sleep fitful and unreliable. All she wanted to do was get in the packed SUV and drive home to southern California and to David.

She left with an SUV containing David's clothes around three-thirty in the morning and was just outside of Sacramento in northern California when Max's call came. He could not find a way into the building.

She told him it probably wasn't anything to worry about as there had been prior break-ins and a lack of security since the new academic building had opened and perhaps the key code access had been changed for that reason. She suggested he come back at seven a.m. when the doors to the building were supposed to automatically open. When Max returned at seven he was able to get through the initial doors, but then his key card did not allow him access to his own lab containing the keys to his own office. While there, he saw that Mikovits's office was unlocked, her drawers were open, and it looked like somebody had rifled through them, plus her flash drives were on her desk.[10]

He called her in a panic. "Now I am really scared. It is not like you to leave your office unlocked."

"Describe what is going on," she asked him in a calm voice.

"Your desk drawers are open," he said.

"Is anything gone?" Mikovits asked.

"I don't know what was in there."

Mikovits conjured up a mental picture of her office. "Which desk drawers?"

"The ones in the main desk."

"Is the file behind the main desk open? The one where the textbooks are?"

"No, it's locked," said Max. "I will take a picture of your desk with my phone and send it to you."

Mikovits looked at the image Max had sent and allowed a sigh of relief. She didn't know what was going on, but the notebooks had to be secured. The desk drawers were open and unlocked but the contents appeared to be fairly undisturbed as each drawer was full to the top. The confrontation with Lombardi had been the final straw in what she suspected were a string of unethical, foolish, and possibly illegal acts at the WPI. She had wrestled with the issue of whether they were simply innocent mistakes or a pattern of something more sinister. What was it the police always asked in a murder investigation? With whom did the victim last have an argument? It was not the final judgment, but it was a good place to start.

She'd had an argument with the Whittemores whom she felt were disgruntled because of the events of the Blood Working Group diagnostic tests and Lombardi's removal from her research program. She had been fired by Annette Whittemore for insolence and insubordination only twelve hours earlier, and now her office had been ransacked: a string of events that surpassed coincidence.

Aside from what looked like evidence of the Whittemores' actions, Mikovits believed the notebooks contained evidence of Lombardi's mistakes and a cover-up of his failure to perform. There was a reason that of all the drawers and files in her two offices only those two drawers had been opened and Mikovits knew what it was. She believed there were a fair number of people at the WPI who would find the information in her notebooks to be at the very least, embarrassing, if not downright incriminating.

"What should I do?" Max asked anxiously.

At the time of her termination, there had been a great deal of controversy over whether contamination could explain the XMRV results. The lab books would be an unvarished record of what had happened.

Max clearly understood their importance, as he later recalled. "We'd always had these lab meetings with the concern of the lab being contaminated. And we pretty much did everything we did to figure out how our lab could be contaminated. In one of the notebooks, I think it was Katy's notebook, there was a lab meeting that Vinnie Lombardi was asking to pull out the plasmids in the bacterial cells, and grow it up. And Judy said no to that. It was in the notebooks. Judy said, 'No, don't do that, it could only get the virus everywhere.' And I think Katy also put an email attachment into her notebook that said, no we're not going to do that because Judy told me not to do that. We were afraid that page or that notebook would just end up missing, and then, I don't know, there would be a lawsuit where everything would get blamed on us."[11]

Mikovits told him he needed to secure the notebooks. "Put them in the lab! Put them in your office! I don't care!"

"I can't get into the lab or my office!" he replied, a little panicked.

"Take them to my condo, then."

"Harvey has the key to your condo, remember?"

"Take them to your mother's place then, for Christ's sake! Just secure them! They have patient data and information. I don't really care what you do, but secure them until I can get back and see what's going on."

"This is my work, too," said Max protectively.

She knew Max well enough that she felt she could almost hear the wheels in his head turning, that he would be appalled at the idea of Lombardi taking credit for work he hadn't done.

She asked him to do one more thing: to go down to her old office in the Applied Research facility and make certain it was secure, and then call back. He called a half hour later and said it was locked and that he did not have the key. "That's okay," Mikovits reassured him. The data in her old office, the data from Peterson and her earlier work were safe because the keys were on the ring she left in the lab with Shanti that day—or were they?

Dr. Mikovits knew she could entrust Max to protect the notebooks. She continued her trek to southern California.

* * *

As the recently fired director of research at the WPI, Mikovits now had a dilemma. She was scheduled to travel to the United Kingdom and Ireland from October 7–16 to give a series of talks in Tulamore, Newry, Belfast, and Sunderland. Should she still go?

Once home in Oxnard with David, she emailed and called the groups in the United Kingdom and Ireland to inform them of recent events. They still wanted her to come and said they would minimize the issue of her recent firing. Mikovits thought that was a good idea, telling her hosts she was sure Annette would get over it, that she was simply mad about the diagnostic test, and that by the time Mikovits returned from abroad the storm would in all likelihood have passed.

She felt a special bond with the UK/Irish sufferers of the disease, and was grateful to turn her attention to something positive and also far from home. The previous October and May she had traveled to the United Kingdom and Ireland for a series of talks. One in particular, her talk in Belfast on May 22, 2011, had touched her. The event was held in the largest meeting room at the Holiday Inn. Most of those in attendance were patients with ME/CFS and their loved ones. A table was set up at the front of the room with the members of the panel, including Dr. William Weir, an infectious disease consultant who chaired the meeting, and Member of Parliament, Basil McCrea.

Joan McParland, who had recently founded a ME/CFS support group, opened the meeting. "I'm finished with potions, lotions, and quacks," she said in her introductory remarks. "It's time for some real science."

Along the wall at the front of the room was a picture display of patients with ME/CFS, listing their names, and how long they'd been ill. To Mikovits, it looked like the sort of memorial display one might have seen twenty-five years earlier during the height of the AIDS epidemic or after the attacks of September 11, 2001. Many of the people in the pictures were not physically present, immobilized by the illness, but Mikovits knew they would be waiting for a detailed report of what had happened at the conference.

As she looked out at the crowd of people, she thought with sadness of the price many would pay in the following days and weeks. The post-extertional malaise, or PEM, of many ME/CFS sufferers meant that any extra energy they expended, such as going to listen to a conference for several hours, would cost them dearly with punishing symptoms. More than a few of the people in the audience would probably spend the next few days or weeks in bed, trying to recover to their prior functioning, just from a trivial outing. Many could not even consider attending: the physical payback would be too severe.

Mikovits was moved by the story told by the mother of a nineteen-year old son with ME/CFS and another family of a cute twelve year old brown-haired boy standing nervously beside his mother, father, and brother. One mother told of how one of the doctors had blamed her son for his condition, claiming he had caused his own joints to swell, a physical impossibility.

Another doctor blamed the mother because she had painted her son's bedroom a color he didn't like. Mikovits believed that future scientists would look back in horror at how these patients had been treated.

At the end of her talk they had presented her with a hurley, a wooden stick that looked like a large paddle, used in the Irish sport of hurling. President Barack Obama was given a similar hurley stick in 2011 by the Prime Minister of Ireland. At the time Obama quipped that he would use it on members of Congress. When the hurley was presented to Mikovits she had wondered aloud, "How am I going to get this past Homeland Security?" On the hurley stick, the members of the committee had engraved the Irish expression *Tiocfaidh ar la*, which means, "Our day will come."

Mikovits loved these patients who could cling to a phrase of uplift in the midst of struggle. She was glad that even though the WPI had fired her, the English and Irish communities still waited with rapt attention for her to discuss her research.

She would not let them down.

* * *

If the Ireland trip was supposed to lift her spirits, the news from Reno, relayed by a harried Max, only deepened her concern.

According to Pfost, after she was fired, the lab had been shut down for seven solid days from September 29 to October 7, in a hasty lockdown.[12] Max had remained locked out of the lab, unable to work. The Whittemores claimed it was a "vacation" for the staff with pay.

When Max was able to get back into the lab on October 8, he found they had quite literally stripped the lab of everything in it. They had confiscated the reagents, the computers, and the notebooks of every worker out of the lab. The Whittemores took Max's notebooks and personal computer.

Pfost said that when they returned his personal computer it had been wiped clean. Given his close relationship to her, Pfost remained a subject of suspicion. As he later related, "I wasn't allowed to do any work. I believe I had my office for a good time. I caught up on three television shows, like five seasons. Just watching in my office because I didn't have any work, but I was still required to show up every day. If I did any work he [Lombardi] would supervise the work over my shoulder."[13]

Max also told Mikovits that while she was in Ireland, they had completely gutted her office. All of this was sounding very suspicious, and Mikovits was relieved Max had gone back to the lab in the early morning hours of September 30 and been able to secure the key notebooks and files from her desk drawers.

Why were they seizing all the data and reagents from the research laboratory? She worried they were planning to manipulate the data in the computer. Mikovits's greatest fear was that the Whittemores were going to create their own narrative of the past five years justifying her censure and demise. As part of this rewritten history, she also feared they were going to claim Lombardi had been working in the research lab, even though he had not, so they could charge the government for part of his salary.

Mikovits was also clear in her belief that Lombardi had not done a single experiment in the research laboratory in all of fiscal year (FY) 2011. And all of that information was in the notebooks of both Mikovits and those collaborators who worked with her since she had moved into the new office in August of 2010. Mikovits strongly believed that Katy Hagen's notebook would show that on March 23rd, 2009, Lombardi had directed Hagen to pull out the patients' cells and culture them along with Silverman's VP62

XMRV expressing LNCaP in Mikovits's own lab, in direct violation of her orders, raising the possibility of contamination.

While this took place after the most critical experiments, it introduced an element of uncertainty in the work, one which would take a great deal of effort to refute. Those two drawers also contained the sequences of viruses generated in Ruscetti's lab. Data supporting evidence of retroviruses other than VP62, as she had shown Harvey in July of 2011.

And if any of this had indeed happened, as Mikovits feared, it would all be documented in the notebooks.

* * *

Mikovits flew into Reno on October 17, 2011, arriving at about one in the morning. Max picked her up in his jeep. He had just turned thirty years old and in the back of his jeep was that orange and yellow celebratory birthday bag. He pointed to the bag and said, "I've got the notebooks."[14]

He told her he had dropped off and then later picked up the notebooks from his mother's house and that they were all secure. Mikovits nodded and they drove to the condo she owned, known as "The Riverwalk" condo.

Once at Mikovits's condo, Max handed her a beer and she stood at the stand-up bar between the kitchen and living room. She took a sip and paused to look out the window at the lights of Reno, the self-proclaimed "Biggest Little City in the World!"

How had it all come to this?

"Okay, Max, here's my plan," she said, pointing to the yellow and orange striped, cheerful birthday bag that now held the notebooks. "When you go to work in the morning, I'm going to go back to the Whittemore condo at the Palladio to get the last of my stuff, then I will take the notebooks to the Kinkos around the corner and copy them. Then I'm going to pack the car and head out of here." Her mind was focused on getting duplicates of the data so that the record of their work would not be destroyed.

Max said he had already started packing her things from the Whittemore-owned condo at the Palladio condo she had been living at for the last 15 months and pointed to some U-Haul boxes scattered about the living room. "I've got your things on that side of the room and mine on the other." She saw how he had tidily sectioned off their possessions in that big open room with the sliding glass doors that framed a sweeping view of the church and Sierra Nevadas.

It was clear to Mikovits that Max felt threatened and perhaps even stalked by Harvey Whittemore, but what he did next completely shocked her.[15]

Max left the room, went down the hallway, and returned with what looked like a small packing box, not much bigger than a shoebox. He opened the box.

Inside was a shining silver Glock 19 handgun.[16] The Glock 19, a petite pistol; looking like it might fit inside a woman's handbag. It's made for concealment, a small weapon that an attacker wouldn't know the victim had until it was too late. Max encouraged her to take the gun for protection.

Mikovits felt a sudden, overwhelming chill: the gun punctuated all of the words that had recently passed between them about the strange goings-on. As a scientist, a person who valued reason above brute force, it looked alien. She felt as if she had slipped into a dangerous netherworld.

"Oh, Max, now you're being a bit paranoid here," she said, hoping to lighten the mood.

When later asked why he even owned the Glock, Max replied, "Because I'm in Nevada and it's fun to shoot guns." He told her not to go to the other condo at the Palladio or Kinkos.[17]

Max continued. "And Harvey's been stalking me the last two or three days."[18] It was clear that Max was very concerned about their safety and felt that he had reason to worry.

"Max, I am not going to use that gun!" she protested.

"Then don't leave this condo," he said, heatedly. "Don't leave for anything. Pack up tomorrow and I'll text you when the meeting starts, so you will know it's safe to leave and pick up the rental car." Lombardi was having his first official meeting the next day as research director. Max was convinced everybody would be in attendance, including Harvey. So he would check to make sure Lombardi's and Harvey and Annette's cars were in the parking lot, and text her then. Once he texted her that it was safe, she could then leave the condo with a baseball cap pulled down over her head and walk to the Silver Legacy casino.

At the casino, she could hop a bus to the airport, pick up a rental car that was rented in the name of a friend, and drive back to the condo to pick up her boxes. It was a stressful, but good plan.

That night, before the next day's meeting at the WPI, Max pointed to a few of the boxes on her side of the room. "Those are David's books," he said. In addition to being an avid reader, David also had a collection of various Bibles, as he often wanted to compare a verse in one translation to the same verse in another. "They're too heavy for you to move. Wait until I come home and I'll help you pack them in the car. I have a dolly and it will be easy for me to move them. But *please don't leave the condo for any reason.*"

"Damn, I really wanted to go to the gym and work out," Mikovits said with frustration.

"You will not be safe if you leave this condo," Max replied sternly.[19] He offered the gun again and again she shrugged it off.

"Fine, I'll just stay here," she replied.

Max didn't leave the gun with Mikovits, but his possession of it would prove to be a pivotal point in coming events. She later noticed the closed box on the desk, and thought the gun was still in it, but she was wrong.

She went over to one of the briefcases she had left on September 29th (as she had planned to return to work at the WPI the following Monday) and opened it. She found her keycard, her parking pass for the WPI, and the institute's laptop computer and spent a moment gazing at them. She left them all after she was fired because she thought Annette would get over her outburst about the cell line and the diagnostic test, and she would begin commuting to direct the grant-funded research. After five years it was difficult to believe their relationship was at an end, as friends and colleagues and—it seemed—groundbreakers. She had loved the Whittemores dearly, and like a suddenly scorned wife, she felt more grief than anger.

She placed everything on the bar. "I don't think I'm going to need these any more," she told Max. "Take them back after I'm out of the state."

* * *

Mikovits went to her room and slipped under the covers wearing her jeans and a T-shirt. She tried to sleep but tossed and turned.

She recalled that she was supposed to have lunch with her former patient coordinator. The Whittemores had promised the woman a full time job after the new building opened, and then hired their niece Kellen Jones instead, without telling anyone. The betrayal of her friend still hurt Mikovits. She texted her former patient coordinator. She hated to lie, but given what Max had said about the dangers of being seen, she did not want to risk the woman's safety. "I missed the plane in Santa Barbara," she wrote. "I will see you in the next couple of weeks when I pick up my things, but I won't be at the Tap House because I'm not in Reno. I don't know when I'm going to get another flight."[20]

The woman texted back, "That's fine. I'll see you when I see you. Safe travels."

Although she drifted off for a few hours, Mikovits woke at five-thirty and went to the living room to begin packing her clothes, towels and linens, glasses, and dishes.

She noticed that Max had rescued her prized possession from the lab, the hurley given to her the previous May on her trip to Ireland. As she held the smooth stick, she let her fingers linger over the inscription, Tiocfaidh ar la.

Our day will come.

She glanced at the front door at the birthday bag that had earlier contained the notebooks when Max had picked her up from the airport. It was now conspicuously empty.

She thought, *Well, that's bizarre.*

But she could not wait to leave that gutted space: like anyone who is packed and ready to hit the road, she figured she would sort out the contents of her boxes later, once she got the heck out of Dodge.

* * *

"Where did you think the notebooks had gone?" Tom Adams asked.

It startled Mikovits to hear him speak. She was talking for more than a half hour, trying to relay the chain of events. As she recounted the story it had all been like a lucid waking dream.

"Well, I'm no rocket scientist, but I thought Max packed them in my stuff to protect me from knowing where they were. I think he was concerned that Harvey Whittemore might stop me somewhere on the road and try to get them from me." She paused. "It was better if I had no idea where the notebooks were."[21]

"But you thought they were there?" Tom asked.

"Yes, I *thought* they were there. But I didn't *know* if they were."

She explained to Tom Adams that after they had gone to their respective rooms to sleep at around 2:30 a.m., Max apparently left in the middle of the night to meet up with a friend, and then went to work at 8 a.m. Around eleven o'clock he texted her that everybody's cars were in the parking lot. Mikovits had gotten up at 5 a.m. to get ready for her trip, so she and Max were both operating on limited sleep.

After Max texted, Mikovits put on her baseball cap, pulled it down, and furtively left the building for the quick walk to the Silver Legacy casino where she caught a bus to the airport. She picked up the rental car, drove into the underground parking, and proceeded to load up her boxes. She worked hurriedly, and soon the only boxes that remained were those containing David's weighty books.

Max texted at about one o'clock and said the lab meeting was starting. The meeting should only last an hour and he would probably get back to the condo by three. When he hadn't arrived or even texted by three fifteen, Mikovits was getting edgy, knowing she had at least an eight hour drive ahead of her. She decided to open David's boxes herself and lighten the load so she could carry them. She placed half of them in a recyclable container. She then put the half-empty box on the U-Haul dolly and took them down to the rental car, returning with the recyclable container. Mikovits loaded the remaining books into the half-empty box, and closed it up.

She was starting to get antsy, thinking that she would not get home until midnight if she left any later. She texted Max and said she wanted to leave. "Hit the road, Jack!" he replied.[22]

Mikovits jumped in the rental car and raced out of dry and dusty Reno, checking her rearview mirror every minute or two to see if she was being followed, smiling when she saw the highway climbing into the cool pine forests of the Sierra Nevada Mountains, and breathing an enormous sigh of relief when the sign on the road announced she had entered California. Harvey Whittemore might be a "Power Ranger" in Nevada, a "king of the legislature," with a close relationship to the leader of the US Senate, but he didn't own California. This was her state and it didn't owe a damn thing to Harvey Whittemore. Stagecoach drivers had once spurred their horses over similar routes West, glancing over their shoulders for lurking, omnipresent dangers. Mikovits revved her rented horsepower toward home.

She thought about Max's text. "Hit the Road, Jack" was of course a song by the legendary Ray Charles. She found herself humming a few of the bars as she made her way south.

* * *

Arriving around midnight, Judy went straight to bed, sleeping weightily until eight o'clock the next morning. David put on a pot of coffee and they started to unload the car. The rental car was due at the Oxnard, California, airport by noon so David drove it back.

At their beach house, Mikovits unloaded all of the boxes with the exception of two boxes of towels and sheets that were supposed to go to their other home, a condo near San Diego. When David returned, he carried the boxes upstairs and put them in the linen closet for the next time they went to San Diego. At the time, as far as she knew, there were no notebooks in any of the boxes.

"You really don't expect me to believe that do you?" Tom interjected, skeptically.

"I wish my attorney trusted my honesty, but that's the truth. Did I have plans to make sure the notebooks were protected? Yes, I did. Did I think they were in the stuff Max packed? Yes I did. But when I unpacked the boxes they were not there. Do you think if I knew where they were when I flew to Washington, DC, on November 15 for my mom's seventy-fifth birthday, I would not have taken them with me for protection given the events of the past week? I wanted to keep copies of the data, the patient information, and the evidence of what the Whittemores and Lombardi had done safe with me. Yet at the end of the day I didn't know where they were, so I didn't ask. We did the best we could to protect the study participants and the taxpayer funded research."[23]

"Okay," said Tom, standing and preparing to leave.

"So, that's it? What's the plan for getting me out of here? When am I going to get bail?"

"I don't know," said Tom. "I'm a civil attorney. You'll have to call your criminal attorney."

And with that, Tom Adams turned and walked out of the jail.

* * *

What Mikovits didn't know while she was in jail, what her attorney didn't know and might possibly have made him reconsider what he thought he knew, was what had happened to Max since she left Nevada.

While Max had played a pivotal role in securing the notebooks, he would also play an equally important role in her false arrest and imprisonment. In the months and years to come she would have her own questions about Max.

Had he sincerely been concerned about her safety when she had arrived back in Reno in mid-October, or had he already been compromised by Harvey Whittemore? Wouldn't it make for an irresistible story to portray her as a gun-toting scientist with stolen notebooks? It would complete the image of a female scientist who had gone completely off the rails. She would eventually come to believe Max was sincere in his actions when she arrived back in Reno after the trip to Ireland, a week after her termination, but it seemed in the XMRV investigation, a little paranoia was almost a necessity.

The extent of Max's interactions with Harvey Whittemore during the following weeks is unclear, but Harvey eventually gained the young man's trust. Max became concerned that the stress was making him unstable and he was concerned about having the gun.[24]

Max recalled, "I asked if I could meet with him. I had my gun in a bag. Harvey picked me up. I talked with him and asked him if he could take my gun for me and hold onto it or take it somewhere so I didn't have it at my disposal. He contacted the university police, I think his name was Todd, and spoke with him. Then we drove to the university police and I gave my gun to them. They took it from me there and held onto it. Then I came back a few weeks later. I think it was Jamie [another university police officer] who returned it to me."[25]

After the gun incident (November 16, 2011), Harvey convinced Max to fill out an affidavit (eventually two affidavits would be created) although at the time Max had "no clue what an affidavit was."[26] As Max recalled, "The first one, I think this is the time I gave the gun deal up, when I got grilled by the lawyer, they had me going to jail, this and that, and I was really blown away because I didn't feel I'd done anything wrong. So then I just explained everything that went on.[27]

"My whole intention was I didn't want those things [the notebooks] disappearing because I really didn't understand what was going on. But it seemed that it was very important to keep those notebooks intact so they didn't end up missing. There was a lot of time and effort put into them. A lot of work and it would be a shame to have it all disappear."[28]

When asked whether Max considered the threat to the notebooks to be coming from the Whittemores or somebody else at the WPI, Max made it clear that he considered the Whittemores to be the threat.[29]

Max also was insistent that Harvey Whittemore had written both of Max's affidavits,[30] an assertion that raises significant conflict of interest questions for a member of the legal profession like Harvey. Was Harvey acting as an attorney for Max? If so, an attorney is not supposed to act in a legal capacity in those situations in which he has a personal interest, just as a surgeon is not supposed to operate on a family member.

Max's exact words recorded in a telephone interview on March 13, 2014 were, "Both affidavits were written up by Harvey."[31] If Max's recollection is accurate and Harvey did assist in the writing of Max's affidavits, it's difficult to imagine how one answers the claim that these actions by Harvey were not a conflict of interest.

These affidavits would be a critical piece of evidence in the decision by the Nevada District Attorney to file a warrant to search the home of Mikovits.

The role of Harvey Whittemore in the management of the WPI has also been a topic of some controversy. In an affidavit filed on November 7, 2011, Harvey described his role:

> "I am an attorney and an unpaid advisor to the Whittemore Peterson Institute for Neuro-Immune Disease ("WPI" or "the Institute), a nonprofit research institute located on the campus of the University of Nevada in Reno with a research team, clinical laboratory, and patient medical clinic devoted to serving people with neuro-immune disease ("NID"). I work closely with the researchers and other employees at WPI. I am also married to Annette Whittemore, the founder of WPI."[32]

This view of Harvey's role at the WPI was further described by Dr. Frank Ruscetti in a declaration he executed on May 17, 2012. Ruscetti wrote:

> "I have had many conversations with Mr. Whittemore and have exchanged emails with him, in which he spoke on behalf of WPI in a managerial or executive capacity. For example, Mr. Whittemore wrote and edited scientific correspondence on behalf of WPI. Exhibit 1 is a May 20, 2011 email exchange between me and Mr. Whittemore about "his suggestion for the cytokine paper response". I also discussed Mr. Whittemore's examination of WPI's scientific data with him before WPI submitted the data to various recipients. Mr. Whittemore was also present with me at WPI's scientific meetings . . . I have participated in conference calls with Annette and Harvey Whittemore, in which Mr. Whittemore spoke on behalf of WPI about this lawsuit. Mrs. Whittemore never once said that her husband was not authorized to make the statements he made."[33]

Max explained his view of the situation and what he thought Harvey was trying to accomplish with his affidavit. "I didn't understand what they were doing was a PR move, plastering them on the internet, trying to make it look like Judy was up to no good. And they were doing that to save face with the patients. I thought it was like I go to court and say what happened. That's what I thought the deal was with the affidavits. Like I'm attesting this is the information I was giving. I tried to go over it with them, and I'm like, you're making it sound like I'm stealing stuff. And that was not my intention, to go in there stealing. That's my research. I paid for my notebooks. They didn't pay for the notebooks of mine."[34]

In summation, Max said, "So I went in there, and every time he [Harvey] word-smithed it to make it look like I was stealing. And as a result, that's hurt my career, labeling me as a thief, and that's not cool. So now my understanding is that the affidavits are no good. I'm not exactly made of money and very quickly I was fighting a legal battle that has nothing to do with whether you're right or wrong, but how much money or power you have. So there was no way I could handle that."[35]

Max later spent time talking to one of the attorneys for Mikovits, Dennis Jones. He was happy to answer questions, but didn't feel comfortable signing an affidavit because of what he had already experienced.[36]

When asked how he felt about the chain of events and what he might have done different, Max said, "I would still continue with what I did, but I would have made a copy of them. I didn't want to fork over the money to copy all of those notebooks, go to Kinkos, because that was like twenty notebooks of 100 pages. I was a little bit cheap there and didn't want to do that. I figured I'd leave Judy to do that."

On November 4, 2011, the WPI filed a complaint against Mikovits saying she had taken documents, computers, and flash drives that did not belong to her. Mikovits officially replied through a lawyer that she had done no such thing. But the legal onslaught continued when on November 9, 2011, a temporary restraining order and a civil lawsuit and were left on her front step, eschewing the normal procedure for how one is legally served. On November 13, 2011, she returned home after a weekend away on a boat with friends and found it there. This prompted a flurry of texts between Max and Mikovits about what this might mean to the WPI's participation in the Lipkin study. Mikovits thought the lawsuit was making Lipkin skittish about the study but felt she could work things out.

On November 15, 2011, Mikovits texted Max that he should feel free to call her attorney if he had any concerns. Max did not reply.

She texted him again on November 17, 2011, but there was no reply.

It was only later that Mikovits would learn that on November 16, 2011, Max had approached Harvey and asked if he could hold his gun for him. Mikovits would come to believe this was when Harvey first learned the notebooks had not been secured by his associates, prompting the mad dash which would would lead a few days later to the search of her house and false arrest.

On November 18, 2011, a combined posse of University of Nevada/Reno campus security and Ventura County sheriffs descended on her house, arrested her for supposedly being a "fugitive of justice," booked her and put her in jail, all without a warrant. This and later events would seem to bear out her fears that somebody, or some group of people, never wanted her or the public to see the contents of those crucial notebooks again.

Mikovits did not see the original version of the WPI lawsuit until March 2014, and it was then when she saw it contained only the affidavits of Harvey Whittemore, Annette Whittemore, and Vincent Lombardi. Surprisingly there was no affidavit from Max Pfost, the one affidavit that was supposed to contain evidence that Mikovits had actually taken the notebooks.

However, Harvey Whittemore's affidavit was of special interest, as it clearly showed his close association with the WPI as well as seemingly corroborated the suspicions Mikovits had that the entire thing had been a set up to frame her and thus keep her quiet regarding the wrongdoing of others. In his affidavit, Whittemore makes clear that he knew exactly in which drawers Mikovits kept the key notebooks, the specific notebooks that would have revealed any errors and indiscretions made by anyone in her program regarding the XMRV study and controversy. So if one wanted to cover up those errors or indiscretions they would first have to acquire those specific notebooks. Whittemore said in his affidavit:

> The last time I was alone with Dr. Mikovits in her office prior to her September 29, 2011 termination, she showed me the locked drawer on the left hand side of her desk where she kept the Mikovits Notebooks and the notebooks of other researchers. On that day, Dr. Mikovits told me that the laboratory notebooks of researchers Pfost, Puccinelli, and Hagen were in her possession and in a locked desk drawer. I observed Mikovits unlocking her desk and pulling out a notebook prepared and used by Ms. Hagen (the "Hagen notebook.") Dr. Mikovits specifically referred me to particular dated references to laboratory activities in the Hagen notebook."[37]

It was the Hagen notebook in which Mikovits had seen potential problems related to the work of Katy Hagen, conducted at the direction of Dr. Lombardi, without Mikovits's approval or knowledge.

Additionally, it was interesting to note that the WPI lawsuit said Mikovits had signed and then breached a contract which stated she would turn in to WPI all copies of all data generated from her. In fact the lawsuit said this supposed contract included anything she had ever written down, regardless of if it was during working hours or not, onsite or not. However, this contract in fact never existed, and when Mikovits later saw it she recognized that the signature page was from another document, now attached to this fabricated contract.

* * *

Around seven-thirty that night Mikovits was lying on her bunk. She kept thinking about the question Tom Adams had asked and now another theory crept in. Could the notebooks be packed inside those two boxes of sheets and towels, so inconspicuous that they had gone unnoticed? She knew she'd have some free time later that evening and could call David.

But she'd have to speak in code since the phone lines were monitored. The bail hearing was set for the next day and there was a good chance that a rational argument might finally prevail. She would ask David, "Honey, would you check the linen closet and the Carol Roman (her friend from New York) bag, because I really need some clean clothes tomorrow when I get out of jail?" Would that be enough to make him open up those two boxes?

She was mulling over a strategy when a call came into the cell. "Mikovits, you have a pastoral visit."

When she looked perplexed he asked, "Didn't you ask for a service?"

Among the slew of questions at her booking, the only religious question she recalled was whether she'd like a Bible during her time at the facility. She said yes, but they never provided a Bible. Mikovits didn't really know what to say, but without a beat replied, "Yes, that's me." She wasn't going to miss a chance to leave her cell.

That seemed to placate the guard, who then asked generously if she thought there was anybody else who might like to attend the service. Mikovits named as many people in the cell-block as she could remember and the staff retrieved them. Eventually they gathered a sizeable number, about thirty women. The guards led the group to a small room where a Catholic priest in a dark robe and white collar and what seemed to be a few lay ministers were setting up for a Catholic mass.

The priest, a husky older man with fine white hair, strolled over. Physically the priest reminded her of Derek Enlander, the doctor and researcher at Mt. Sinai Hospital in New York she'd been planning to see before she'd been arrested. "You're Judy Mikovits, aren't you?" he asked.

Mikovits couldn't get over how much the priest resembled Enlander. At this very moment she should have been sitting with Enlander in New York, but yet here she was with what seemed to be his costumed double, dressed in a black robe and white collar.

She was so dumbfounded she took a minute to answer. The other members of the priest's group, mostly women of about her age, hovered around and took great interest in whether or not she was Mikovits. "Yes, I'm Judy Mikovits," she finally said, feeling conspicuous.

"Then we're in the right place," he replied with a twinkle in his eye.

The other group members gave her warm smiles then went back to setting up.

The inmates were told to find a seat. One of the lay ministers motioned for Mikovits to take a place at the front of the room, which she reluctantly did. She still felt confused as to why they were singling her out.

The opening passage was a tale of jealousy, betrayal, and justice from the Book of Daniel, the account of Daniel being thrown into a lion's den. Mikovits remembered the historical context of how a Jewish man had found himself at the court of the Persian emperor. Jerusalem had been sacked years earlier by another Persian king and a number of Israelites were taken prisoner in what was referred to as the "Babylonian captivity." Daniel was hand-picked to receive special training as a court servant, which attracted the favorable notice of King Darius.

From Daniel, chapter 6, 1–3.

> *Then this Daniel was preferred above the presidents and princes, because an excellent spirit was in him; and the king thought to set him over the whole realm.*

What were the chances of the two men she was supposed to be spending this Monday night with, one a researcher in New York, and the other a priest visiting a jail in southern California, looking so alike? The question of this serendipity kept intruding her thoughts. Had God reached into her prison cell to let her know she was not forgotten?

The priest recounted how favoritism shown by the Persian king towards David didn't sit well with the other princes and members of the court. Knowing Daniel was an observant Jew who according to the dictates of his faith, prayed three times every day in the direction of Jerusalem, they concocted a hateful plan. They maneuvered Darius into signing a decree, which forbid citizens from asking any requests of God or any man other than the king for a period of thirty days. Failure to follow this decree would result in a defiant person being thrown into a den of lions. Daniel continued to pray three times every day in the direction of Jerusalem. He was seized by his enemies who brought him before the king. Darius was distraught over the arrest of his loyal appointee, but it was his own decree and he had to follow it.

Mikovits could not help but reflect that even after thousands of years, when powerful people wanted to do evil deeds they still followed the same playbook: twist the law to make a virtuous person into a criminal.

In the morning, King Darius arrived at the lion's den to ask if Daniel's god had protected him. Daniel loudly declared:

My God hath sent his angel, and hath shut the lions' mouths, that they have not hurt me; forasmuch as before him innocence was found in me; and also before thee, O king, have I done no hurt.

Daniel was removed, found uninjured because of his righteousness, and the tables then turned:

And the king commanded, and they brought those men which had accused Daniel, and they cast them into the den of lions.

The harsh Old Testament justice of the story made Mikovits chuckle to herself.

Even if unjustly imprisoned, she certainly did not want her friends-turned-foes devoured by lions.

The priest gave his homily, stressing the need for people to act with conscience, even if they may be attacked. They must hold tight to faith that if they do the right thing, God will protect them. Those words felt personally directed, resonating deeply with Mikovits.

At communion time, most of the thirty women got up to receive the sacrament. Mikovits had grown up in the Catholic Church and thus knew one is not supposed to take communion until given confession or unless one's soul is free from sin. She had switched from Catholicism to a Protestant denomination prior to marrying David so it had been more than a decade since her last confession. The pastor for the facility noticed she was not taking the sacrament and walked over. "You're not Catholic?" she queried.

"No, not lately. I'm a devout *Christian,* but I'm a Presbyterian."

The pastor was confused. "We got an email from Ireland that said it was *imperative* we have a Catholic mass for you because you're a devout Catholic and this was very important to you."

The Irish ME/CFS patients must have set this up. Mikovits was touched by their efforts to comfort her in this darkest hour. "Well, that's okay," she smiled. "I still appreciate it very much." Mikovits thought of the hurley stick and its inscription.

Tiocfaidh ar la.

Our day will come.

CHAPTER TEN

The Autism Question

"You don't talk about autism in the US—it's too politically charged," Mikovits claims Coffin told him. She believes Coffin turned against her that very day. Coffin confirms he was upset that Lombardi presented such preliminary data on such a fraught topic but says, "I did not 'turn against' Judy at that or any other point."

—Jon Cohen and Martin Enserink, "False Positive," *Science*[1]

On October 14, 2009, Mikovits and Annette Whittemore surprised the ME/CFS community when they appeared on a show called *Nevada Newsmakers* and introduced the compelling possibility on air that there was a connection between not just XMRV and ME/CFS, but also autism.[2]

The WPI had been interested in both autism and Gulf War Syndrome as apparent neuroimmune diseases. Mikovits believed that autism involved a dysregulation of the immune and nervous systems. Despite the dissent of one of the WPI advisors, Mikovits kept pursuing the idea. She then chanced upon a pertinent article from the Department of Medical Microbiology and Immunology at the University of California at Davis.

A study overseen by Dr. Judy Van de Water of the M.I.N.D. Institute (Medical Investigation of Neurodevelopmental Disorders) investigated one of Mikovits's prime subjects, natural killer cells. The title of her paper, "Altered Gene Expression and Function of Peripheral Blood Natural Killer

Cells in Children with Autism,"³ jumped out to Mikovits like a lighted billboard on a quiet road, flashing the message, "Autoimmune problem! Check for viruses and other pathogens!"

Natural killer cells (NK) are white blood cells with a vital role in killing tumor and virally infected cells, and they were thought to be centrally involved in early cases in ME/CFS, as measured in by researchers at the NCI in a paper coauthored by Cheney and Peterson in 1987.⁴⁵ The killer cell maneuvers itself into close proximity to a cell that it has identified as being cancerous or virally infected, and then releases special proteins such as *perforin* and proteases known as *granzymes*. The perforin opens pores in the cell membrane ("perforating" it) and this allows the granzymes to enter. The granzymes then induce *apoptosis* (cell suicide), destroying the cell without releasing infectious viral particles, or virions.

A body fighting off cancer or viruses desperately needs its natural killer cells to function properly. Some scientists have suggested that natural killer cells have other indispensible purposes like modulating the immune system.

Van de Water found significant differences in natural killer cell function between normal children and those with autism spectrum disorder (ASD). As she stated:

> "This present study shows distinct and significant physiological differences in NK cell responses in children with ASD . . . An altered NK cell population may have several consequences which could impact upon immune function in ASD and could explain some of the immune findings previously observed in ASD.⁶

This work by Van de Water clearly revealed that mainstream researchers were investigating autism as a potential *neuroimmune* disorder, very much in line with the central mission statement of the WPI.

The UC Davis group had found abnormal autoantibodies in the mothers of children with autism, specifically a 37 kilo-dalton protein that they could not identify,⁷ as well as autoantibodies in the systems of the children. Autoantibodies were already on Mikovits's radar. This expression "autoantibody" comes from the Greek words, "auto" meaning "self," "anti" meaning "against," and body, the self turning internally against itself. The immune system manufactures a protein directed against one or more of a person's own proteins. Some autoantibody production is thought to be due to a combination of genetics and environmental triggers, such as a viral illness or exposure to certain toxic chemicals.

Mikovits and Peterson thought that a protein of 37,000 molecular weight the UC Davis researchers had discovered might be directed against a part of the RNase L enzyme, indicating another weakness in the body's antiviral defense system. The enzyme was likely amorphous and broken, and the body had decided these fragments were invaders and that it should mount a defense. What could have possibly triggered such a breakdown in the RNase L enzyme?

It could be caused by viral or bacterial infection, but chemical or environmental exposures were also possible. As Van de Water later wrote in an article entitled, "Autoantibodies to Cerebellum in Children with Autism Associated with Behavior":

> "Collectively, these data suggest that autoantibodies towards brain proteins in children are associated with lower adaptive and cognitive function as well as core behaviors associated with autism. It is unclear whether these antibodies have direct pathogenic significance, or if they are merely a response to a previous injury. Future studies are needed to determine the identities of the protein targets and explore their significance in autism."[8]

Mikovits and Peterson traveled to visit Van de Water and others at the M.I.N.D. Institute to discuss these intriguing findings. At the time, Mikovits's team wasn't as focused on retroviruses, but rather on general patterns of immune system regulation suggested by the research. Overall immune dysregulation might explain the pathogens associated with neuroimmune disorders, such as HHV-6 or the Epstein-Barr virus, argued as both causal and passenger viruses in ME/CFS.

If a retrovirus was at the heart of various neuroimmune problems, the situation could be one of a retroviral "driver" and other "passengers," the way the Kaposi's Sarcoma herpes virus (KSHV or HHV8) was a "passenger" in AIDS patients.[9] While the herpes virus does not technically cause Kaposi sarcoma (a type of cancer) in healthy patients, the HIV retrovirus knocks out a part of the immune system, which allows for KS to reactivate.

The M.I.N.D. Institute meetings were fruitful and it became clear to Mikovits that Van De Water, her staff, and the head of the M.I.N.D. Institute, David Amaral, were interested in further research.[10] Specifically, they wanted to examine the natural killer cell defect and were intrigued by the possibility that the 37 kilo-dalton protein might be an antibody directed against a broken fragment of the RNase L enzyme.

After an ongoing back-and-forth via telephone, Annette, Mikovits and Peterson even donned formal clothes and drove down to Davis, CA, for the M.I.N.D. Institute Ball in September of 2008. The three of them were disappointed to find the ball wasn't raising funds for the biological investigations of Van De Water and other researchers, but for behavior modification.

"They just want to build a better wheelchair," said Annette. Annette had often used this expression to describe useless research into ME/CFS that had seemingly given up on being curative. Annette wanted to spearhead a dramatic change for these patients, not only make tiny improvements in their quality of life.

* * *

One of the first pediatric cases the WPI investigated involved a family with two kids who suffered from a genetic mutation-associated disease called Niemann Pick Type C, often referred to as a kind of childhood Alzheimers disease due to progressive deterioration of the brain and organs caused by a build-up of lipids, which results in dementia, seizures, tremors, and other progressive symptoms.[11]

While both parents tested negative for any active XMRV virus, they tested positive for the antibody to the envelope of XMRV. The children also tested positive for XMRV envelope protein. The "envelope" of a virus is often used as a means of detecting the virus, although this line of testing can be imprecise because related viruses can react to the detecting reagent (ie, the antibody to SFFV envelope detects all members of the xenotropic murine leukemia viruses including XMRV.) If a virus is wearing a different envelope protein, like a suspect in a borrowed coat and hat, identification may be confusing.

Mikovits and her team also found compelling overlap with other disorders. Of three patients they tested with atypical multiple sclerosis, all three were positive for antibody to XMRV envelope protein and gag DNA, (partial sequence of a structural protein of the virus).

Twelve out of twenty patients (60 percent) with fibromyalgia were also positive for XMRV gag DNA.

The autism cohort was the largest non-ME/CFS group they investigated and the results varied depending on which test was used. Twelve out of thirty children with autism were positive for XMRV gag DNA, or 40 percent positive. Twenty-three out of thirty (76 percent) were positive for XMRV envelope by Western blot. Sixteen out of twenty-eight (57 percent) were positive for serum antibody to the XMRV envelope. Depending on the test used, the autism kids showed some XMRV viral evidence from a low of 40 percent to a high of 76 percent.

As they were working on more than three hundred samples from ME/CFS patients, the other disorders they were investigating were more marginal interests, yet notable given the shocking data versus controls.

Another paper done earlier by Dr. Paul Levine and his colleagues informed much of Mikovits's thinking. Levine had worked at the NCI at the same time as Mikovits, but they never interacted until she arrived at the WPI. Levine's study looked at several generations in a family in which one of the members had ME/CFS, and delved deeply into the question of the natural killer cell activity of specific individuals. Levine found, remarkably, that the NK cell activity level correlated with the overall *severity* of disease.[12]

Mikovits extrapolated from Levine's work that if a person's natural killer cell activity was *normal,* no disease would be present. If the NK cell activity was around 50 percent, indications of subclinical disease would appear. NK cell activity of less than 50 percent meant that in all likelihood the person had a chronic illness.

As Mikovits, Ruscetti, Silverman, and Lombardi began to develop and research their idea of a retrovirus at the heart of ME/CFS, the illness no longer looked like a "catchall" but more like a cellular diaspora—with everything fanning out from the central element of retroviral infection and NK cell problems. If a person's NK cells functioned normally, the retrovirus would stay suppressed. This pattern was also observed in HTLV-1, the retrovirus found mainly in Asian countries that caused leukemia or HTLV-associated myelopathy in about 5 percent of the patients who carried it.

Retroviruses didn't have to be as fast and furious as HIV, directing a person toward certain death if left untreated. But such drama tended to grip public attention.

Although people liked to focus on vicious-seeming viruses like Ebola, those didn't actually present a long-term threat. Their very lethality meant that they would burn quickly through a population, killing swaths of people but eventually extinguishing themselves. HIV had been a cataclysmic scenario, showing that a retrovirus could coexist with its victims for many years, eluding the attention of the immune system, and yet still result in lethal chronic diseases. Mikovits thought of the admonition by famed scientist James Lovelock: *an inefficient virus kills its host. A clever virus stays with it.* HIV was not just a "driver" but a hijacker.

If a retrovirus was at the heart of ME/CFS and autism, it seemed to lower immune functions of the patient enough to allow the proliferation of other pathogens, but not completely sabotage immunity to cause imminent death, dialing down functionality just enough so the pathogen could survive, but making the daily life of the host much more difficult.

Those with the virus suffered greatly, but since the medical profession couldn't get a clear read on the problem because of conflicting signals from other pathogens and biomarkers, as well as the baffling array of symptoms, patients were disbelieved or ignored.

The patients had to linger in their own private hell with this "clever virus" directing the course of their lives.

* * *

Mikovits and Frank and Sandy Ruscetti first presented their evidence to the federal government that a retrovirus might underlie many childhood neuro-developmental disorders in early September 2009 at the NCI.[13]

Besides the detailed July 22 meeting, two other special XMRV conferences happened that involved Mikovits and the public health community prior to the publication of their article in the journal *Science* on October 8, 2009. Each meeting gave public health authorities time to review the work and identify any weaknesses in the original data or concerns they might have had about the research conclusions, before to the publication of *any* data.

The level of caution seemed failsafe.

Frank Ruscetti, as a government employee, was slated to present the data on the two children with Niemann-Pick Type C disorder. When the two children's results for natural killer cells and antibodies were matched against the results from ME/CFS patients, they looked remarkably similar. Frank also presented evidence from several children with autism, showing abnormal *gag* sequences from viral proteins thought to be associated with XMRV, as well as data showing that the childrens' plasma could transmit what appeared to be a retrovirus into LNCaP cell line. He included data from the family members, showing an apparent familial pattern of both immune system abnormalities and neuroimmune disorders.

What impressed Mikovits the most about the September meeting was how seriously the public health establishment responded to this information.[14] A number of researchers from the NCI had attended. John Coffin was there as well as a smattering of career retrovirologists. Also in attendance at the meeting was an HIV researcher and antiretroviral drug specialist who had once worked with Mikovits. XMRV and the apparently-associated diseases created a loud rallying cry in the research community, convening great minds and significant forces to combat it.

In that air of optimism, Annette Whittemore also had a visible agenda. She proposed to the scientists in charge of the research funding to the extramural community at the NCI and the National Institute of Allergy and Infectious Disease NIAID (the institute in charge of overseeing infectious disease research) that the WPI be made a center of excellence and secure separate funding to study these diseases. While the attendees seemed to comprehend that a timely reaction to the growing epidemic was imperative, they emphasized that there were channels to go through to secure research funding and resented, like most bureaucrats, a nonscientist attempting to circumvent the process.

Annette Whittemore hit a wall with a senior grants administrator of the National Institute of Allergy and Infectious Dieseases (NIAID) that was directed by Dr. Anthony Fauci. Fauci was a notorious foe-turned-friend to many people with AIDS in the early years of the pandemic, when his later-friend Larry Kramer (the well-known activist and writer with AIDS) called him a *murderer*.[15]

Fauci had his own critics in the ME/CFS community, especially over his 2002 decision to move research for ME/CFS from the National Institutes of Health to the Office of Research on Women's Health[16]. One activist wrote of the decision, "A major concern of CFS sufferers has been the tendency of society to quickly classify emerging illnesses which predominantly affect women as psychiatric, stress-related, behavioral, or emotional dysfunction."[17]

Mikovits felt that the senior grants administrator spoke in a seemingly condescending tone when she explained that Annette didn't understand how the American research culture worked.[18]

The National Institutes of Health would put out a call for proposals.

It would then take about three or four years for the proposals to come back from the various researchers in the field and institutions.

After such time, the medical intelligentsia would decide which submissions should be recommended to the appropriate research agencies.

From the picture presented by the senior grants administrator, it appeared the federal government would jump right on this potential public health crisis in about five years.

This meeting, in addition to some of the early critical response to the *Science* paper, drove Annette and Mikovits to make a brash move by appearing on the *Nevada Newsmakers* show.

* * *

Nevada Newsmakers opened with punchy music, colorful graphics, and the announcer promising a "No-holds-barred political forum!"

Host Sam Shad spoke in a soothing voice as he began. "I can't think of a more exciting program I've done because what we're about to talk about is world-shattering news."[19] Shad recounted the recent publication of the *Science* article and the association between a newly discovered retrovirus, XMRV, and patients with ME/CFS.

Mikovits began in the confident tones of a professional who had spent decades in medical research.

> "There are only two other human retroviruses known—human T-cell leukemia virus and HIV, the retrovirus that causes AIDS. Both of these viruses cause cancers, neurological disorders, and inflammatory diseases in man."[20]

Shad asked what had prompted them to look for this retrovirus in patients with ME/CFS, when it had previously only been reported in men with prostate cancer: a seeming leap. Mikovits noted that the hereditary prostate cancer gene #1 (HPC-1) coded for an antiviral enzyme RNase L, and that this enzyme *was also* deficient in patients with ME/CFS.

The host solicited Annette Whittemore's thoughts when the research veered down this unexpected path. Annette replied:

> "I knew the one thing we had to do was look for answers. And I believed all along we were going to find an underlying pathogen. It didn't make sense that so many different viruses had been claimed to cause this disorder. I was absolutely thrilled. This is exactly what we started out to do."[21]

Then Shad broached a hot button issue, addressing the oft-mistreatment of ME/CFS in the medical community. Annette fielded the question.

> "The exciting thing is that Judy can tell you how the virus works and every one of those symptoms makes sense. It's wonderful to be able to have this puzzle—just all the pieces come together and fit so well."

Mikovits also launched into the science:

> "It's a simple retrovirus, which means its expression, the on or off switch, is controlled by just—and we've learned this—it's unpublished data, by just three things. The response to hormones and the response to inflammatory cytokines. It's called NF Kappa B element. Cortisol, which is the stress

hormone, turns on the virus very rapidly, and continues to have it expressed. So do inflammatory events as caused by other pathogens. And so do hormones like androgens and progesterone, which also make sense with regard to prostate cancer."[22]

Shad then asked trenchantly: *how* could the retrovirus be transmitted? Annette replied the other two known human retroviruses were unable to become airborne and were transmitted through blood and other bodily fluids. Then the conversation turned into interesting territory.

* * *

Shad observed that the station had been vigorous supporters of the autism community and asked how this research might effect autism.

Mikovits replied. "It's not in the paper and it's not reported," she said, "but we've actually done some of these studies and found the virus present in a number, a significant number, of autistic samples that we've tested so far."

Shad said the news had tremendous potential for parents looking for treatments or cures for their autistic children. Mikovits replied by saying that XMRV:

> "might be linked to a number of neuro-immune diseases, including autism. It certainly won't be all, because there are genetic defects that result in autism, but there are also the environmental effects."

Then with barely taking a breath she crossed the Rubicon.

> "There's always the hypothesis that my child was fine, then they got sick, and then they got autism. Interestingly, on that note, if I might speculate a little bit. This might explain why vaccines lead to autism in some children because these viruses live and divide and grow in the lymphocytes, the immune response cells, the B and T cells. So when you give a vaccine you send your B and T cells into overdrive. That's its job.
>
> Well, if you're harboring one virus, and you replicate it a whole bunch, you've now broken the balance between the immune response and the virus. So you could have had the underlying virus and then amplified it with the vaccine and then set off the disease, such that your immune system could no longer control the infections and create an immune deficiency."[23]

The segment broke for commercial.

* * *

After break, Shad announced that Annette and Mikovits wanted to clear up an issue about vaccines. Annette spoke up:

> "I just wanted to say that we are certainly advocating vaccinations and understand how important they are to the well-being of the children. But what we're hoping for is by finding out whether or not somebody is positive for XMRV, whether it's one family member or another, and then looking for it in children, you could alter the immune response in such a way that you can protect the child and still be able to vaccinate, and avoid autism in these kids.
>
> And again, I don't think either one of us are sitting here and saying vaccinations cause autism, rather a number of factors—the genetic susceptibility to the illness, the infection itself, and then on top of that you're adding something to that mix that takes the child over the top."[24]

Shad then jumped to the hopeful topic of drug treatments, stating that HIV/AIDS was once a death sentence that was overturned once current treatments allowed people with HIV/AIDS to live long and healthy lives. Mikovits quickly interjected that there were existing FDA approved drugs that might also be effective in combating the XMRV retrovirus and believed the time was right to begin "rational clinical trials."

Shad said this was going to be an exciting global breakthrough, and he was thrilled by the potential as he had a family member with the excruciating nerve pain of fibromyalgia. Annette added that they had been studying fibromyalgia and found those patients were also testing positive for XMRV in high numbers.

The segment ended.

* * *

On November 9–10, 2009, Silverman hosted a workshop on XMRV at the prestigious Cleveland Clinic.[25] It was the first workshop on XMRV at the facility internationally-renowned for its innovative, forward-thinking approaches and therapies.

In an air of levity and sportsmanship, he even had black baseball caps with XMRV emblazoned across the front made for the attendees. Silverman had been one of the lead researchers on the team that first discovered pieces of the retrovirus in prostate cancer tissue and the Cleveland Clinic could reasonably lay claim to a fair share of the credit for the discovery of XMRV in ME/CFS, even though others like Joseph DeRisi had contributed with the use of his ViroChip diagnostic test.

Silverman and Lombardi had been professional pals since that meeting the previous year at the Michael Milken Prostate Cancer event when they hiked the Sierra Nevadas. Mikovits considered it a positive sign that even though their work had cast considerable doubt on Silverman's assertion of a link between a specific genetic variation in the RNase L gene and infection with the XMRV retrovirus, it had not damaged their collegial relationship.

The association between a defect in the RNase L antiviral enzyme and XMRV infection in ME/CFS was still very robust. They simply hadn't identified the genetic variation, which was responsible for the defect in the enzyme. Maybe it wasn't even a problem that could be blamed on genetics.

Despite some unanswered questions, Silverman invited Lombardi to give a presentation on the RNase L research and the immunology in XMRV-infected ME/CFS patients. He also asked Mikovits to present on the virology in the *Science* paper and she prepared both of their talks. At the conference, she and Lombardi met up with Sandy Ruscetti, who had coauthored the paper. Mikovits felt confident that Lombardi would expound on his research well, and she was excited for the young postdoctoral fellow presenting data from his first publication.

But Mikovits came down with severe laryngitis the day they arrived and couldn't talk above a whisper. So the night before her own presentation, Mikovits reviewed it slide by slide with Lombardi, explaining what to say about each, the transitions, and the questions he might be asked. Frank Ruscetti had contributed essentially all of the virology in the paper and Lombardi the immunology.

In retrospect, instead of simply naming Frank Ruscetti co-first author, the team probably should have revisited the decision to name Lombardi as a first author on the *Science* paper after Dean and Gould found that there was no association between the RNase L genetic variation and infection with the XMRV retrovirus. But Mikovits had been in the middle of credit disputes before and didn't have the stomach for them.

In her mind, once the decision had been made around authorship, it was settled.

* * *

John Coffin presided over the event from the stage, sitting behind the table with his arms folded, his wild white hair and thick beard framing his face. Mikovits sat in the front of the audience, her throat still raw from laryngitis, a rare transient mutism for the woman nicknamed "tsunami Judy" by her friends for the torrent of words that poured out when she had a pressing thought.[26]

In his fifteen minute presentation, Lombardi seemed to have little understanding of the virology. He stumbled over the slides and scrambled important information. It was a disaster. Mikovits sat tensely in the audience, trying to hold herself back.

The presentation also included a slide on the children with Niemann-Pick Syndrome. Including data on children had been a strategic decision. Because ME/CFS sufferers were mostly women, disbelievers and deniers could easily build on a long history of supposed female hysteria.

It was one thing, Mikovits and Ruscetti reasoned, to ignore sick women, but who would stomp on sick children?

During the question and answer session, an audience member noted that Lombardi had mentioned vaccines, and asked if these kids had started falling apart after a vaccination. Mikovits thought she had been abundantly clear with Lombardi about answering this question.

Although she did not fully understand how starkly the lines were drawn on the question of vaccines, she labored to develop an answer that would acknowledge the importance of vaccines to public health, while at the same time being consistent with the existing evidence that among some children those same vaccines could cause enormous damage.

She had instructed Lombardi to say that they might be dealing with some children who have an inborn genetic defect, which would make it difficult, if not impossible for them to respond to the immune challenge of a vaccine.

Lombardi needed to be clear that they were not talking about any fear that XMRV might be *contained* in a vaccine or that anything was wrong with the vaccine. The hypothesis he was supposed to advance was that if these children *were harboring a retrovirus*, possibly as a result of a genetic defect in their antiviral defense system, the retrovirus would most likely be hiding in the B and T cells of their immune system.

The intended effect of a vaccination is to expand the B and T cells of a person's immune system to fight the weakened or killed virus, thus creating antibodies that can protect a person in the event of an actual infection.

But in the case of a child who had the retrovirus in their B and T cells, the very immune activation that was supposed to confer immunity for one virus would end up causing this newly discovered retrovirus to replicate. It would look to a parent like they took their healthy child in for a vaccination and afterwards their child started to develop problems. There was also the very real possibility that as a child's immune system developed, it would control the retrovirus on its own, and the developmental or neurological problems might never occur.

The vaccine *could* be the trigger that tipped the child into immune and neurological problems, but there were other issues. In other words, the argument had a lot of nuances.

But Lombardi gave a pat answer: yes, these children began falling apart after a vaccination. He began to fully explain the hypothesis, but was cut off by hostility from the rear of the room. "You can't say anything about autism!" one researcher shouted from the back.

"Why are you talking about vaccines?" yelled another.

Lombardi responded, "I didn't say it was in the vaccines, I said the vaccines were stimulating the very cells which might be harboring the retrovirus."

Mikovits could see Coffin, now on his feet, glaring at Lombardi.[27] Somebody yelled, "Are you saying these retroviruses are in the vaccines?"

Lombardi tried to settle the fracas. Researchers typically did not shout over each other like Wall Street traders, but now they were doing just that.

This just agitated the attendees. The audience members, particularly from the back, continued to shout. Mikovits tried to defend her collaborator, but her laryngitis prevented her from speaking in anything louder than a whisper.

Finally, Silverman calmed the crowd and Lombardi left the stage.

Mikovits thought that Coffin's eyes were staring daggers at the young postdoc.[28]

* * *

A scientist present at the conference confirmed much of what Mikovits reported, but put a different spin on the controversy. It's left to the reader to determine whether this opinion reflects proper scientific humility in the face of processes which are still unclear, or something which would have been foreign to scientists like Galileo or Darwin who chose to follow a hypothesis to its conclusion, regardless of who it bothered.

About Coffin, the scientist said, "I think he was just concerned. I think all of us were concerned. We didn't want anybody to get upset about something. Or think that we had found the answer to something before it was really well investigated. I'm trying to think of the right word. I think he was sensitive to the fact that there's been a lot of controversy about studies on ME/CFS and autism. And I think that's part of the reason we still haven't figured it out. There's so much controversy and the scientists don't want to get involved in the controversy and do the research. And it's left to people who are maybe not the best researchers to do it. But I think he was concerned, like don't go there."[29]

Indeed, "there" seemed to be a minefield when it came to autism and what role, if any, vaccines might play in the disorder.

* * *

Mikovits and Frank Ruscetti debriefed Lombardi's discussion, as well as as a separate set of problems raised by Dr. Brigitte Huber, namely her inability to detect XMRV in a ME/CFS cohort.

Huber had done some very preliminary work in a handful of ME/CFS samples, using primers and assays quite different from those used by Mikovits, and had been unable to find XMRV: data she brought up pre-publication. Mikovits believed the poor results were because Huber was focusing on Silverman's sequences and PCR conditions, not starting with the spleen-focus forming virus assay which seemed to her to be the closest match to the new retrovirus. Mikovits also believed Huber was getting funding from the HHV-6 Foundation and often scientists will have an unconscious bias against research which conflicts with their own. If XMRV turned out to be a significant pathogen in the development of ME/CFS and autism, it would displace HHV-6 as a possible causal candidate.

From an email Mikovits later sent to Silverman discussing what had happened at the conference:

> . . . I am sorry that we were both sick and that Vinnie said some inappropriate things, he is young and trying his best. That said, I am disappointed that Brigette Huber was allowed to show very preliminary data and turn our presentation into a circus. . .[30]

Silverman responded by saying that he had declined Huber's request to speak, but then she talked to some other people who agreed to let her speak. Then he talked about the autism data.

> The autism data has rubbed a lot of people the wrong way since it is out there on the internet without a peer-reviewed publication and uninformed people are jumping to conclusions . . .[31]

Mikovits thought it was a little strange to say that the data on autism, the most common childhood developmental disorder "rubbed a lot of the people the wrong way," but she wanted to maintain the forward momentum. Shouldn't this be exciting news?

It concerned her that Huber had presented such preliminary data (also unpublished) even after being told no by Silverman. Mikovits responded:

> We are going to get all of the autism off of the sites possible today and do damage control as possible. The intent of the slide was as should have been stated and I did last week say we found it in families. Vinnie missed the subtleties and Frank had felt that these data needed to be presented to scientists as they do provide clues to transmission so that studies could start as soon as possible . . . point well taken . . . the slide has been "burned."[32]

Even at this early stage, the outlines of a two-front war were starting to take shape. The first front was on XMRV itself, and how Mikovits believed researchers were choosing to ask a different question than she had done in the *Science* article. She had found something *close* to Silverman's XMRV, and it might be exactly that, or it could be something closely related.

They were hunting an elusive prey, a virus that might spend most of the time hiding in tissues, and not usually be present in the bloodstream. That was why she had taken *multiple* samples from the patients, looking for that time when the virus could be detected in the blood. It wasn't a one-look proposition. This fight would be very public.

The second front, almost a secret war, the one thing nobody seemed to want to talk about, is what this might mean for those with autism. Mikovits would come to believe that powerful forces were fighting on the first front so hard, because they didn't want to reveal what this second front meant to a generation of children, and their own culpability.

* * *

After the Cleveland Clinic seminar, Mikovits phoned Frank Ruscetti to tell him what a disaster the conference had been.[33] Frank tried to calm her down, telling her it couldn't have been that bad, but he didn't convince her.

When Mikovits returned to Nevada and met with Harvey and Annette Whittemore, she told them that in order to protect Lombardi's professional future, they would need to remove him from the research. After the dust-up at the Cleveland Clinic she'd heard murmurings that Lombardi would not get any other government grants. And Mikovits knew how clannish the research community could be.

Lombardi hadn't been working at the research lab much anyway; he spent most of his time in the clinical lab. Mikovits suggested that Lombardi could head up the clinical lab and work part-time in the research lab. If he could not get funding, he could not teach students whose careers depended on academic reputation and grant support. Mikovits saw no other viable solution to keep Lombardi safe. Lombardi would not be allowed to be an academic research scientist, at least for several years.

But Lombardi was still interested in doing research. He would work on her grants as a co-investigator and continue to submit research grant proposals to the government with the support of Mikovits and Ruscetti. None would be funded.

If the presentation had been about a less charged scientific subject, Lombardi might not have received such hostile treatment. But in his first outing he had grabbed the third rail of western science, and Mikovits doubted that her colleagues would forgive the heresy.

* * *

Harvey Whittemore made several phone calls urging Frank Ruscetti to meet with Senator Harry Reid to discuss the research and control the damage done by Lombardi's presentation.[34] Despite his misgivings (federal employees are not supposed to meet with congressional representatives unless they get approval), Ruscetti eventually accepted.

After all, Reid, the Senate Majority leader, wanted to meet with *him* and surely had the authority to approve the meeting. Reid noticed Ruscetti's Boston accent, asked about his background, how he got into science, his military service, and engaged in other basic small talk. Ruscetti did his best to remain charming and engaged. Eventually, Reid asked Ruscetti if he thought the research into XMRV was important.

Ruscetti believed Reid just wanted to know if Ruscetti thought if XMRV and ME/CFS research was worthy of more research funding. The autism-vaccine question never came up. A photographer came in at one point to take a picture of Senator Reid, Ruscetti, and Harvey Whittemore. Harvey and Frank and then Annette and Frank posed behind the majority leader's chair with gleaming grins.

After the meeting, Harvey and Frank had lunch without Annette who had some political meeting to attend. During lunch, Frank told Harvey not to sell a diagnostic kit for two reasons. First, the science paper needed to be independently reproduced first, and second, even if it was right, the self-righteous scientific leaders would think it unseemly to make money directly from an individual's discovery.

As far as vaccines and how they might contribute to the number one childhood developmental disorder, neither the scientific nor political establishment wanted to "go there." ME/CFS was already well-branded, and the patients were by and large tucked away in rooms, but autism was far too much of a wild card.

Trying to find answers for the millions of children with autism apparently "rubbed a lot of people the wrong way." Given the lack of interest, and even outright hostility expressed by those at the Cleveland Clinic conference on XMRV, it's difficult to understand what part of the "public health" these scientists were interested in protecting.

CHAPTER ELEVEN

Science and Sensibility

Here, we identified XMRV nucleic acids in the peripheral blood mononuclear cells (PBMC) of 68 of 101 (67%) of CFS patients, whereas only 8 of 218 (3.7%) of regional, healthy controls contained XMRV DNA. Furthermore, infectious virus was transmitted from activated PBMC as well from purified B and T cell cultures and plasma derived from CFS patients by establishing a secondary infection in uninfected primary lymphocytes and indicator cell lines.

Full length sequencing revealed 99% homology with previously isolated strains of XMRV. These results were supported by the observation of type C retrovirus particles in patient PBMCs using transmission electron microscopy.

Taken together, these data demonstrate the first direct isolation of infectious XMRV from humans and implicate a role for XMRV infection in the pathogenesis of CFS.

—The original abstract of the *Science,* article, October, 8, 2009[1]

Otto Von Bismarck once said, "No one should see how laws or sausages are made." In this century, Bismark may have added scientific publications to that list.

To Mikovits, publication in the top journal *Science* certainly involved sausage factory antics, with some offcuts from her original article never reaching the public eye. To her, the decision over which passages were included in her team's published *Science* article called into question the stringency around protecting public health. She felt that the steps leading to the identification of a new retrovirus in ME/CFS patients were as important as the actual discovery, and should be elucidated.

People needed to know the *why* and the *how* to find out why these patients were so sick. So in the July 13, 2009, version of her team's *Science* submission, which the referees approved for publication before some of it ended up on *Science's* cutting floor, her team detailed its rationale for seeking out the retrovirus in CFS:

> Chronic fatigue syndrome (CFS) is a multi-system disorder, manifested by inflammatory sequelae including innate immune activation, antiviral enzyme (RNase L) dysfunction and low natural killer cell numbers and function. The recent discovery of the gammaretrovirus, XMRV, in men with familial prostate cancer harboring a reduced variant of RNase L prompted us to test the hypothesis that XMRV might be associated with CFS.[2]

The RNase L commonality was a crucial part of why the team had gone hunting a retrovirus in ME/CFS, an initial linkage between seemingly disparate men with bad prostate exams, and fainéant-looking women wrinkling the paper of exam tables while doctors dismissed them with lines like, "just try exercising more." Given the melee that followed the *Science* publication, it is notable that Mikovits and her coauthors also did launch into a detailed description of their sampling and testing methodology, as well as stressing their safeguards against contamination:

> To investigate possible routes of transmission, we next examined plasma from heparinized blood that had been frozen in liquid nitrogen within one hour of blood draw and shipped on dry ice to the NCI. To determine if infectious XMRV could be detected in plasma, we employed a virus isolation spinning protocol, which had previously been shown to greatly enhance the ability of retroviruses and Kaposi's sarcoma-associated herpes virus, to infect different cell types in vitro.
>
> After co-culture with plasma, the LNCaP cells were sub-cultured for 2–4 passages. Both XMRV gp70 Env and p30 GAG were abundantly expressed in 10 out of 12 plasmas (6 of which are shown in Fig. 4A).

While the details about spinning protocols and liquid nitrogen may make non-scientist eyes glaze, they are important because liquid nitrogen can nearly fix something in time. The expedient freezing of samples in liquid nitrogen (a range between -346 and -320 degrees Fahrenheit) is known to preserve evidence of retroviral activity that could otherwise quickly degrade.

So if the samples in Mikovits's study were properly processed after being taken, labs could also later use existing procedures to stimulate the replication of retroviruses to a high enough level for detection. This was, in part, why Mikovits's restructuring of Peterson's sampling methods had been imperative groundwork for their research. The team went on to explain:

> After 10–14 days, LNCaP cells tested positive by intracellular flow cytometry for the presence of XMRV p30 Gag proteins (Fig. 4B). We then confirmed cell free transmission to the T-Cell line SupT1 by placing cultures of patient PMBC in the top of 0.22 uM trans-wells in 6 well plates. After three days, the trans-wells were removed and the Sup T1 cells were positive for XMRV *gag* sequences by RT-PCR. (Fig. 4C).
>
> We next showed secondary infection of cell-free virus could also be demonstrated, since virions from patients' activated PBMC were transmitted to normal CD4+ T-cells and from those T-cells to secondary normal CD4+ T-cells (Fig. 4D). These results strongly support routes of both cell-associated and cell-free transmission of XMRV in this patient population.[3]

Mikovits and her team then addressed the genetic identification of sequences of the retrovirus and its *homology*, or similarity, to known samples of XMRV, to demonstrate a direct linkage between this pathogen and the one in human prostate cancer samples, thus distinguishing it from its mouse counterpart (and lab contamination):

> Sequences of viral isolates from the peripheral blood of CFS patients were highly homologous to the XMRV clones constructed from the patients with prostate cancer and share homologies with other well-documented gammaretroviruses.[4]

The previous passage is of vital importance. The "sequences" of virus they isolated were closely related to the "XMRV clones" (VP-62) that Silverman had constructed.

Mikovits and her team were comparing the sequences that they had taken from actual patients with "XMRV clones." But these XMRV clones could be compared to a bad copy machine, producing a replicate, which was of poor quality and significantly different from the original. It wasn't just a matter of comparing apples and oranges, it was like comparing an apple to a Xerox of an apple. Mikovits would later argue that this was the *critical* mistake made by many researchers in their work. They were calling this synthesized XMRV clone, the natural virus, keeping the standards for calling a match extremely narrow and in the process missing anything that might be closely related.

They then outlined how the detection of the virus in T and B lymphocytes made it consistent with the "tropism" (the growth or movement of a biological organism) of a retrovirus:

> In addition, our demonstration that infectious virus is present in both T and B lymphocytes is consistent with the tropism of other well-documented targets of human retroviral infection.[5]

Other researchers had for decades studied similar retroviruses and their disease associations in *animal* models, as retroviruses can cause both cancer and neurological disease in their animal hosts. Some of the symptoms in animals with retroviral infection look intriguingly like those in known human diseases, and the conditions have suggestive names such as "retroviral spongiform polioencephalomyelopathy" and "simian acquired immunodeficiency syndrome." The first sounds a little like myalgic encephalomyelitis (with a bit of the spongiform Creutzfeldt-Jacob disease, or human "mad cow disease," thrown in), the second if one drops the "simian" turns into Acquired ImmoDeficiency Syndrome: "AIDS":

> Several retroviruses such as the MuLVs [mouse leukemia retroviruses], primate retroviruses and HTLV-1 are not only associated with cancer but also associated with neurological diseases. Investigation of the molecular mechanism of retroviral induced neurodegeneration in rodent models revealed vascular and inflammatory changes mediated by cytokines and chemokines and these changes were observed prior to any neurological pathology.
>
> Neurological maladies and upregulation of inflammatory cytokines and chemokines are some of the most commonly reported observations associated with CFS; the involvement of XMRV may account for some of these observations.[6]

Mouse and monkey retroviruses also induce inflammation and changes in blood flow, a correlative to the inflammatory markers Mikovits and her team had found in human patients, the measurable chemokines and cytokines.

> Retroviral involvement has long been suspected not only for CFS but also for other neurological diseases such as multiple sclerosis (MS) and Amyotrophic Lateral Sclerosis (ALS). McCormick et al recently explored the candidacy of XMRV in ALS; however, they did not find XMRV in the blood or CSF [cerebral spinal fluid] of the 25 ALS patients where reverse transcriptase (RT) was detected.[7]

In retroviral research, reverse transcriptase was a smoking gun: an indication that a *retrovirus* was on the scene. To deny the significance of reverse transcriptase in a disease, even if the retrovirus was yet to be determined, was much like ceasing police questioning after ruling out just one suspect in a crime. But a crime had been committed and a suspect was still at large.

While Mikovits and her team cautioned that "The causality of XMRV in CFS is probable but not definitive at this time," they left little doubt that the path of future research for CFS might resemble that of AIDS research, where antiretroviral treatment became the primary focus after HIV was found, and AIDS-associated cancers and opportunistic infections were addressed aggressively, but secondarily to the retrovirus.

> Retroviruses are strongly associated with malignant transformation and neurological disorders. The higher incidence of neoplasia has been associated with outbreaks of CFS; investigation of the role of XMRV infection in cancer development of this population warrants further study.
>
> In HIV infection, co-infection with other pathogens, e.g., herpes viruses, contribute to the disease progression and outcome, thus the association of HHV-6 and HHV-7 with CFS (although in this study we found two CFS patients positive for XMRV but negative for CMV, EBV, HHV-6 and HHV-7) may not be fortuitous, but a strong cofactor.[8]

The researchers' mention of the two herpes virus-negative CFS patients would support the idea that herpes viruses were passengers and not drivers in ME/CFS. In AIDS, the presence of HIV opened the door to the bus, but pathogenic "passengers" were varied characters, bringing their own traits that might cause one patient's skin to bloom with purple lesions and another to have a throat lined with white thrush. Secondary pathogens in ME/CFS, or their reactivation after a period of dormancy, might also be enough to spur the replication of B and T cells as the vaccines did with children, leading the production of more B and T cell-embedded XMRV and then overwhelming the immune system.

Mikovits believed that her team's revised submission of July 13, 2009, to *Science* had addressed the relevant concerns of the referees, hence their subsequent approval of the publication. But after *Science* more formally accepted the paper on August 31, 2009, its editors wanted a full one-third cut from the article. Mikovits did not have a pleasant experience working with the senior editor. She would email a question and wouldn't get an answer for weeks, and then be sent a caustic response along the lines of, "Oh, you're the worst writer I've ever seen. You've got all this jargon in there. Eliminate it. All we care about is the virology."[9]

What the senior editor considered chaff was the immunology and scientific precedent that gave background into hunting a retrovirus in ME/CFS. Mikovits felt every sentence buttressed the research, and with cuts, there were too many pieces missing.

Even with the immunological data, the entire paper was less than eight and a half pages in length, double-spaced, not counting the supplemental material, which mostly contained references and pictures of *gag* sequences from gels, electron micrographs, other figures, and a description of the methods and materials used.

Mikovits made the changes to the *Science* article, trying as best she could to preserve the most critical information. But while she edited it, she thought: given the seventeen million people estimated to suffer from ME/CFS worldwide, the 1 in 100 children with autism (increased in 2013 to 1 in 50), not to mention the multitudes afflicted with multiple sclerosis and ALS, was it *really* too much to ask the world's leading scientists to read just eight and a half pages?

* * *

Fortunately, although John Coffin later became Mikovits's fiercest critic, in that shining hour he was her champion, publishing supportive commentary in conjunction with her article in *Science*. After attending the meeting of July 22, 2009, at the National Cancer Institute, he understood the centrality of her research in the current scientific dialogue.

His seven paragraph article coauthored by Jonathon Stoye, "A New Virus for Old Diseases," was a clarion call to medical researchers about the appearance of a retrovirus in ME/CFS and its alarming implication for other diseases. Coffin seemed like a researcher wrestling with the results of new data, but sincerely concerned about the impact of the new information on millions of people with chronic disease. He described the results as showing about two-thirds of ME/CFS patients as testing positive for XMRV.

> Both laboratory and epidemiological studies are now needed to determine
> whether the virus has a causative role, not only in this disease, but perhaps in
> others as well.[10]

While Mikovits had cut what she considered critical parts of their paper at the editor's request, Coffin's article bolstered her research with excellent background on the history of the disease, the retrovirus, and its possible importance to further research. He recounted that the virus had first been detected three years earlier in a few prostate cancer patients, and had recently been discovered in about a quarter of all prostate cancer biopsies. He also went onto say:

> It has been isolated from both prostate cancer and chronic fatigue syndrome
> patients, and is similar to a group of endogenous murine leukemia viruses
> (MLVs) found in the genomes of inbred and related wild mice.[11]

Coffin had noted the extensive research into gammaretrovirues via animal studies, and accurately reported that fifty years of investigation into retroviruses had not shown a link to any *human* diseases other than HIV/AIDS and human T-cell leukemia virus. He then raised the point that retroviruses can become integrated in the DNA of an individual, passing from mother to child.

> Endogenous viruses, such as xenotropic MLV, arise when retroviruses infect
> germline cells. The integrated viral DNA, or provirus, is passed on to offspring
> as part of the host genome. Endogenous proviruses form a large part of the
> genetic complement of modern mammals-about 8% of the human genome,
> for example.[12]

Then Coffin got to the hot button issues, transmission and the possibility of lab contamination, which would consume so much of Mikovits's attention in the years to come.

> The propensity of xenotropic MLVs to infect rapidly dividing human cells
> has made it a common contaminant in cultured cells, particularly in human
> tumor cell lines.[13]

Coffin identified a theoretical danger posed by the retroviruses, which was their ability to infect germ-line cells, those passed from mother to child. He also explained that the retrovirus did not seem to present a real problem to mice because their cells generally lacked the proper receptors that allowed the retrovirus to gain entry.

Just like mice, humans already harbored a large amount of genetic material (8 percent according to the latest estimates) that appeared to be viral in origin, meaning that the genetic material of viruses had been fused with human DNA. While mice generally lacked the cellular receptors that would allow these viruses to enter the cell, humans were not so lucky: they had receptors that would allow these same viruses to gain entry into cells. It was known that these types of retroviruses could easily infect human cells cultured in a laboratory, especially cancer cells since they were rapidly dividing cells.

Given the propensity of these viruses to contaminate cell cultures, particularly of cancer cell lines, Coffin next attacked the issue of whether this new discovery might conceivably be the result of contamination. His opinion, in what seemed like a pre-emptive strike given the events to come, was *no*:

> There is more than 90% DNA sequence identity between XMRV and xenotropic MLV, and their biological properties are virtually indistinguishable, leaving little doubt that the former is derived from the latter by one or more cross-species transmission events. There are several lines of evidence that transmission happened in the outside world and was not a laboratory contaminant.[14]

While Coffin would later become the fiercest proponent of the idea that the XMRV findings were the result of contamination, it is revealing to read his own words at the time of the announcement of the discovery. It is clear that at this time Coffin believed there were multiple lines of evidence showing this virus to be a threat to the public health, and he detailed them.

> One is that XMRVs from disparate locations and from both chronic fatigue syndrome and prostate cancer patients are nearly identical; The viral genomes differ by only a few nucleotides, whereas there are hundreds of sequence differences between XMRVs and xenotropic murine leukemia proviruses of laboratory mice. Other evidence includes the presence of XMRV and high amounts of antibodies to XMRV and other MLVs in chronic fatigue syndrome and prostate cancer patients.[15]

Coffin argued cogently that XMRV found in these human samples could *not* be a laboratory contaminant, because of the pace of its evolutionary advancement. As a natural consequence of evolution, a virus needed to undergo significant changes to adapt to a new host, but these varied in mice and humans.

There were hundreds of nucleotide sequence differences between the XMRV in the human population and similar retroviruses among mice, which implied a distinct evolutionary leap between species. Coffin also noted the presence of antibodies to XMRV. Human biological materials in a test tube do *not* generate antibodies. The production of antibodies can only take place in a living organism.

Coffin stated that much was still unknown about XMRV—whether it was limited by geographic region, whether it was a consequence rather than a cause of an impaired immune system, and other concerns.

> We do not know how the virus is transmitted, and the suggestion, based on indirect evidence, that there is sexual transmission is premature. Given that infectious virus is present in plasma and in blood cells, blood-borne transmission is a possibility. Furthermore, we do not know the prevalence or distribution of this virus in either human or animal populations, and animal models for infection and pathogenesis are badly needed.[16]

Coffin then focused on some important attributes of XMRV—the near-identity of viruses recovered from individuals in different diseases, and in different geographic reasons. This did not seem to bother him, as it meant XMRV was less like HIV and more like the lesser known HTLVs.

> In this respect, XMRV more closely resembles human T cell lymphotropic viruses (HTLVs) isolated from the same geographic region. As in the case with HTLV, the lack of diversity implies that XMRV recently descended from a common ancestor, and that the number of replication cycles within one infected individual is limited.[17]

Genetic drift is an important concept. Among higher organisms, it's relatively simple. The more closely related organisms are (say, two siblings), the greater the degree of genetic similarity. There are fewer genetic (or nucleotide) differences. However, if one selects a random human being from a country thousands of miles away from where one's ancestors originated, there will be more genetic or nucleotide differences, simply because so many generations have passed since early ancestors wandered out of Africa to fan out over the globe.

As Mikovits had been thinking for some time, XMRV looked more like the human T-cell leukemia virus (HTLV) co-discovered by Frank Ruscetti, and less like HIV. The lack of diversity among the XMRV isolated from patients meant it had in all likelihood crossed over into human beings in recent genetic history, and that it did not appear to go through many cycles of replication once established in a person (a trait that might help survival, since pathogens are often killed during a point in a replication cycle, which is why one class of AIDS drugs is called "reverse-transcriptase inhibitors"). This might explain why before the 1934 epidemic among doctors and nurses at Los Angeles County General Hospital this disease had never before been observed, and why autism seemingly emerged during a similar time frame.

Mikovits later recalled how when she showed the evidence of immune and other biological abnormalities in ME/CFS patients to a group of doctors at the Defeat Autism Now Conference (DAN) in January of 2010, they had observed the similarities to their patient population and made the declaration, "We think autism is simply ME/CFS in kids."[18]

While this statement seemed to leap several fences, researchers such as Dr. Rich Van Konynenburg had previously observed common methylation defects in both populations, as well as other overlaps such as specific nutritional deficiencies, neurological hypersensitivities, and mitochondrial defects. Researcher Dr. Cecilia Giulivi from the University of California, Davis, whose work was reported in *JAMA: The Journal of the American Medical Association*, stated that "Children with mitochondrial diseases may present exercise intolerance, seizures and cognitive decline, among other conditions . . . Many of these characteristics are shared by children with autism."

Just as Dr. Sarah Myhill reported on dramatic mitochondria-based energy defects in ME/CFS patients as measured in blood, Giulivi and her team found in blood sampling "that mitochondria from children with autism consumed far less oxygen than mitochondria from the group of control children, a sign of lowered mitochondrial activity."[19] Myhill, similarly, found in her published research on ME/CFS patients that "Only 1 of the 71 patients overlaps the normal region of mitochondrial functioning."[20] These parallel findings are significant when one understands that mitochondria are like gas stations for the tissues and organs, producing energetic "fuel" called ATP, and the heart and brain use up the largest percentage of this fuel hence they are most vulnerable to mitochondrial defects and sometimes show the most obvious symptoms (such as neurological hypersensitivity, or post-exertional fatigue).

Underscoring the possibility that XMRV could be a ticking time bomb in massive numbers of people, Coffin ended his 2009 *Science* commentary discussing what Mikovits considered one of the most disturbing findings of their research, the prevalence of the retrovirus in the general population.

About 4 percent of the healthy population showed evidence of having this virus.

> If these figures are borne out in larger studies, it would mean that perhaps 10 million people in the United States and hundreds of millions worldwide are infected with a virus whose pathogenic potential for humans is still unknown. However, it is clear that closely related viruses cause a variety of major diseases, including cancer, in many other mammals.[21]

Ten million individuals in the United States and *hundreds of millions worldwide* might be infected with this virus. If the virus was hiding out in the B and T cells of the immune systems of people that meant that *any* immune challenge, a fever, a cold, an accident, and yes, even a vaccination, might be enough to make the virus come out of its hiding place and rampage through the immune system. And it was clear that Coffin believed the implications might go far beyond prostate cancer and ME/CFS.

> Further study may reveal XMRV as a cause of more than one well-known "old" disease with potentially important implications for diagnosis, prevention, and therapy.[22]

Mikovits thought Coffin's commentary elucidated the most important issues. The medical community agreed that closely related viruses were responsible for "a variety of major diseases" and that it was possible that many more diseases would be found in association with XMRV.

In the days leading up to October 8, 2009, patients in the ME/CFS buzzed that *Science* was going to publish groundbreaking research on the illness. The news spread like wildfire, lighting up blogs and phone lines. Patients had grown wary of hope, yet knew that being represented in the hallowed pages of *Science* with auspicious backing seemed hugely promising. Patients with shifted circadian rhythms awoke early, or just stayed up all night, and then October 8, 2009 dawned.

The scientific community was looking at the largest retroviral threat to the world since HIV/AIDS.

CHAPTER TWELVE

Day Five and Freedom?

On November 15, 2011, Dr. Lipkin emailed that Dr. Mikovits would be the study principal investigator in my lab at National Cancer Institute. Her participation was considered essential for the accuracy of the replication study. On November 17, 2011, I was notified of an official internal review of my laboratory conduct of the XMRV affair. On November 30, Lipkin notified me that NIH [National Institutes of Health] supervisors had banned her from the NIH and security would uphold the ban . . . The failure to have Dr. Mikovits's full participation meant to me that it was not a full replication study that the NIH wanted.

—Frank Ruscetti[1]

Tuesday, November 22, 2011
The wakeup call came during the hour of farm reports and hot cups of coffee somewhere on the "outside." Marie slept like a pile of sandbags, while Mikovits waited by the small window in the door as pre-dawn blackness became light. She could only think that soon she would again enjoy the predawn cup of coffee David made her each day. This was the day of her arraignment and she expected to make bail.

At 9:00 a.m., a guard told her to get ready to move out. She rolled her mattress and stuffed her blanket and papers in the jail-issued box. Then she was taken to a group of more seasoned inmates in A3 block. They were housed in an overload space in the common room. The guards directed Mikovits to set her box on one of the bunks and wait with the group.

From there, the guards would call up individuals, escort them to a holding cell where they waited for transport for several hours before being shackled to another prisoner, and then loaded onto the bus for the trip to the courthouse.[2] When Mikovits was called up, the guard told her to leave her box, which meant she would have to return even if she made bail. It was extremely discouraging news.

The repeat offenders were rowdy: one flirted with the male prisoners, peeling off her shirt to flash them. Drugs like heroin had made a comeback in southern California, so there was a different kind of "combination drug therapy" making rounds. A few inmates were pretty rough-cut, while some were just big talkers.

There was one nicknamed "Crazy Mary," who seemed almost giddy. She had been picked up for public drinking but was homeless, living in a small encampment on a walking trail behind a grocery store in Ventura. Even in southern California, November and December were tough months to be on the streets, rainy and chilly. Mary was skinny as a rail and had the rough, leathery skin of sun-baked Californians.

The guards seemed to like Mary. Mikovits gathered from their playful jabs that Mary got herself picked up a few times a year, normally when the temperature dropped or she needed reliable meals. Mary prattled about current events, somebody's hair, past music festivals. She was fun and light, but Mikovits still hoped to be shackled to another low-key woman she recognized from before, Karen.

Instead, as luck would have it, Mikovits was paired with Crazy Mary.

For the shackling, one prisoner put her hands behind her back to be handcuffed, and then the shackles and chains were attached to those of another prisoner. The male prisoners were led out first and then the guards returned for the women.

Mikovits and Crazy Mary were the last to get onto the big white bus with black letters identifying Ventura County Sheriff's Office. Every two seats of the interior were housed in an individually locked cage, and they were placed right behind the driver. For the fifteen-minute ride, they were shackled to each other, locked to the cage, and then locked inside the cage.

Chatting up the whole bus, Crazy Mary zeroed in on Mikovits. Mary asked what kind of music she liked and did she have any brothers or sisters? Mikovits had been staring out the window like someone trapped on a long flight beside a toddler. But Mary turned out to be friendly and funny, so Mikovits dropped her guard: "I've got an older brother, a twin sister, and a younger sister" she replied.

"You got a twin sister?" Crazy Mary yelled incredulously over sudden traffic noise. "Well, is she just like you?"

"Sometimes we're almost the same, sometimes we're so different," Mikovits replied, with Mary giving a serious nod.

The bus pulled up to the Hall of Justice, a drab, four-story southern California office building with politely-placed fountains and palm trees. It was a civic structure of facile appearances, an illusion of order and security. But instead of stopping at the public entrances, the bus drove to the cavernous underground garage, pulling up to an area often referred to as "the dungeons," where prisoners were held before court appearances.

They were led toward a holding cell where the group of about eight women shared a single toilet and a few benches. On the way there, one guard offered a paper bag lunch of two pieces of white bread and a slice of bologna. Mikovits was about to hand it back when Crazy Mary protested, "Don't!" and snatched the crinkly brown paper from Mikovits's hand. Undernourished and obviously not sure when she was going to eat again, Mary wolfed down both sandwiches.

After what seemed like an incredibly long time but was probably not much longer than two hours, they were escorted from their cells, up a back elevator, and were placed in a specially designed cage at the back of the courtroom.

From a prisoners' cage behind the judge's bench, Mikovits could see David and some of her friends in the spectator's gallery. Her lawyer, a prominent criminal defense attorney, stood a slim six foot three with a smooth shaved head, glasses, and a sharp professional air.

"It's all right," were the first words he said to her, adding that he knew she didn't want to be extradited to Reno and would not let it happen. She was greatly relieved.

"But there's another issue," said her attorney. "There's a reporter here from *Science*, Jon Cohen. He wants to take a picture of you in court."

Mikovits recalled how Cohen had been the one who included the comment from Coffin in the *Science* article the previous September, comparing her to Joan of Arc. Cohen was also the writer Coffin had designated at the "Invitation Only" meeting in July of 2009 at the National Cancer Institute, three months prior to the publication of the XMRV article, to explain the story to the public. Mikovits couldn't help but believe that Cohen was playing for the other team.

Mikovits told her lawyer she did not want her picture taken.

Cohen painted the scene for the readers of *Science Insider* in his November 22, 2011, piece entitled, "Inmate Mikovits Meets Judge."[3] The article described her entering Room 13 of the Superior Court in a "prison-issued blue jump suit with an orange T-shirt underneath," the area for the inmates in the courtroom was a "room-within-the-room that had white metal bars for walls," the "four bailiffs with Taser guns," and the other inmates who were "heavily muscled and tattoed men and street-tough women." Cohen noted that the fifty-three-year-old Mikovits "appeared composed, but wildly out of place."[4]

It seemed to Mikovits that as she was talking to her attorney, Cohen was looking for an opportunity to photograph her disheveled state. She told this to her attorney, and he obliged by blocking her from Cohen's view. After blocking Cohen from taking any possible shot, he told Mikovits that she would probably be released on bail to her own recognizance. What she did not know at the time was that after the call from Max and Harvey, David had found the notebooks in the box of towels and linens and surrendered them to the police only a few hours earlier. After the judge heard about a dozen other cases, Mikovits pled not guilty and waived extradition to Reno. The judge then asked about Cohen's photo request. Mikovits and her attorney vehemently answered "No."

The judge asked Cohen to speak about why he wanted the picture. Cohen asserted that a photograph could let others in the science world see what was possible in these situations. Mikovits felt that she was being displayed as a warning for anyone in science who dared venture into controversial areas. The judge noted that Mikovits had not been convicted of any crime and concluded, "I can see no benefit to the scientific community in putting the picture of a woman presumed innocent in a professional journal."

An extradition specialist then took the podium to detail why extradition to Reno was not required. Her attorney joined him in explaining to the judge that Mikovits had owned her home in Ventura County since July of 2003; that she'd had a normal homing instinct right after being fired and returned to her home, and she was no "flight risk."

The two of them explained how Mikovits had been welcomed back to her community, a fact supported by the many friends from her church and yacht club who were in the courtroom.

After confirming that she could post bail, the judge ordered that she did not have to appear in Reno and that she be released immediately from custody.

* * *

Unbeknownst to Mikovits a very different chain of events had been taking place in Reno, Nevada, New York City, and Bethesda, Maryland. A conference call was held on Monday, November 21, 2011, the day before her arraignment, between Ruscetti, Lipkin, and Harvey and Annette Whittemore to discuss securing Mikovits's release from jail and settling the civil lawsuit against her.[5]

Ruscetti recalled, "Mr. Whittemore started the conversation by saying: 'We told Judy that she could have a copy of the notebooks.' I replied 'I have read the application for TRO [Temporary Restraining Order] and that is a bold face lie.' Before a verbal fight could ensure, Lipkin interrupted and emphatically told Whittemore that the culture of scientific community demands that scientists have access to their data and that of their students, for multiple reasons. Whittemore was insistent that the notebooks had to be returned before Mikovits (we always referred to her as "Judy") could post bail and be released from jail."[6]

When the conference call was over, Ruscetti and Lipkin talked privately about the situation, and Ruscetti sent a follow-up email to Harvey in which he wrote that many of Judy's relatives believed Harvey was "capable of doing her physical harm."[7]

Harvey emailed back that "we would never do anyone physical harm. That is just simply insane," then advised Frank not to "become an accessory after the fact" and suggested "her lawyers should immediately call the Washoe County District Attorney."[8]

Ruscetti then called Whittemore "and told him I didn't appreciate being threatened. Harvey replied that his response was not a threat, just friendly advice."[9]

The notebooks were returned to the Ventura County Sheriff's office in southern California first thing in the morning of Tuesday November 22, 2011, by Mikovits's husband, David Nolde.[10] Later that morning, Harvey sent the following email to Ruscetti. "Frank, the first positive step has taken place, but the damage continues because Judy's lawyers failed to provide a statement to her 'bloggers' to stop the stupid attacks on the WPI and Annette. If David can help Judy understand this and direct these people to simply stop we will have a very good chance of a positive outcome. I will do everything I can to help the authorities get her out of jail today in time for Thanksgiving with her family. This is not my decision, but I will help as is appropriate."[11]

Frank discussed the situation some more with Harvey Whittemore that day. Whittemore told Ruscetti that Mikovits needed to pay the Whittemore Peterson Institute $200,000 in restitution for their investigation, even though he knew she did not have that kind of money.[12]

Whittemore also wanted Mikovits to sign an apology letter so that "the internet bloggers would stop their attacks on the WPI and the Whittemores."[13] Ruscetti then "told Mr. Whittemore that the stimulus for the bloggers' attack was WPI's illegal firing of Judy, the legal attack by WPI against Judy, and WPI's arrangement to have her thrown in jail."[14]

Frank eventually tried to edit an "apology letter" drafted by Harvey and Annette Whittemore. He was aided in this effort by Ian Lipkin, but the two of them became convinced that the apology as written by the Whittemores was "an admission of guilt and had criminal consequences."[15] A selection of some of the passages, which they found to be problematic:

"I am deeply and honestly sorry from the depths of my heart and soul for causing great harm to the Whittemore Peterson Institute, patients around the world who counted on me, and all those who work at the Institute . . .

"I deeply regret that I am responsible for allowing those closest to me to create a cesspool of mistrust about the Whittemores, telling lies about them, and not stopping the malicious rumors from spreading out of control when I could have. The Whittemores have done nothing to deserve this horrible treatment by me, my friends and those who think they are acting on my behalf . . .

"I promise to restore the faith that all of you had in me by dedicating my life to honest and careful scientific discovery. I want to make full restitution to those I have hurt by having my apology posted on all blogs, forums, and websites by those who have inaccurately reported stories adverse to the interests of the Institute and those associated with it. . ."[16]

But the effort foundered when Ruscetti became convinced that the Whittemores were going to use the apology against her in a civil hearing, scheduled in Reno, Nevada, on November 22, 2011, the same day Judy was scheduled for her arraignment in southern California. Ruscetti could not understand how a person who was in jail in one state could be expected to prepare her defense for a case in another state.

For Ruscetti, it all looked like a rigged game and he didn't want any part of it.

When this became clear to Ruscetti after Harvey's urgent texts that the apology needed to be signed that day, he recalled, "I did not call Mr. Whittemore because I knew that Dr. Lipkin and I had been duped and Harvey was planning to use the apology as a confession against Judy at the hearing."[17]

* * *

After her termination from the WPI, but before her arrest, Mikovits wanted to continue her work with Lipkin on the XMRV replication study. On November 15, 2011, she sent him an email which seemed to set the ground rules for the study.

She wrote to Lipkin, "Thank you for giving me the opportunity to do the replication study in Frank Ruscetti's lab at the NCI. I understand that no money can be given for these studies but that you can provide resources, such as PCR and culture supplies and sequencing or other services (contracted through a third party). Frank will need to get permission for me to work as a special volunteer from Drs. Fauci and Varmus in order for me to work in his lab."[18] (Anthony Fauci was head of the National Institute for Allergy and Infectious Diseases and Harold Varmus was head of the National Cancer Institute.)

As things eventually shaped up, the agreement for Mikovits's participation in the Lipkin multi-center study must surely rank as one of the most unusual in all of science.

In an email sent on December 1, 2011, to Varmus and copied to Fauci and Ruscetti, Lipkin wrote that Mikovits would be associated with the National Cancer Institute for purposes of the replication study, but she could not step foot on the NCI campus, a prohibition which would be enforced by security.[19]

Ruscetti would later call this banning of a scientist from working in a lab and yet wanting to have her name on the ensuing paper, "unprecedented in contemporary American science."[20]

Mikovits strongly believed that her questions about XMRV, ME/CFS, autism, and the possible role of vaccines had caused consternation at the highest levels of the scientific community. But in her wildest dreams she'd never imagined a scientific study of international importance would contain a clause which held that a scientist would be designated as working at a specific lab, (the National Cancer Institute where she'd been employed for more than twenty years), but also provide that the scientist could not step foot on the grounds of that lab or would be escorted away by security.

It was a little like the Catholic Church allowing Galileo to participate in a "replication study," but then denying him the use of his telescope.

* * *

Mikovits thought she'd be released soon after the hearing, but she and Karen were ordered to get back on the bus. Mikovits protested but a guard said there was no paperwork authorizing her release.

As they were processed at Todd Road, the receiving guard saw the names and exclaimed, "You're not supposed to be here!" Within an hour both Karen and Judy were in a van heading back to the Ventura County Courthouse.

Back in the catacombs, Mikovits joined the prisoners who had made bail and were slowly transitioning back to freedom. Their clothes and personal belongings were returned: for Mikovits this meant gym clothes, flip-flops, and her baseball cap. She had also surrendered her wedding ring after the arrest, slipping it into her gym shorts pocket. It had not been off since she married David in 2000, so she had the thought that he should be the one to put it back on.

The guards offered another bag lunch and Mikovits had the good sense to take it, handing it off to Crazy Mary. Mary was still chatting, occasionally flitting over to Mikovits to ask a question.

At one point she asked, "What were you in here for?" Mikovits explained her story. Mary listened in rapt attention then shouted, "This is a doctor! This is a PhD! She's in here because she stole her own brain!" Mikovits couldn't help wondering why it seemed her story only made sense to a homeless woman with something of a sketchy grasp of reality.

Mary's shouts attracted two sergeants. One of them asked Mikovits, "So, you're a PhD?"

"Yeah," said Mikovits. "When I came in they only asked me if I'd graduated from high school and I said yes. I figured I shouldn't answer questions you didn't ask."

The guards started laughing. Since cancer is such a universal disease, a few of them asked if she might have anything to offer their family members or friends with cancer. Mikovits said sure and collected a few of their cards.

Wherever Mikovits went, it seemed like she was picking up patients.

Mikovits and Karen were ushered out at the same time and ushered out into the nearly empty lobby of the Ventura County Courthouse at the same time, but neither of their husbands were there. The two of them walked out into the large parking lots, looking around in vain. Mikovits remembered a nearby restaurant and thought they might let her use the phone to call David. But Karen wanted to continue looking in the parking lot.

"I hope you get your brain back!" she said wryly, in parting, though Mikovits was more concerned about getting her reputation and credibility back.

* * *

At the Hill Street Café Mikovits asked to use their phone to call bail bondsman, Bill Burns to arrange a meeting. She also called David.

When Bill arrived, he found Mikovits a little dazed, drinking a beer at the counter. Burns would later recall she had "just this sense of wonder about what had happened to her."[21]

It was Bill's experience that among people who had spent their whole lives outside of the criminal justice system and suddenly found themselves at its mercy, there was often a "wide-eyed amazement at what can happen to a person." This was often coupled with a profound relief at leaving jail.

David walked in soon, and they headed over to Bill's office, signed the papers, copied their drivers licenses on the Xerox, and wished each other a happy Thanksgiving.

* * *

The Nolde clan had planned to spend the holiday week at the Sonoma Mission Inn, starting the previous Sunday, but David had gotten them to hold the reservation.

Mikovits said she didn't want to go back to their house because it would only serve as a reminder of her upsetting arrest. She asked that he simply pack a suitcase of their things and they start driving north from the café. They planned to have lunch with one of David's sons in San Francisco the next day, then make the final hour and a half drive from the city to the pastoral Sonoma Wine Country dappled with gnarled old grapevines.

"Try not to speed," Bill Burns said as he left them.

They headed up north in their 2003 blue Audi A6 and it was clear to Judy that her husband was rattled. He relayed how he'd been on the phone constantly with Frank and Harvey over the past few days, trying to figure out a solution to everything. He mentioned Harvey's "apology letter" and showed it to her.

She started to read it with trepidation and finally said, "I'd never sign this."

David said nothing; he stared ahead with tears of frustration and relief welling.

"We just wanted to get you out before Thanksgiving."

*   *   *

Judy also learned what had happened to her friend, Lilly Meehan, on the day of her arrest.

Lilly had first read about Mikovits shortly after the October 8, 2009, publication of her XMRV article in the journal *Science*. When she noticed Mikovits's email address as the senior author, she jotted off a message about her close family member who'd been struck by ME/CFS at the age of fifteen and was completely disabled. Lilly was surprised when Mikovits speedily wrote back and found that the scientist often spent weekends and vacations at her Oxnard, California, home—as luck would have it, not far from where Lilly lived.

They made plans to have lunch when Mikovits came home for the Christmas holidays. Mikovits told her to be sure to bring her family member with ME/CFS if that person could manage it. When they sat down for lunch Mikovits quickly asked the family member about her symptoms, such as light sensitivity, nausea, post-exertional fatigue; the kind of questions the other doctors almost never asked that Lilly knew as central features.

The women became fast-friends, and when the unpleasantness started with the Whittemores, Lilly had launched a public outcry. As the situation deteriorated from a wrongful termination into a civil lawsuit on November 9, 2011, Lilly publicly posted a response from Mikovits's attorney at the time.

If she had understood more about legal culture, she would have fully comprehended the bad strategy of publishing it, but since Annette was posting accusations against Mikovits on the WPI's Facebook page, it was already public.

Shortly before her arrest, Mikovits was convinced she was being followed.

Lilly believed it was a credible threat. Soon after, from the second floor of her house, Lilly saw a man walking along the sidewalk and suspected he was a process-server for the Whittemores. Moments later, he rang her doorbell. Lilly froze and stayed quiet in her home, and then did a few chores.

She soon heard men's voices in her backyard and then three men entered her house through the back door.

Within a minute, Lilly counted a total of nine police officers in her home. They sat her in the recliner in her living room while the police officers searched her house, handing her an odd search warrant. It listed her address, the issuing judge (the Honorable R. Wright), and the case agent supervisor (Detective Horigan), but not any items for which they were searching.

Lilly later showed the search warrant to her sister-in-law, a paralegal. Because of the lack of specificity in the text, Lilly's sister-in-law simply stated, "Something powerful is behind this." With Lily in her recliner and two policemen standing guard, the rest of the officers searched her house.

For Lilly, who considered herself a model citizen, having been married thirty-six years and never having gotten so much as a traffic ticket, it was both surreal and traumatizing. She heard police officers rifling through boxes that had been in the attic for years. "I kept wondering how this could happen," she later explained.

Mikovits called Lilly at 8:15 the morning after her release to apologize for what she had endured.

"Those were *her* notebooks," Lilly later said, with outrage. "She was the principal investigator. They couldn't deny them to her. She was supposed to have them."

* * *

The next morning while David was showering, his cell-phone rang. Judy picked it up. "Hello?" she said.

"Is David there?" asked the familiar voice of Harvey Whittemore.

"He's in the shower."

"I shouldn't be talking to you, but have David call me," said Harvey brusquely. There was a pause in his voice, something that Mikovits hoped was a small shred of humanity from a man she'd considered family for the past five years. "I'm really glad you're out of jail," he said gruffly.

"Yeah, right," said Mikovits hanging up.

She didn't even tell David about the call.

* * *

Later, they were driving up to San Francisco to visit David's son for lunch, when a call came in from her civil attorney, Dennis Neil Jones.

"You need to turn yourself in at the Reno courthouse on Monday morning," he said. Mikovits said the judge in Ventura ruled she was free on bail so she wasn't going. But Jones insisted it was mandatory. "You'll be in and out in a couple hours," he promised. He'd found criminal attorneys for her in Reno. They were now handling the criminal case.

A preliminary injunction had been issued from the Reno court on November 22, 2011 (the day she'd been arraigned in southern California), and it ordered her to make copies of all her personal emails to provide them to the court.

When Mikovits phoned, the attorney answered and seemed to know her case. Mikovits told the attorney she had been released on bail and was not turning herself in to the court in Reno as she never possessed any property of the WPI. The attorney insisted that if Mikovits did not turn herself in, she would be picked up and jailed.

Mikovits unleashed her frustration. She had paid the bail, and besides that, she had not seen the notebooks or the birthday bag since October 17. The attorney repeated that she must be there by 11:00 a.m. on Monday or police would pick her up.

On their "vacation" at the Sonoma Mission Inn, David and Judy went to an Apple store and bought an iPad and tried to download all 22,000 emails from Mikovits's personal Gmail accounts, excluding the ones that contained classified patient information or the ones irrelevant to her WPI work. Mikovits and her husband drove directly from Sacramento to the Reno office of her new criminal attorneys, Scott Freeman and Tammy Riggs. The attorneys escorted them to the courthouse where they were besieged by news reporters.

Mikovits was booked, had her mug shot taken, and told she was free to go. But someone else was waiting in Reno: Jon Cohen of *Science*.

And he finally got the mug shot denied him by the California judge. A mug shot to adorn his article on Mikovits.

"Dispute over Lab Notebooks Lands Researcher in Jail" was the headline in the December 2, 2011, issue of *Science*, complete with the mug shot.[22]

* * *

One issue that needs to be addressed is how Frank Ruscetti's name came to appear on a paper authored by John Coffin entitled, "Multiple Sources of Contamination in Samples from Patients Reported to Have XMRV Infection" published in February of 2012.[23] For many following the XMRV story, it seemed a significant change in Ruscetti's support of Mikovits and her investigation, especially since he was Mikovits's mentor and one of her most prominent and respected supporters.

In July of 2014, Ruscetti explained how his name came to be on the paper:

> In August 2010, blood samples drawn from the homes of patients and controls were sent to my lab at NCI where cells and plasma were separated and sent to the Drug Resistance program (DRP)(directed by Coffin). They reported that on August 23 that all four ME/CFS patient plasma had MLV and mouse DNA, but no XMRV, while 2/5 normal had XMRV, but no MLV or mouse DNA. These patients had not been found to be XMRV positive.[24]
>
> Nothing was said about this data, for almost a year, after the SOK [State of Knowledge, 2011] conference where Alter stated there was no evidence of contamination at the WPI. I was never allowed to see any of the data before my bosses. It smelled like an inquisition.
>
> I was aghast when I saw a draft of the paper. The original 2010 data was in the paper without any replication and the tone of the paper placed most of the blame for multiple contamination on the WPI and Dr. Mikovits. Since there was no evidence in this paper supporting that contamination and my belief the XMRV presence in the normal donors came from making of the DNA in the Coffin lab, I declined to be an author. On October 6, 2011, [a week after Dr. Mikovits had been terminated from the WPI] I was told it would be good for my career if I was a coauthor.
>
> Since there was an internal review of my program looming and a reevaluation of funding in two years instead of the usual four, I relented and consented.[25]

* * *

As a scientist, Mikovits was trained to consider data points to determine if any larger patterns could be found.

First, she'd been arrested at her home on Friday, November 18, 2011, without an arrest warrant, and held without bail until Tuesday, November 22, 2011.

Second, because she'd been in jail in southern California she couldn't prepare for or appear in Reno, Nevada, on Tuesday, November 22, 2011, for the civil case brought by the Whittemores. At that hearing, the judge issued a preliminary order that she turn over all of her emails, something she could not do according to privacy laws regarding study participants.

Third, the most powerful scientific authorities in her field, Anthony Fauci of the National Institute for Allergy and Infectious Diseases and Harold Varmus of the National Cancer Institute, wanted her name on the Lipkin multi-center study, but didn't want her setting foot in a lab, an edict they were willing to back up with security.

Fourth, at her southern California hearing, the judge had denied Jon Cohen's request for a picture of her in her prison garb, and denied that she had to appear in Reno.

Fifth, after being released, it was suddenly mandatory that she appear in Reno, and when she arrived, Jon Cohen was waiting.

"You must really have pissed off someone important," the bail bondsman, Bill Burns, had told Mikovits after he first investigated her case.

Mikovits was starting to believe she must have pissed off a lot of important people.

CHAPTER THIRTEEN

The Counter Offensive

"We and others are looking at our own specimens and trying to confirm it," he said, adding, "If we validate it, great. My expectation is that we will not."[1]

—Dr. William Reeves

Four days after the *Science* article lit up the media, the ME/CFS director at the CDC—William Reeves—publicly dismissed the idea of a retroviral association with ME/CFS and broadcast a damning prophesy in the *New York Times*, saying his team did not expect to validate the XMRV findings. They had not performed a single experiment, and moreover had attended the July 22 meeting, worked with the samples, and found no issues with the Mikovits team's research findings. Patients with ME/CFS had grown weary of William Reeves's decades-long claim that their symptoms were psychological.

Around the time of the October 2009 *Science* publication, Reeves may have been working on his research that was later published the following July in his coauthored paper, "Personality Features and Personality Disorders in Chronic Fatigue Syndrome: A Population-Based Study," which concluded that, "CFS is associated with an increased prevalence of maladaptive personality features and personality disorders. This might be associated with being noncompliant with treatment suggestions, displaying unhealthy behavioral strategies and lacking a stable social environment."[2]

Reeves had taken over the NIH's ME/CFS program from Dr. Stephen Strauss (a physician turned virologist), who was also quoted in the *New York Times* in 2001 (eight years before the *Science* publication) on his belief that "What's important about CFS is many people get over it. Individuals who have it for many years lose hope. They then take on a series of maladaptive behaviors, which sustain their illness because they become so focused and so phobic: they avoid exercise, disrupt their sleep patterns. It gets harder and harder for them to regain normalcy."[3]

Patients had to wonder what these government heads would consider an "adaptive" as opposed to a "maladaptive" set of illness "behaviors," when most of them expressed a relentless wish to deal positively with hardship, by seeking out social networks, downgrading to new hobbies that matched their limited abilities, and pursuing medical care that might help them get their lives back.

One patient in an ME/CFS support group was told by her admiring doctor, "You're not depressed *enough* given what you're dealing with!"

* * *

Andrea Whittemore's picture appeared with the *New York Times* article, quoting Reeves. She was leaning on a couch, still youthful but with a drawn, pale face a pretty blue blouse hugging her thin frame, her blonde hair and make-up styled, with her right arm crooked across a pillow and her head almost leaning against her hand. The last detail was not lost on ME/CFS patients, who joked about their alternate reality to that of "The Uprights"—people who could stand up easily without propping themselves on something. An oxygen tube curled from the lower edge of the picture up to the nasal cannula peeking out from Andrea's pert nose. At thirty-one, she had been sick for nineteen years. She looked directly into the camera as if to say, *I'm here, I exist.*

The *New York Times* summarized the history of the disorder well, but impersonally. It seemed to describe a natural oddity, rather than conveying the horror of a tragic, devastating affliction for millions around the world. ME/CFS was so destructive that patient communities had gone decades without the physical ability to arrange large protest marches on the Washington Mall. It was hard to stand up and demand your rights when standing itself often presented many challenges.

The journal *Nature* gave the isolation and association of XMRV with ME/CFS some positive press, although it also covered the position of

William Reeves. *Nature* published their XMRV article October 8, 2009, in coordination with the article in *Science*, adding additional information on further research efforts that had gone on since the May 23, 2009, submission.[4] They noted that the authors detected XMRV in the immune cells of 67 percent of the ME/CFS patients but in only 3.7 percent of healthy controls, and that the authors showed that the virus was able to spread from infected immune cells to cultured prostate cancer cells. They also noted the authors' finding that the virus' DNA sequence was more than 99 percent similar to the sequence of the virus associated with prostate cancer.

Mikovits believed the association between XMRV and ME/CFS may have been even stronger than her team's work indicated. She wanted to study ME/CFS exposure to the virus more broadly. In an unpublished investigation, she and her colleagues analyzed blood cells in about 330 ME/CFS patients and found that more than 95 percent had evidence of XMRV infection, whereas only about 4 percent of healthy controls did.[5]

While the *Nature* article quoted the negative prediction from Reeves, it also ended with comments from Coffin, who was still then very supportive of the XMRV research. The Tufts University virologist who had studied MLV (mouse leukemia virus) pointed out that the XMRV virus' prevalence in healthy controls "is in some ways, an equally striking result. It's highly preliminary, but if it's in fact representative, then there are 10 million Americans with this infection, which is very similar to MLV and is now linked to two important diseases."

Coffin added: "There's a lot we don't know, including whether XMRV causes disease, but that's always the case when the first paper, like this one, comes out." Overall, the early press coverage was positive, but there were hints that the battle ahead would be more brutal than anyone imagined.

* * *

The night of the publication, Mikovits and her team gathered at the Whittemore mansion.[6] It was fall in the Sierras, with mild sunny Reno days dropping to near-freezing at night. The first winter storms expected to dump snow on the high mountain passes were only a few weeks away.

But the Whittemore mansion was warm and inviting, and Harvey and Annette had—in their usual gracious style of hosting—made sure there was enough food and drink for everybody. Mikovits, Peterson, Lombardi, University President Milt Glick, and their research staff including Katy Hagen and Max Pfost were present along with other dignitaries. Harvey knew which news shows would be covering the XMRV story, and the group watched in rapt silence as the screen displayed pictures of their lab and interviews with Mikovits and other scientists commenting on the work. The attendees felt like they were part of a pivotal moment in science history, like NASA employees at a private showing of the moon landing.

The British Broadcasting Corporation (BBC) wanted to hold the first live interview with the researchers for their morning radio show UK Radio 5 Live—with Nicky Campbell and Shelagh Fogarty—on October 9, 2009.[7] Since the interview would start at around midnight Nevada time due to the time difference, the group was tired but a little wired. England was home to several academic researchers who believed ME/CFS was nothing more than "mass hysteria" including psychiatrist Simon Wessely, who had taken to calling the disorder a "functional somatic syndrome," which was simply "mass hysteria" under another name.[8] It was a little like calling a woman "a daily serial monogamist with a rotating list of intimate partners" rather than a "tramp." They were still calling the patients crazy, but with just a bit more discretion.

The derogatory view of the illness, typically called myalgic encephalomyelitis (ME) in England, had apparently infected the BBC radio hosts and their tone became patronizing. The host mocked patients, and by extension the researchers, as mentally unstable.

Hagen retorted by addressing them in a faux British accent like a chiding nanny, and when the BBC team realized they were being laughed at, they cut away to the weather. Mikovits observed that Annette seemed to disapprove of Hagen's antics, but the rest of the team enjoyed the comic relief.

* * *

The Mikovits team found rousing support at home. The Whittemores hosted Ruscetti and Mikovits and her fellow researchers at their skybox suite at the University of Nevada, Reno's Mackay Stadium for a football game by the team and its fanatical fans known as the "Wolf Pack."[9] Mikovits enjoyed watching the Wolf Pack take on adversaries from the awesome view in the skybox. University President Milt Glick often dropped by to say hello.

"Hey, Judy, turn around," said Harvey eagerly, during the half-time show. On the Jumbotron was a tribute to Mikovits and Annette Whittemore for the publication in *Science*. The scrawl noted that this was only the second time in the university's history that an article from their school had been published in the prestigious journal.

Thirty thousand fans of the Wolf Pack roared their approval, sports fans cheering on *science*. Glick smiled and shook Mikovits's hand, offering congratulations.

It was a banner moment for all of them: a good game, good company, and football fans unexpectedly showering love for their scientific accomplishments.

* * *

Within a few weeks of publication, the research team began receiving letters forwarded by *Science* from researchers, including one from a group at the *British Medical Journal* inquiring about their patient selection. The group seemed critical of the attempt to find a cause behind the disorder, without having a good handle on the virology.

The editors of *Science* often gave the Mikovits team just twenty-four hours to draft a response, which meant everybody had to drop everything they were doing to respond.

An email exchange between the senior editor at *Science* and Mikovits on January 5, 2010, revealed the decaying nature of their relationship. The letter referred to their response as "carelessly prepared" and that "The reader should come away knowing exactly what criteria you used for CFS diagnosis and who exactly were studied as healthy controls (and how many controls were studied)." The senior editor added, "If you cannot send us a clear, informative and polished Response, I'm afraid we will have no choice but to publish the Tecchnical Comments on their own without a Response from you."[10]

The "carelessly prepared" document and accompanying material sent to *Science* in response to questions raised by Lloyd, Sudlow, and van de Meer is reproduced in part below:

> In our recent publication in Science (1) we analyzed blood cells from a patient population with a clinical diagnosis of Chronic Fatigue Syndrome (CFS) and found that sequences encoding the human retrovirus XMRV were present in 67% (68 of 101) of the patients but in only 3.7% (8 of 218) of healthy individuals. Our observation of actively replicating virus as well as antibodies directed against XMRV in this CFS patient population strengthened the hypothesis that this virus, which has also been detected in a subset of prostate cancers (2, 3), is a human pathogen . . .
>
> . . . Clinically, CFS is a heterogenous syndrome and its diagnosis is based on the exclusion of other diseases. Thus, the criteria used for the diagnosis of

CFS varies greatly. All patient samples included in our study were diagnosed by a licensed physician, using the Centers for Disease Control (CDC) criteria that diagnosis of CFS requires that an individual have persistent and relapsing fatigue for six months. All other known medical conditions had been excluded.

The patients included in our study all met at least four of the eight criteria defined by the (CDC) (4) as a prerequisite for CFS diagnosis; tender axillary lymph nodes, sore throat, significantly impaired cognition and short term memory, muscle pain, multi-joint pain, new headaches, unrefreshing sleep and post-exertional malaise lasting more than 23 hours. The patient samples used in the present study were not selected with any bias other than a clinical diagnosis of CFS.

The CFS patients were from the United States. Blood samples were drawn from patients in 2006–2009 from 35 patients identified in the 1984–1988 CFS outbreak in Incline Village, Nevada; 15 sporadic cases came from upstate New York, 15 sporadic cases from North Carolina and Florida, 36 diverse cases from western United States, Canada, Australia, and Ireland. Inclusion in the study was based on a diagnosis of CFS which met the CDC and Canadian consensus diagnosis of CFS regardless of severity.

The study population was 67% female, reflecting the gender incidence of CFS, and the age distribution was 19–75 years of age with a mean of 55. The control population of 320 was from samples drawn from healthy people visiting doctor's offices in the western United States. These samples were from used as controls for the PCR studies. The healthy controls used in the study were representative of the at-risk population (that is, zip code matched as well as age and gender matched).

For both patients and controls, peripheral blood mononuclear cells were collected from heparinized blood and aliquoted into four samples: RNA, DNA, protein plasma and DMSO freeze for culture later. Samples were prepared within 6 hours of blood draw and frozen immediately in -80 C or liquid nitrogen. Blood was drawn from the patients and controls at at least two different time points within three months of each other . . ."[11]

For a twenty-four hour turn-around requested by *Science*, the researchers believed it was a thorough job, with very minor editing mistakes.

In the third paragraph the team should have written that, "The CFS patients were mainly from the United States," although the sentences following that one detailed the country of origin for each sample. There was also some confusion over the number of control samples, whether it was 320 or 218. The quick answer was that the original number of controls provided by the lab was 320, but when trying to age and gender match the samples, this number was winnowed down to 218.

These were not grievous errors given their quick reply, and easily correctable.

The increasingly-scrutinizing attitude by the senior editor at *Science* was all the more puzzling to Mikovits and her team, since their paper was probably one of the most thoroughly vetted papers ever published by the journal. It had not only undergone peer-review, but was the subject of an all-day pre-publication conference of the world's leading experts on retrovirology as well as the focus of another, smaller conference attended by Frank Ruscetti a month earlier, almost a scientific hazing atop a peer review.

Mikovits reached out to a leading immunologist and advisor for the WPI on the paper and the antibody assay. She'd relayed her suspicion that *Science* would begin attacking the research because of the way they were handling the letters of complaint, so when the critical letter came from *Science*, Dr. Mikovits forwarded it to the advisor with the declaration, "Right on cue!"

He emailed back barely an hour after Mikovits had received the email from *Science* and told her to "[C]larify the writing and numbers; stick to what is in the paper."[12] It was good advice and Mikovits followed it.

* * *

The first negative study was published in the open-access Journal *PLOS-One* (Public Library of Science) and involved researchers from England long-associated with a psychiatric explanation for ME/CFS such as the previously-mentioned Simon Wessely, who preferred to call ME/CFS a functional somatic syndrome, rather than "mass hysteria."

The Wessely research was even partially funded by the South London and Maudsley NHS Foundation Trust/Institute of Psychiatry National Institute of Health Biomedical Research Centre. Compared to those for *Science*, the standards for publication in *PLOS-One* were relaxed. The work of Mikovits and her team was submitted to *Science* on May 4, 2009, the referee comments were received June 6, a resubmission made in early July, an NIH Conference held to discuss the findings in late July, an official acceptance issued on August 31, followed by more than a month of editing during which a full third of the paper was cut, and it was not published until October 8, 2009, five months after the initial submission.

In stark contrast to this lengthy review process, the first *negative* study on XMRV was submitted on Tuesday, December 1, 2009, and accepted and published by *PLOS-One* on Friday, December 4, 2009,[13] somewhere around seventy-two hours.

Within just a couple more months, another negative paper on XMRV and ME/CFS was published in *Retrovirology* in February 2010. It was also from the United Kingdom, from the laboratory of Dr. Kate Bishop.

Bishop and her team studied two cohorts of ME/CFS patients from the UK to look for the presence of viral nucleic acids by using a PCR assay for XMRV *gag* gene and the presence of serological responses using a viral neutralization. No association was found in that study between XMRV and CFS, while considerable cross reactivity was found in the serological assay.

The next negative paper came from the Netherlands and used samples that had been gathered from patients between December 1991 and April 1992: it was published in the *British Medical Journal*.[14] The method the Dutch used was a "real time polymerase chain reaction assay targeting the XMRV integrase gene and/or a nested polymerase chain reaction targeting the XMRV gag gene."

The researchers found "no XMRV sequences in any of the patients or controls in either of the assays." The authors noted in their conclusion, "we found no evidence for a role of XMRV in the cause of CFS in Dutch patients. Over the past decades we have seen a series of papers prematurely claiming the discovery of the microbial cause of CFS. Regrettably, thus far none of these claims have been substantiated."[15]

Mikovits was particularly concerned about the last paper since her team had established a collaborative research agreement with the Netherlands researchers. The Netherlands group had in fact *found positives*, and Mikovits's team *confirmed* the positives, but when the paper was published there was no mention of the positive samples.

The researchers never reported their positive findings because they had altered their PCR procedures to make the assay more stringent (less likely to pick up closely-related virus strains or non-specific results, thus narrowing the probability of any positives).

Mikovits considered this scientific selectivity to be unfair. It was the relaxed stringency, which had allowed the retrovirus to be identified in the first place. They didn't have a good handle on what it was, but they knew what was close to it.

* * *

Mikovits believed best way to characterize these three studies was to say that they were unable to find XMRV in those with ME/CFS or healthy controls using Silverman's VP-62 plasmid as the gold standard. This was frustrating to Mikovits as she had given several slide presentations around the world and said specifically that researchers needed to be looking for more than the VP-62 sequence, because there could be others.

It was time for Mikovits and her team to mount their response.

* * *

Mikovits and her team addressed the challenge of responding to the negative studies as well as some positive findings that had come out in the research literature.

Their response paper was entitled, "Distribution of Xenotropic Murine Leukemia Virus-Related Virus (XMRV) Infection in Chronic Fatigue Syndrome and Prostate Cancer" and was soon published in the *AIDS Review*,[16] a venue where researchers had already endured a complicated and conflict-laden history around HIV.

The explanation for the negative studies revolved around four potential arguments.

First, there might be differences in the geographical distribution of XMRV. Ruscetti's human T-lymphotropic virus type 1 (HTLV-1) exhibited was relatively easy to find among Japanese populations, but less so in other Asian countries, and exceedingly rare in regions outside of Asia. Maybe XMRV was largely confined to the United States, the way *borrelia burgdorferi* was the commonly accepted cause of Lyme disease in the US, whereas *borrelia garinii* was more common in certain parts of Europe.

Second, PCR data (which was used by all three negative studies) is *particularly* sensitive to sequence variation. Therefore, if the viral sequence was well-characterized and the regions which were conserved and did not change markedly as the virus evolved, *then* PCR was a good diagnostic tool. At the time, however, the genetic variation between different XMRV sequences was about 0.03 percent, even though they were obtained from geographically distant areas. This variation was even smaller than the differences observed in HTLV-1, a virus that did not appear to exhibit much sequence difference. Thus, there was not enough known about the sequence variation to determine whether these regions were well-conserved or not, so PCR was not necessarily reliable.

Third, the question of whether individuals infected with XMRV had the ability to form antibodies to the viral proteins was still unanswered. While their research had suggested patients did form antibodies to XMRV, it was still too early to use this as a reliable test (and it required the use of the spleen-focus forming antibody test as an analog). As previously noted, an antibody response could be blunted by a compromised immunity.

Fourth, and perhaps most significant, the clinical criteria used by the various researchers varied dramatically, a major point given the vast arguments over the vagueness of some case definitions of ME/CFS, which easily could include cases of (extremely common) idiopathic "chronic fatigue" caused by unrelated issues such as heart failure or overwork.

Patients with ME/CFS had argued for years to judge them with more stringent diagnostic standards, so they would not be lumped in with people who were simply overtired who suffered from an obviously different ailment.

The Canadian Consensus Criteria—after much lobbying from patients and doctors—had thus become the gold international standard for defining the ME/CFS population, as it relied not just on physical symptoms, but also highly measurable and clinical and immunological data.[17]

Mikovits and her team had used the Canadian Consensus Criteria as well as abnormal chemokine and cytokine markers to more clearly define the population and weed out unrelated illness, while other studies had seemed carelessly unconcerned with distinguishing between ME/CFS and generalized, idiopathic fatigue.

While the negative studies were getting the most press, other research had also been conducted during this time that supported the findings of Mikovits et al.

Researchers from Germany had found XMRV sequences in the respiratory tract secretions of 2–3 percent of healthy donors and 10 percent in immune compromised donors.[18] The clear implication from the German research was that the immune status of individuals might be an important factor in XMRV detection.

The argument Mikovits and her team made was that their paper was not simply "a survey of CFS using PCR or a search for antibodies to XMRV proteins" in a poorly defined group of patients. They had used several different assays for the presence of XMRV nucleic acids, proteins, and infectious virus.

The research might still be challenged, but if disproving the findings involved a direct replication of the testing done by the Mikovits team, it had not been accomplished.

* * *

In February 2010, Mikovits flew with the Whittemores to Washington, DC, to visit with Senator Harry Reid.[19] They spent the night at the historic Hay-Adams Hotel, just across from Lafayette Square, with a view of the White House. The hotel was founded by John Hay, a former secretary to Abraham Lincoln, and Henry Adams, an acclaimed author and descendent of father and son US presidents, John Adams and John Quincy Adams.

The Whittemores and Mikovits managed to check in just before an ice storm hit the region. The storm left the trees latticed with ice, and when Mikovits stepped outside into the cold the next morning, she noted the absolute quiet that had blanketed the capitol city.

The ice storm had closed most of official Washington, but that didn't deter the group from making their way to the Office of the Senate Majority Leader. They were on a mission to get funding for the planned institute and see if they could get it designated as a "national center of excellence," which would make it much easier to obtain federal grants.

The Whittemore's lobbyist, Alan Freemeyer, drove them to meet with Harry Reid. On the way, Freemeyer had a special stop planned for them. As they drove, Mikovits thought the neighborhood looked familiar. They pulled up to a house, and it clicked.

"This is my grandfather's house!" she exclaimed happily, suddenly flooded with nostalgic memories. "Where my mother was born and where my grandfather died of cancer!" Freemeyer pulled the car over and Dr. Mikovits stepped out onto the sidewalk.

She remembered all the times she'd gone there, the times she'd listened to baseball on the radio with her grandfather, and when he'd been too sick from cancer to listen. In that very house, she had made the decision to fight the disease that was stealing her beloved grandfather away. She knew she had mentioned the address, 417 Constitution Avenue, to the Whittemores in her stories about her grandfather, but didn't think they'd been listening that closely.

But they had, and they were honoring her by remembering it, and taking her to the location so she could recall the young girl who so desperately wanted to fend off her grandfather's death. That young girl still lived inside Mikovits, and now she was going to visit one of the most powerful men in the country. She had come full circle, fulfilling the promise she'd made in that very house.

The Office of the Senate Majority Leader was almost overwhelming to Mikovits. It was elegant and radiated centuries of power. Portraits of John F. Kennedy, Andrew Jackson, and Mark Twain (whose wife "Livy" Clemens had suffered from "neurasthenia," prompting Twain to write about his beloved, "She was *always* frail in body and she lived upon her spirit, whose hopefulness and courage were indestructible"[20]) hung on the walls.

The office was appointed with large wingback leather chairs, dark mahogany tables, the seal of the Senate, and pictures of the Senator with various world dignitaries. As the group sat down to speak with the Senator, a large fire crackled in his private fireplace.

Reid inquired after Andrea's health with his typical concern, before they all turned to Mikovits. She began by talking about her background, her experience culturing HTLV-1, her doctoral work on HIV latency, the years she'd spent at the NCI biological response modifiers program and in the screening technologies branch of the division of cancer treatment, and determining which compounds and drugs might be helpful in treatment for AIDS associated malignancies.

The recent publication of the *Science* article was what really held his interest, and how the WPI could be designated a center of excellence, to make it easier to obtain federal funding.

The momentum seemed swifter than it had ever been.

Mikovits mentioned how she had become aware of ME/CFS, the estimated seventeen million people worldwide and one million in the United States who suffered from the condition, and the hellish day-to-day experience they endured. Senator Reid eased back into his chair, a finger at his lips, his eyes watching intently.

She knew the look.

It was one of active interest; and while he might not be following all of her points, she was convincing him of her expertise for the task of conquering this illness that was known to be more common than HIV, lung cancer, and breast cancer in women (with no small number in men and children as well): a national epidemic.

When she was finished, Senator Reid's eyes darted to Harvey, then Annette. He said he saw why this was so important and wanted to help in any way he could. Reid's staff could help with any details. What more could any researcher want than to have the support of the most powerful man in Congress?

* * *

Mikovits was about to enter the most perilous phase of her professional career, and by late May and early June of 2010 it would come to be viewed as the time when angels and devils revealed themselves.

The publication of her article in *Science* put Mikovits at the top of the list of speakers for the fifth *Invest in ME Conference* in London on May 24, 2010, and she was slated to be the final speaker for the day.[21] When she had attended the year before, she'd been blown away by the keynote address of journalist Hillary Johnson with her speech, "The Why," that recounted in vivid detail not just what it felt like to have ME/CFS, but the manipulation and mismanagement of the disease by the CDC starting with the 1984–1985 Incline Village outbreak.

A year after that conference, Mikovits had become friends with Johnson and she shared with the journalist much of the inside drama of what had taken place since the publication of the *Science* paper. The paper had come under some attack from the negative studies, but researchers at the conference were still enthusiastic about XMRV and what it might mean for the treatment of ME/CFS.

During the conference Mikovits and Annette Whittemore got the opportunity to have several long talks with Cheney about the research. Cheney had more of a research bent than Peterson and the theory of a retroviral origin for ME/CFS could reasonably be credited to him during his early presence at the 1984–1985 Incline Village outbreak. He had also worked with Dr. Elaine DeFreitas, who in the 1990s had published evidence of a retrovirus in the disease, but then had come under vicious attack.

A 1996 *Newsweek* review of Johnson's book *Osler's Web* depicted the eerie déjà vu of an "arrogant" CDC that seemed capable of unfairly skewing retroviral replication studies on ME/CFS again in the future: ". . . when the CDC publishes a paper saying it has been unable to replicate her findings, her support evaporates. By early 1995, the saga has cost [Dr. Paul] Cheney and [Dr. David] Bell their marriages, and a regretful DeFreitas fears her career as a scientist is finished. The book closes with the image of an infectious disease spreading unchecked as an arrogant medical establishment looks the other way."[22]

Cheney had long ago parted ways with Peterson and moved to North Carolina to open his own practice, while also devoting a significant amount of time to research. Mikovits was impressed with Cheney's intelligence. The WPI was looking for a medical director, as they'd long planned to open a treatment center in August of 2010. Peterson had resigned earlier that spring.

That left an opening for a medical director, one who could step in and immediately understand the complexities of the disease as well as the patient population. Cheney expressed interest in the position, assuming they were able to get the clinic up and running.

For Mikovits, landing somebody with Cheney's background, intelligence, and credentials was a coup after what had been a rough couple of months. They were going to win this fight and offer hope to beleaguered patients. Cheney had been there for the beginning of the battle, and they all planned to keep fighting.

* * *

A reporter for the *Chicago Tribune,* Trine Tsouderos, contacted the WPI after learning that Mikovits was scheduled to speak at the AutismOne Conference in Chicago. This was the same event Nobel Prize winner, Luc Montagnier, would attend two years later, leading to a scientific reprimand by his old rival, Robert Gallo.

The Whittemores discussed the merits of Mikovits responding to Tsouderos with the WPI public relations company. They decided that Tsouderos could submit her questions in writing, Mikovits would answer them, and then the public relations consultants would review them before sending them back to Tsouderos.

But after hearing that Paul Cheney was willing to come on board as medical director, and having spent her life as a scientist telling the truth as best she saw it, Mikovits decided to email Tsouderos directly from the plane from London.

Trine Tsouderos was not held in high regard among many of the regular attendees of the annual AutismOne Convention in Chicago. Many years around the time of the conference articles by her would be published, questioning some new therapy or research being investigated by the autism parents. The larger questions of what might be causing the disease or whether the federal response to the epidemic was adequate, rarely found their way into her articles.

Prior to working for the *Chicago Tribune*, Tsouderos wrote for *People* magazine. She later left the *Chicago Tribune* in September of 2012 to become the Healthcare Media Director for Golin Harris International, a communications firm, serving such clients as Dow Chemical, GlaxoSmithKline, and Merck Pharmaceuticals.[23]

To Dr. Mikovits it seemed as if Tsouderos had already written the article in her head prior to even speaking with her. The article ran in the June 7, 2010, edition of the *Tribune*. Entitled "Hope Outrunning Science on Chronic Fatigue Syndrome," it painted a picture of Mikovits as an out-of-control scientist:

> Mikovits said she accepted an invitation to speak at AutismOne to help sound the alarm. She accused researchers and government agencies of being more

interested in previously published research linking XMRV and a form of prostate cancer than in her work . . .

So research dollars will go for XMRV infected men with cancer but not women with CFS," she wrote in an email. "(This) left me no recourse but to play the autism card! Will they ignore the children too?[24]

Mikovits was known for not mincing words, but she'd never perceived herself as a public figure before. In the right context, such outspokenness could shake people up and mobilize them to create change.

But in the fray of a politically sensitive scientific discovery, with ramifications that reverberated out to cancer, ME/CFS, multiple sclerosis, and the nuclear issue of autism and vaccines, such outspokenness could cause a cataclysm.

* * *

While in Chicago, Mikovits and Harvey Whittemore met with Dr. Marcus Conant, a well-known AIDS researcher and Dr. Joseph Burrascano, a Lyme disease specialist, to discuss putting them in touch with a wealthy benefactor who owned a diagnostic company and was interested in developing an XMRV diagnostic test.[25] Conant was a legend in the AIDS community, having first identified Kaposi's Sarcoma in AIDS patients as an associate professor at UCSF, then developing some of the most used HIV medications and helping to found the San Francisco AIDS Foundation which had served as a steady anchor through the darkest years of the AIDS epidemic.

As an advocate, he was the successful lead plaintiff in a case that claimed the First Amendment protected the right of physicians to recommend the use of medical marijuana to people living with HIV and AIDS. Burrascano, as an expert on Lyme and tick-borne coinfections, had testified before Congress on his areas of expertise.

At one point in the meeting, a call came in on Harvey's cell phone. He listened for a few moments and then exclaimed, "She did what?" Tsouderos had apparently shared with the WPI what she was planning to publish and the WPI was going into overdrive to try and limit the damage.

Though apologetic, Mikovits felt the entire situation was unfair, that she should be able to talk freely, that it was an absolute necessity for scientific investigation and most scientists did not come under this level of scrutiny. Yet she also knew that it was reasonable to vet public comments through their PR firm.

Harvey said, "We love you, warts and all, Judy. But don't talk to anybody without our permission."

* * *

When Mikovits first made her plans to speak at AutismOne, shortly after she'd spoken at the Defeat Autism Now! conference in January of 2010, there had not been much media attention on her. As the spotlight increased, Annette expressed misgivings about Mikovits going to AutismOne. They discussed it and Mikovits said she was going because she had given her word.

Even though Mikovits planned to go with the full knowledge of Harvey and Annette, they tried to limit her exposure there. Normally at AutismOne, every talk was recorded and the entire conference then placed on a DVD so that parents and professionals could review it. Her talk would not be recorded.

Hillary Johnson accompanied her to the conference and together they walked the halls and commented on what they saw. They both felt that the community of parents in autism were much better organized than the ME/ CFS patient community. But they also agreed the research conducted on autism was woefully lacking.

Mikovits's talk was on the program at the same time as another presentation by the editors and writers for the website *Age of Autism*, founded by Dan Olmsted, who had been an editor for twenty-five years with *United Press International* as well as *USA Today*, and Mark Blaxill, a Harvard MBA who ran his own consulting company.

One of the writers for *Age of Autism*, this book's coauthor Kent Heckenlively, had become interested in the XMRV story from the autism angle and had talked to Mikovits several times.

* * *

"Welcome to the Rebel Alliance," Dan Olmsted said in his opening remarks, referencing *Star Wars*, then introduced the various editors and writers for *Age of Autism*. The small room was packed, so after he stood up in front of the group of about three hundred, Heckenlively raced out of the room to Mikovits's presentation, where the audience of some thirty spectators was dwarfed by a cavernous ballroom.

Heckenlively listened to the talk, then joined the small group of people huddled around Mikovits to ask questions. As he waited for his chance to talk to Mikovits, Heckenlively saw something that set off warning bells in his head.

Hillary Johnson was wearing a press badge for *Discover* magazine. In January the magazine had published its "100 Top Stories of 2009." The research linking XMRV to ME/CFS was ranked a respectable #55.[26] The #1 science news story of 2009, according to *Discover* magazine—ahead of stem cells, swine flu, or even the discovery of the first Earth-like planet orbiting another star—was "Vaccine Phobia Becomes a Public Health Threat."[27]

For parents like Heckenlively, who had seen their children regress into autism after a vaccination, the wholesale dismissal of their observations as a "phobia" was like waving a red cape in front of an angry bull. He immediately felt combative toward Johnson, briefly considering grabbing her by the scruff of the neck and tossing her out of the nearest exit, but when he introduced himself, she surprised him saying, "Oh, you're one of the people I'd been hoping to meet here!"

She said before the conference she had reviewed the articles on *Age of Autism* and had come away thinking that the two people she wanted to meet were Katie Wright, the woman whose parents had founded Autism Speaks, and him.

"I thought you'd be the best person to understand XMRV in the autism community," Johnson told the stunned Heckenlively.

* * *

"You know they're going to come after you, don't you?" Heckenlively said to Mikovits at dinner that night with Johnson.

Mikovits explained that she had taken precautions by gathering strong scientific and political support and that new findings that upset old paradigms were often the subject of heated attacks. She felt safe keeping her nose to the grindstone and focusing on science.

"But you talked about autism and vaccines," he answered. "You've got yet another bull's-eye on your back. No offense to your XMRV discovery, but it was story #55 in *Discover* magazine's top 100 stories of 2009 and told with neutral language. The anti-vaccine parents were #1. In their eyes we're the same enemy."

"I really haven't joined the anti-vaccine campaign," she protested. "I think vaccines have by and large been a good thing, it's just that in some people they may have negative effects."

"That's what we all say at first," continued Heckenlively. "We say we understand that vaccines have done wonderful things, but may cause problems in a 'sub-group' of people. We politely ask, 'won't you please look at this question so we can continue to vaccinate safely?' Then we got attacked for being 'anti-science' when all we wanted was *more* science for our kids. I have a child who can't speak and I want answers."

Heckenlively told the story of his own children. His daughter, Jacqueline, had been born without complications and developed normally for the first six months of life. She'd gone off of breast milk at six months, starting a cow's milk based formula and a few weeks later got her six-month series of shots. During this time she also had a series of ear infections and received six rounds of antibiotic treatments. The decline in her had been gradual, and she wasn't diagnosed with a seizure disorder until about four months later.

"I'd heard about the vaccine theory for problems with these kids around that time," Heckenlivey continued, "It just seemed so unbelievable. I actually remember telling people, 'shit happens' and that unfortunately something terrible had simply happened to my daughter with her seizures and autism. Then came my son, Ben, and he made me a believer."

Ben was born without complications and from the very start was so active he was nicknamed, "Busy Ben." Heckenlively's wife's family had a history of milk allergies and amidst all of the problems with Jacqueline they uncovered a milk allergy.

When Ben went off breast milk, his parents gave him a hypoallergenic milk formula until he was about fifteen months old. But it was expensive and since their son was doing so well, they figured it didn't justify the extra cost. Heckenlively's voice became softer and more intense as he continued Ben's story.

"I took Ben to his eighteen month checkup in early January. I'd given the pediatrician hell, because I thought she might have missed something at my daughter's six-month checkup years earlier. They were on the lookout for problems with my kids, so they put Ben through an entire developmental milestones series of tests. He passed with flying colors. He had fifteen or twenty words, looked at the doctor, and was very engaged. They gave him his shots and we left."

Within a few days of getting vaccinated, Ben became mute.

On Martin Luther King Day in 2001, Kent was in a bookstore and ran across the book, *Unraveling the Mystery of Autism and Pervasive Developmental Disorder* by Karyn Seroussi. The book presented the theory that a diet free of wheat and dairy (gluten and casein) might help children with autism and other developmental disorders. Both Ben and Jacqueline went on the diet the following day. For Jacqueline it did little, but for Ben it had a profound effect. Ben started the diet within five days of going mute, and after seven days of the diet he spoke his first word again.

"I'll never forget it as long as I live," Heckenlively told the two women. "We were outside the Nordstrom's department store in Walnut Creek, where there's a small fountain and a pillar with a lion's head on each side with water coming out of the lion's mouth into the small pool. Ben was fascinated by this stream of water going into the pool. He looked like he was focusing all of his attention, everything he had inside of him, and then he just said, 'Bubbles!' He kept saying it again and again, and my wife and I were laughing and encouraging him to keep saying it and he did."

Heckenlively related how his wife, Linda, a speech therapist trained in the steps of language acquisition, observed in the months that followed as Ben slowly regained the language he had lost. It took about a year for him to catch up with his developmental age for speech, and about two years for what seemed to be a painful sound sensitivity to subside, but by the time he started kindergarten he was just like any neurotypical kid.

"I was able to reverse something in my son that I wasn't able to do with my daughter. She's twelve years-old now and can't speak. Someday I want to have a conversation with her. That's why I'm doing all of this."

"You will have that conversation someday," Mikovits reassured him.

"Promise?" said Heckenlively.

"Yes."

"Okay, I'm going to hold you to it, doc. Someday I will have a conversation with my daughter."

For Heckenlively it was all so clear. Even though she didn't fully realize it, Mikovits had already joined the Rebel Alliance.

* * *

The next day, Hillary Johnson and Kent Heckenlively met for breakfast. Heckenlively thought it hilarious that the same *Discover* magazine that had "Vaccine-Phobia" as its number one science story of the previous year would send such a sympathetic journalist to the AutismOne Conference who would seek him out.

Johnson told Heckenlively that some people with ME/CFS also traced the onset of their symptoms to a vaccination, just as Andrea Whittemore had traced her relapse to the MMR shot required before she could start at the University of Nevada, Reno.

Heckenlively wondered how he could talk so candidly about these issues with a world-famous scientist like Mikovits and a top-notch journalist like Johnson, while the press seemed to take an almost ghoulish delight in painting the parents as fanatics and the scientists as the voices of reason?

* * *

While Mikovits understood why Heckenlively might be worried about her future, there were a few things he didn't know that made her believe things would soon move in a more positive direction. Something that could be cited as evidence of possible contamination in the XMRV study was the lack of sequence diversity between known isolates of the virus.

The less diversity in the virus, the more likely studies using the Silverman strain were subject to contamination. But Frank Ruscetti's collaborator at the NCI, Dr. Kathy Jones, the scientist who replaced Mikovits when she left Ruscetti's lab to head the Lab Of Antiviral Drug Mechanisms, had been doing work on the sequence diversity of the virus. They were finding different strains of the virus and greater diversity than they had seen previously. "Diversity is our friend," Mikovits and Ruscetti would often say to each other.

The same weekend Mikovits attended the AutismOne Conference, the Cold Springs Harbor Conference on Retrovirology, where presentations on XMRV a year earlier had caused the NCI to hold the invitation-only meeting to discuss public health threat of XMRV, was held.

At the Cold Springs Harbor conference, Jones discovered another group, led by Dr. Harvey Alter, chief of the Infectious Disease Section in the Department of Infectious Disease at the National Institutes of Health, and his collaborator, Dr. Shyh-Ching Lo, had been doing similar research. Alter was best known for his discovery of the hepatitis C virus, for which he received both the Distinguished Service Medal (the highest award that can be given to civilians working in the US government), and the Lasker Award for clinical medical research. The Alter/Lo paper showed significantly more sequence variation in XMRV than had previously been reported, thus lowering the possibility that the findings of Mikovits and her team were due to contamination. They were finally being backed by solid supporters.

Kathy Jones learned that Alter and Lo's study had already been submitted to the Proceedings of the National Academy of Sciences. Although Jones felt her own paper had more data about the sequence variation, it would be good politics to let Alter and Lo publish their findings first since they had no ties to the original XMRV study.

If all things went as planned, Alter and Lo's paper would come out within a few weeks of the Cold Springs Harbor meeting.

Back at the WPI, Mikovits hosted Marcus Conant, the AIDS doctor. As an AIDS veteran, Conant was known for forceful advocacy and would later tell those working in ME/CFS that they needed to proactively create their own luck without waiting for either the press or the government to understand their work.

Conant shared with Mikovits his hope and determination that XMRV would not be subject to the same government indifference that held back HIV/AIDS research for several years. Conant strongly believed that Mikovits should get a lawyer to protect herself and the WPI against legal challenges, which were likely to start flying. Those who had been on the front lines of the AIDS fight had come to expect heavy adversity.

Conant had an attorney in mind, a man around seventy years old like himself, and he picked up the phone with Mikovits present. "Hey buddy!" Conant said. "What if I were to tell you this scenario? Some researchers found a virus and the government says it doesn't exist, yet they have a serology test that can distinguish it and prove it's there. What would that sound like to you?"

Even though she couldn't hear his answer, the bemused expression on Conant's face told her what the answer had been.

The new discovery sounded just like HIV/AIDS.

CHAPTER FOURTEEN

The Long, Hot Confirmation Summer of 2010

The FDA and the NIH have independently confirmed the XMRV findings as published in Science, *October[2009]. The confirmation was issued by Dr. Harvey Alter of the NIH during a closed workshop on blood transfusion held on May 26–27 in Zagreb. . . .*
The association with CFS is very strong, but causality not proved. XMRV and related MLVs are in the donor supply with a prevalence of 3% and 7%. We (FDA & NIH) have independently confirmed the Lombardi group findings.[1]

—*ORTHO*, a Dutch magazine for health professionals.

"Your lives are about to become more interesting," Frank Ruscetti wrote to study authors Drs. Harvey Alter and Shyh-Ching Lo on June 23, 2010.[2]
It's difficult to overstate the importance of a confirmatory study by highly qualified and well-respected leaders in the field, such as Alter of the NIH and Lo of the Food and Drug Administration. A positive confirmatory study simply replicated original findings, therefore bolstering their legitimacy. In this case, it would mean that the scientific community would accept that XMRV or a related retrovirus had a natural history of infection in humans, with greater numbers among people with ME/CFS as

compared to healthy controls. Even if there were further negative studies, the scientific community at large would accept an association. The more specific issue of causation would be a question for future studies.

Ruscetti, Lo, Alter, and Mikovits would be stunned at the twists and turns that the confirmation study would take in the next few months, politely summarized by Lo's rather cryptic response to Ruscetti a few weeks later: "We thought our results very much support your earlier findings, although there were apparent variations in sequences. Unfortunately, our situation has become more complicated. We are facing various challenges and uncertainties."[3]

* * *

Within a few days of the *Science* publication, Mikovits received a call from Dr. Michael Busch.[4] He was professional, calm, and courteous, and said it was of critical importance that the blood supply be protected. The disaster of HIV/AIDS in the blood supply infecting thousands of individuals who received blood transfusions was still a vivid memory for many scientists.

Busch, an MD with a PhD in experimental pathology,[5] was a professor of laboratory medicine at the University of California, San Francisco, as well as the head of the Blood Systems Research Institute (BSRI), which in 2012 *The Scientist* listed as one of the "Top 10 Best Places to Work in Academia."

Thus they quickly assembled the Blood XMRV Scientific Research Working Group (SRWG, in this text called the "Blood Working Group" for clarity), which included Mikovits, Ruscetti, Busch, John Coffin, as well as William Switzer from the CDC, determined that their work should go in phases. They needed to design assays for XMRV capable of testing for its presence in the blood supply. Typically these assays would use PCR (polymerase chain reaction testing) and be conducted on serum or plasma, rather than using the laborious testing done by the Mikovits team.

Mikovits worried about viral latency. Other viruses, including herpes viruses and HTLV-1, would *not* appear in blood tests (such as PCR) on infected individuals, becoming detectable only after the patient developed cold sores or immune deficiencies. Mikovits was concerned that XMRV might fly under the radar in a similar way.

She and the team came up with additional techniques that made it possible to more easily find the virus: if they let the blood set inside a biological containment hood for a day or two, the cells would die and release nucleic acids into the plasma, which could be more easily extracted, allowing them to test more accurately for the virus. To address the concern more thoroughly, she and another member of the BSRI and the Blood Working Group (BWG) wrote a grant proposal to study latency in the virus.

Among the concerns Mikovits had about the Blood Working Group was the growing size of the committee due to the inclusion of some members for political reasons, even though they might not have the requisite expertise. This was not about XMRV in ME/CFS, it was about XMRV in the greater blood supply.

After the Blood Working Group decided on the assays that would be used, they would set to work determining whether the virus could be detected in blood supply. What happened next would create serious doubt in the minds of Mikovits and Ruscetti as to whether those in charge at the CDC were truly interested in unbiased testing of XMRV in the blood supply, even with public health at stake.

* * *

After the publication of their paper by *Science*, the laboratories of Mikovits and Ruscetti kept working on sequencing isolates of XMRVs, revealing strain differences, independent of the Silverman lab.

The initial sequencing done by the Silverman lab indicated that various isolates didn't vary more than 0.03 percent, a curious result and one which could indicate a lab contaminant rather than a virus with a history of natural infection in humans. The antibody and protein findings still argued that the virus was a human infection, but genetic diversity greater than 0.03 percent would have been stronger evidence of independent infections.

Ruscetti's lab at the NCI kept pursuing sequencing work, though, and eventually identified three genetically distinct strains of the virus, which according to standard classification practice, Mikovits had named strains A, B, and C. The work suggested a genetic difference of about 1–2 percent between the strains, but even that was a well-informed guess.[6]

Of the less than 8,000 estimated base pairs in the virus, they'd only been able to sequence about 300 base pairs of the *gag* protein and 600 base pairs of the envelope, or about 10 percent of the complete virus. Still, a 1–2 percent genetic difference between strains of XMRV is what they could expect for a mouse related-retrovirus, which generally would not show much more genetic diversity than that. These sequences were similar to those Lo/Alter were about to publish.

Mikovits was in charge of sample selection for BSRI phases two and three. Mikovits had arranged for the WPI study coordinator and phlebotomist to draw three fresh samples from subjects who had one of the various strains, and a negative control sample, and send them off to BSRI for distribution to the testing sites, including to the WPI, to Bill Switzer of the CDC, and to the NCI lab directed by John Coffin.

Mikovits chose the study participants with great care. Two were from patients who were local to the Reno area and tested positive in multiple assays on multiple occasions. The third positive sample was taken from a subject whose blood had been put under an electron microscope to reveal a clear picture of a gamma-retroviral particle budding from the cellular membrane. Study participants with variant sequences were included. The fourth sample was from a lab worker who had never been in Reno and had not worked with any of the samples.

The tests done at the CDC and WPI confirmed their research: all three patient samples, positive, negative control sample, negative.

When John Coffin heard this, on a conference call with other members of the Blood Working Group, he went ballistic. "That's not a well-done study! You're simply contaminating everything! We need to do a phase two (b)!" he thundered.[7]

Mikovits directly challenged Switzer in an email she sent on June 27, 2010, which also went to the other members of the Blood Working Group, stating

"Please also note that I sent 8–10 confirmed XMRV positive patients plasma to Bill Switzer in September 09 and he failed to detect XMRV in any of them while Ila Singh correctly identified XMRV in the exact same sample set. Clearly the clinical sensitivities of the assays were the difference."[8]

Instead of using less stringent PCR conditions, which had allowed Mikovits to find sequences closely related to Silverman's XMRV sequences, the CDC switched to more stringent PCR conditions for calling a sequence a positive, and also added the requirement of finding sequences for envelope and polymerase, other viral proteins.

Switzer served on the Blood Working Group, but was concurrently doing his own independent research for the CDC trying to show that the *absence* of XMRV in ME/CFS was correct. For the CDC, the virus had to be the sequence of XMRV Silverman, the VP62 plasmid, or else there was no infection. Four days after Mikovits's email, Switzer would publish his negative XMRV study in *Retrovirology,* while the positive study of Drs.

Alter and Lo would be held up by the *Proceedings of the National Academy of Sciences* for six more weeks.

* * *

"Although we will almost certainly continue to accumulate low-hanging fruit, where simple relationships will be found between the presence of a cultivable agent and a disease, these successes will be increasingly infrequent. The future of the field rests instead in our ability to follow footprints of infectious agents that cannot be characterized using classical microbiological techniques and to develop the laboratory and computational infrastructure required to dissect complex host-microbe interactions."[9]

This was how in 2011 Ian Lipkin described the problems inherent in discovering and identifying new pathogens with established techniques. There was a need to branch out to methods that involved more sleuthing, "following footprints," and using different methods for catching elusive prey.

Mikovits and her team had been quite adept at identifying the "footprints," and through their electron micrograph they'd even captured a picture of the beast.

But for Coffin and the CDC, it wasn't enough. They wanted the footprints, the beast, and its DNA, which identified only one very dubious species. Even if they discovered the body of a slightly different beast, it would be as if no beast existed.

Unbeknown to all at this time, this beast was quite unusual. What Mikovits would later come to refer to as XMRV Silverman, (or, more precisely, the VP62 clone he had stitched together from various sequences), was a laboratory recombination, a virus which didn't actually exist in nature, but would produce viral RNA to higher titers than HIV in human cell lines in vitro and it could jump to cell lines in tissue culture hoods.

A virus with *no natural history of infection* would be their standard. This was not a virus that had been cultured from an actual human subject, as Mikovits has no doubt she had done, but one which had been created in a lab, and that they hoped was the same as was found in nature.

* * *

The major problem, which Mikovits had started to publicly identify as early as the May 2010 *Invest in ME* Conference in London, was that Silverman had not isolated the virus as done in classic retrovirology, but had simply created an infectious molecular clone (a copy of the complete viral genome made by laboratory techniques).

To many, this was the same thing as isolating an actual virus from a patient, but for those trained in more classical virology, the absence of confirmation using other retrovirological techniques could be a recipe for disaster. Mikovits knew about the process of creating an infectious molecular clone, having done some of the pioneering work in the field while a postdoc at the Derse lab and creating the first infectious molecular clone for Ruscetti and Poiesz's HTLV-1.

In Mikovits's mind, relying on an infectious molecular clone rather than an isolated and sequenced virus from an actual human subject was simply poor science. The more traditional investigative techniques for retroviruses involved determining whether reverse transcriptase (the enzyme by which an RNA virus converts its genetic code to DNA and inserts itself into the genome) was present, then culturing and transmitting the suspected infectious virus, then using PCR to get the sequence data.

But mistakes and misconceptions are a common part of science. Some of the greatest discoveries in science had started out as errors due to misconceptions and unknown exotic biology which presented curious and unsuspected aspects to the research.

The question few seemed willing to consider was whether something very similar but not identical to XMRV Silverman actually did exist in patients with ME/CFS and other neuroimmune disorders, and whether it could be properly hunted. Although it may seem somewhat ironic given Lipkin's need for scientists to move beyond "classical microbiological techniques," Mikovits and her team had always been taught to use that approach. Mikovits had approached the question by combining classic microbiological techniques of viral isolation and transmission, electron microscopy, serology, protein detection, and PCR, as a way to support each other. She treated each of these methods as complementary, revealing different and changing facets of an immune system under apparent attack.

Mikovits believed that Coffin and the CDC seemed to be using these techniques in an exclusionary manner: any sequence variation was deemed to be a contamination. In the view of many working on the issue and the patients, it seemed almost calculated to *avoid* finding evidence of a retrovirus in the patient population.

Yet, as Lipkin argued, using the old techniques in the old ways meant the chances for finding something new would be "increasingly infrequent."

* * *

The disagreements between members of the Blood Working Group over which assays to use did not appear to lessen the enthusiasm of other government entities to take the issue of XMRV and the blood supply quite seriously.

On May 10 of 2010, shortly before Mikovits boarded a plane for Europe, the American Association of Blood Banks released a bulletin that read: "The AABB Inter-organizational task Force on Xenotropic Murine Leukemia Virus-Related Virus reviewed the risk of transfusion transmission of XMRV by individuals with CFS. The task force presented its recommendations to the AABB board of directors, which approved an interim measure intended to prevent patients with a current or past diagnosis of CFS from donating blood or blood components."[10]

Even ten months later the question of whether XMRV posed a risk to the blood supply was still not answered. A committee report from the Blood Working Group published in the journal *Transfusion* asked: "Does XMRV pose a risk to transfusion recipients?

> Given that XMRV is a retrovirus and appears to be present in PBMCs (peripheral mononuclear blood cells) and in plasma, blood transmission, although undocumented, is possible. This possibility is supported by the demonstration that human cells have been infected in the laboratory by virus found in human specimens and by the transfusion transmissibility of other retroviruses (e.g., HIV and HTLV). Related viruses are oncogenic and immunosuppressive in different animal species . . ."[11]

The Blood Working Group considered new data as it appeared, specifically the confirmatory study of Lo and Alter as well as research using rhesus macaque monkeys.

The study involving rhesus macaque monkeys was particularly valuable as it showed that the virus quickly disappeared from the blood stream, presumably going into tissues. As with HIV/AIDS, immune stimulation caused the virus to reappear in the blood, where it could be detected by standard clinical methods.

In addition, the parents of children with autism would probably have found a single fact interesting about the rhesus macaque study of XMRV latency. What was the scientists' immune stimulant of choice to provoke the virus and cause it to replicate to detectable levels, and presumably cause disease?

A vaccination.

* * *

When the Dutch magazine *ORTHO* broke the story that Drs. Alter and Lo had confirmed the findings of Mikovits and her team regarding XMRV, the patient community buzzed with excitement over this development. Public health authorities reacted very differently.

In an email to Frank Ruscetti on June 10, 2010, Lo explained the very curious chain of events which had cause the *Proceedings of the National Academy of Sciences* (*PNAS*) to put a hold on their confirmatory study. Lo explained that when their paper was submitted for review they were asked to retest and document their PCR assay system for any trace amount of mouse DNA. They did and found no mouse DNA.

The paper was accepted for publication, and the galley proof was ready, when they learned the CDC also had a paper in press, which did not find any evidence of XMRV in ME/CFS patients. Although the Alter/Lo group had briefed the CDC three months earlier, they had never mentioned that they had a negative study in press.

The CDC group as well as Lo/Alter were asked to put a hold on both papers, as it might be confusing for the public that two government labs were reporting different results. Lo did not think this was a good scientific precedent and confessed at the end of the email that he was tired of this runaround but still hoped that he and Ruscetti might work together on the problem.[12]

The email from Lo to Ruscetti was revealing, and bolstered Mikovits and Ruscetti's suspicions about the CDC's investigation into XMRV.

From the manner in which Reeves had told the *New York Times* that he did not expect that his team would validate the findings, to the CDC's narrowing of the search parameters for finding XMRV or a related murine leukemia virus (MLV), to their insistence on using Silverman's VP62 XMRV plasmid as the control standard while the Lo/Alter results confirmed Mikovits's assertions about going beyond Silverman XMRV, it was becoming difficult to believe that the CDC was interested in doing an objective scientific investigation of the XMRV question.

The question, which still remains unanswered, is why research into XMRV seemed to cause public health authorities to make such an exceptional departure from standard scientific research protocol.

* * *

One would have expected that the editor-in-chief of the *Proceedings of the National Academy of Sciences*, Randy Schekman, could easily explain the scientific reasoning behind the hold on the Lo/Alter confirmatory paper.

But if anything, he only complicated the matter, raising far more questions than he answered. In an interview published by *The Scientist* entitled "Why I Delayed the XMRV Paper,"[13] Schekman tried to answer the question of how a thoroughly vetted paper by a figure like Harvey Alter, (recipient of a Lasker Award for clinical research and the Distinguished Service Medal), could be held up in the galleys.

Interviewer Cristina Luiggi explained, "According to Schekman, the paper had gone through the proper review at their journal, and was ready to be published when he was 'called by someone in the retrovirus field who expressed concerns.' This person remains unnamed, as does the 'established person in the field' with 'very high standards,' not 'involved in the controversy' who would conduct 'independent review of the work.'"

Who was this mysterious figure in the retroviral field who trumped the designated referees of the paper? This "someone in the retroviral field who expressed concern" must have been of great renown in the medical community.

Mikovits wondered if it was Coffin, who since the July 22, 2009, meeting on XMRV at the NIH, three months before the publication of her research in *Science*, seemed to have set himself up as the arbiter of any research into XMRV?

Did *PNAS* have such little faith in the quality of its peer reviewers that a call from an outside scientist could make them completely abandon their established procedures? Luiggi asked how the process could have gone smoother:

> RS: . . . I feel we might have done a better job was in identifying how this particular paper might have caused [a controversy]. I was not aware of the paper until I got a call from this outside retrovirologist. If we're going to be faulted at all at PNAS is that we weren't aware of the implications of this work until after it was already accepted. And we try.
>
> The staff at Washington, they're not trained scientists, but we try to examine the papers that we're reviewing to see if they have anything unusual about them that would cause additional publicity . . .[14]

According to Schekman, the Lo/Alter paper was sent out for review to "two well-known retrovirus experts" who had reported favorably on it. Then the *PNAS* Washington staff of "not trained scientists" which would normally sniff out anything that would draw undue media attention or controversy also gave the green light.

Despite the positive reviews of "two well-known retrovirus experts" and no warnings from the Washington staff of "not trained scientists" it wasn't until Schekman got a call from this "outside retrovirologist" that he realized the journal might reasonably be faulted for not being "aware of the implications of the work until after it was already accepted.

The delay of the Lo/Alter confirmatory study did not appear to be based on any scientific concerns that Schekman wanted to share. Who was this mystery "outside retrovirologist" who was so well-known that he could make Schekman override the established procedures of the prestigious *Proceedings of the National Academy of Sciences*?

And maybe more importantly, what did he say?

* * *

If sins exist in the world of scientific research, non-collaboration and unwillingness to share data or samples is without doubt thought a mortal sin, far more egregious than its venial cousin, refusal to share cookies in the cafeteria. Shared progress, the health of the world, and the improvement of the human condition depend on conscientious sharing of all relevant information.

The scientist who doesn't share his data isn't just undercutting a rival, but sacrificing the health of others for personal gain.

As Mikovits and Ruscetti became aware that William Switzer, (not an MD or PhD) and others at the CDC were planning to release a negative study about XMRV, the question came up as to whether the WPI had shared samples with the CDC. Mikovits sent out an email to the group to document her growing fears.

> I have just confirmed from my notebook codes and discussion with Frank Ruscetti that all 20 plasma samples sent to you (CDC) on 9/30/09 were confirmed positive in 3 independent laboratories including the subset of 12 sent to Ila (Singh).
>
> Not all assays were done at all laboratories on all samples, due to limitations of quantity for some of the samples. XMRV was detected by DNA, PCR, CDNA PCR, WB using monoclonocal antibodies for gag P30 and Env, transmissible virus to LNCaP, direct immune precipitation of virus and direct PCR detection of XMRV in the plasma . . .

Patient samples from which XMRV was cloned and sequenced by Jaydip
Das Gupta in Lombardi et al. were all in the samples set sent.

Judy and (Frank)[15]

What would strike Mikovits as so unusual, given the dissent in the Blood Working Group over whether Switzer and his colleagues had been given positive XMRV samples as found by the Mikovits group and others, as well as the conflict between the CDC study and that of Lo and Alter, was that the CDC would be allowed to go ahead unilaterally in the manner that it did.

Even accepting the premise that two different agencies of the public health establishment should not come to differing conclusions on an important question, the standard expectation would be that both papers should be held until they were reconciled.

In fact, that is what people anticipated would happen, as detailed in an article by Amy Dockser Marcus of the *Wall Street Journal,* who wrote about the controversy in an article published on June 30, 2010:

> It is unusual for a paper to be held after it has already gone through the formal peer-review process and been accepted for publication, say scientists who publish frequently.
>
> "It's fair to say it's not a usual kind of thing," said John M. Coffin, a special adviser to the National Cancer Institute and a professor at Tufts University in Boston who wrote an editorial alongside the *Science* report in October. Dr. Coffin said he couldn't comment specifically on the XMRV papers, but that scientists often come up with conflicting data, especially when a virus is new and not well understood as in the case with XMRV.[16]

It's probably reasonable to say that a casual observer of the unfolding XMRV saga would believe that the two papers would be compared, coordinated, and released at the same time so that some clarity might be brought to the issue.

But that's not what happened.

Two days later, on July 1, 2010, the Switzer/CDC negative paper was published in the journal *Retrovirology.* Lo and Alter's paper would not be published until August 16, 2010, six weeks later. If the intention behind holding the Lo/Alter paper had been that the public health establishment didn't want to confuse the public, wouldn't it make more sense to coordinate the release of both papers rather than staggering the publications so that one of them was read by the public first?

What message did it send to the public to release a negative study first, and then wait six weeks to release a confirmatory study in August, the month when even doctors often take long vacations and when most people are distracted by summer's final events, moving into a dorm room, or getting the kids ready for school?

* * *

The CDC's XMRV study would come in for strong criticism based on the way they diagnosed patients suffering from CFS. From the CDC's own publication of July 1, 2010:

> Briefly, between 2002 and 2003 we sampled adults 18 to 59 years old from Wichita, Kansas and between 2008 and 2009 we sampled adults 18 to 59 years old from metropolitan, urban, and rural Georgia. In both studies, we used random digit-dial screening interviews to classify household residents as either well or having symptoms of CFS. A follow-up detailed telephone interview was administered to all individuals with symptoms and to a probability sample of those without symptoms.[17]

"They just called people up on the phone and asked if they were tired!" was the way Frank Ruscetti later characterized it.[18]

If ME/CFS was the result of a viral infection, similar in many ways to other outbreaks, it made sense to go to those areas of the country that were reported to have experienced such outbreaks, such as Incline Village, Nevada, or Lyndonville, New York. Nobody could recall an outbreak or even a cluster of ME/CFS in Kansas or Georgia and yet the CDC had used samples from these two groups (the "phone book" cohorts) exclusively in studies for decades, infuriating Mikovits as much as it had her team.

The lack of any clinical diagnosis of ME/CFS, either through the exclusion of other conditions, or findings of abnormal cytokines/chemokines, other inflammatory markers, NK cells or RNase L data, also made the study group suspect.

It was the computer programmer's classic dilemma of "garbage in, garbage out."

If one chose a population of people unlikely to have the disease, it followed that one would be unlikely to find any pathogens linked to the disease. Why were so many intelligent scientists missing these very elementary concerns?

As Mikovits and Ruscetti often said to each other when these issues came up, "They find what they want to find!"

* * *

Although the publication of the CDC paper in early July while the Lo/Alter research was still held from publication concerned Mikovits and her team, they had other issues to occupy their attention; namely, the grand opening of the $77 million dollar facility at the Center for Molecular Medicine on the grounds of the University of Nevada at Reno.

When fully up and running, the WPI would have a state-of-the-art patient medical clinic, neuro-imaging through SPECT scans, and a clinical laboratory that would provide testing for patients, support physician studies, and develop diagnostic markers of disease.

There was a reception for donors, government officials, and personal friends of the Institute on Monday, August 16, 2010. The speakers included US Senator Harry Reid, US Senator John Ensign (a Republican), members of the US Congress, Shelley Berkley and Dean Heller, State Senator William Raggio, and University President Milton Glick. Hillary Johnson and Amy Dockser Marcus of *the Wall Street Journal* were also in attendance.

During Annette Whittemore's remarks she announced that the WPI's basic research laboratory would be named the Judy Mikovits Laboratory.

Fourteen months later, Mikovits would be thrown in jail and held there five days without the opportunity for bail and without an arrest warrant, and to this day has not had a single day in court.

The next day there was an invitation-only scientific symposium, which, as described in a WPI press release, brought together "leading researchers, clinicians, and key collaborators from the medical and research community to discuss the latest XMRV research findings, medical research data and evaluate treatment outcomes since the publication of their 2009 XMRV study. The symposium organized by the WPI, entitled *XMRV in Human Disease*, was led by WPI Director of Research, Judy Mikovits."

From a press release of August 16, 2010:

> The completion of this building is the realization of a dream and a beacon of hope for patients afflicted with neuroimmune diseases," said Annette Whittemore, president and founder of the WPI. "Our strategic partnership with the University, combined with this dynamic setting gives us confidence that the research done here will translate into innovative medical care for patients around the globe. We are proud that our work will soon lead to new avenues of research and treatment for these diseases.[19]

On the following Saturday, August 21, the WPI had an open house for the public that included tours of the new facilities as well as researchers on hand to discuss the new facility and its plan for further investigations.

The entry to the building had a large, open atrium and large panes of glass, which let in the rays of the summer sun. Against the white stone planters which dotted the space, many ME/CFS patients rested, often squinting in the harsh light. Some were in wheelchairs, some had oxygen tanks with the accompanying nasal cannulas snaking into their noses.

It was sobering to realize these were the "healthy" sufferers of the disease, those who could even travel to the WPI. In the large theater, which probably seated close to three hundred people, there was a slide show of those who were more severely affected than the ones who could get there.

The silent slide show presented pictures of many patients before they had become ill, often showing them engaged in athletic activities like running or hiking, then followed with pictures of them confined to their beds for most of the day, the shades drawn to keep out the sun because of their sensitivity to neurological stimuli like light.

These were people for whom the idea of spending an afternoon at an "open house" was akin to getting on a rocket and traveling to the moon. But for those who suffered from ME/CFS, the opening of the Whittemore Peterson Institute seemed like one giant leap for all of them.

* * *

With the publication of the Alter/Lo study, there was the inevitable flurry of interest from the press, so on August 23, 2010, a question and answer teleconference was held with Alter and Lo. Alter began the teleconference by recapping what had happened over the past several months.

Alter began by talking about his coauthor, Shyh-Ching Lo, who had a number of samples from ME/CFS patients which had been gathered in 1990 by Anthony Komaroff from Harvard University. The samples were "well-pedigreed" and had been stored "in pristine fashion in a frozen repository." When the XMRV story broke, Lo became interested in testing the samples.

Alter and Lo found sequences which were closely related to what had been designated XMRV by Silverman, and noted that the Whittemore-Peterson group led by Mikovits had discussed with them how they'd also found a more diverse group of retroviruses than had originally been reported. Alter ended the teleconference by saying:

So it's really probably a better term is murine leukemia virus-related viruses which encompasses XMRV so we found this in a very high percentage of the chronic fatigue patients that Dr. Komaroff had sent to us—about 86 percent—and simultaneously found that in about 6.6 percent of our healthy blood donors.

So there was a dramatic association with chronic fatigue syndrome, with the syndrome of chronic fatigue but that's all it is . . . we think basically it confirms the findings of the Whittemore Peterson group.[20]

Now with the findings of a combined team from the NIH and FDA, led by a world famous scientist, the results of the WPI had been confirmed. In most scientific investigations, that would have been the end of the story.

Findings confirmed, proceed to the next stage.

But the investigation into XMRV wasn't playing out like any other story in science.

CHAPTER FIFTEEN

The December 19 Hearing and Damages

In our opinion, Dr. Mikovits has been "home-towned" from the outset of the criminal and civil cases, by the Washoe County criminal justice system, as the result of undue influence exerted by WPI's owners. . . .
But issuance of the warrants pales in comparison with the bail hold placed on Dr. Mikovits, which prevented her from bailing out of jail in Ventura County for the better part of a week and severely limited her ability to participate in her defense to WPI's motion for preliminary injunction. . . .
Dennis Neil Jones, Dr. Mikovits's lead attorney in the civil case, asked the court for permission to appear at the preliminary injunction by telephone, because he was recovering from knee replacement surgery and he was physically unable to attend the hearing. The Court refused his request, forcing Jones to arrange a last minute replacement unfamiliar with the case to attend the hearing.[1]

—From a planned, but unreleased, press release
by Mikovits's attorney, Dennis Neil Jones

December 15, 2011

Judy Mikovits was sitting in the lobby of the Lucerne Hotel in New York City when she got a call from her attorney, Dennis Jones.[2]

She and Frank Ruscetti had traveled to New York to work with Ian Lipkin of Columbia University, who had been given the task by NIH of determining why some groups had been unable to find XMRV among the ME/CFS patient population while others like Lo and Alter had found it. Maureen Hanson of Cornell University and journalist Hillary Johnson were also present for the meetings.

Since her termination only three months earlier and arrest on November 18, Mikovits had flown largely under the radar, avoiding the press and publicity.

After a few words, her attorney got to the point and told her she needed to come back and appear at a hearing in Reno on December 19, four days away. Mikovits told her attorney that he was out of his mind, that she was in New York City, and not going to any court controlled by the Whittemores. She then used some colorful language common to many New Yorkers about what the Whittemores should do to themselves.

While the language she used may have passed without much notice to the majority of the denizens of the Big Apple, the staff at the upscale Lucerne Hotel expected such language to be left on the street.

By their disapproving looks, they were telling her to keep her voice down.

Jones tried to convince Mikovits that she would get an unbiased hearing before Judge Adams and that she needed to go. "This is ridiculous!" she shouted into the phone. "You don't understand! There is no way you'll get a fair shake in that courtroom!"

Jones attempted to soothe her. "We decided you're going to take the Fifth Amendment, and I'm still comfortable with that. But you've got to produce your personal emails, all 22,000 of them. We really don't have any leeway on that."

In the preliminary issued while Mikovits was held on in a Ventura county jail, Judge Adams had given Mikovits two weeks to sort through all 22,000 emails and decide which were personal and which were professional. Under normal circumstances this would be a difficult task with an unfair time frame, but since the police had taken all her electronics, leaving her without a computer, this was an especially impossible task.

"Over my dead body," said Mikovits. She told Jones that when she signed on as the principal investigator that she was required to protect the identity of the more than 800 subjects who had signed consent forms in the various studies.

Mikovits thought her attorney really didn't comprehend the responsibilities of a principal investigator on a government grant. The participants had to know that when they handed their blood or other samples over to the study, their information would never find its way into the hands of an insurance company that might deny them coverage or physicians who might restrict or deny treatment.

Her frustration was at a fever pitch.

Jones again tried to convince her that she needed to go to Reno.

"You can hold your kangaroo court, but I won't be there! And you will not get my emails! Your attorney promised my personal email! I did not! I was in jail! I told you no. That's not how a principal investigator works!"

Back in her hotel room, Hillary Johnson, Maureen Hanson, and Frank Ruscetti begged Judy to just hand over the emails, fearing that she would be jailed again. But she was firm about not handing over documents containing study subject names.

* * *

Her plane flew back to Santa Barbara in the early morning hours of December 18. Mikovits, David, and Dennis Jones went to Reno together later that day. This was the first time she actually laid eyes on Dennis Jones and the condition of the man was startling, even for someone accustomed to the worst ravages of cancer.

Because of the knee replacement he'd undergone in November, he was using two canes to walk with great difficulty. This manner of getting about made it difficult for Jones to schlep his laptop around, so he simply brought his files.

This was their champion, teetering about on his canes with his paper files.

* * *

In the car on the way to the airport, David drove and Dennis sat in the front with him while Judy sat in the back. Jones wanted to discuss a contract Mikovits had allegedly signed on February 22, 2009, claiming that in the event she was fired she would return all copies of all data to the WPI. Dr. Mikovits looked at the paper and could not recall ever having seen it.

She believed the signature page had been taken from another document. February 22, 2009, was almost three years after she had begun working to develop the research program of the WPI. The idea that an academic researcher and professor would sign such a clumsily-worded agreement that effectively turned her from a professional scientist with rights and responsibilities into an employee-for-hire was ludicrous. There were other reasons to believe she had never signed it.

On January 20 of 2009, she'd signed a joint agreement with the Cleveland Clinic and the NCI for their investigation of XMRV and ME/CFS. The information developed in that joint investigation belonged to all three institutions. This new document claimed it superseded three years of collaborative research agreements and that all intellectual property would be owned solely by the WPI.

Dennis Jones was astute, but Mikovits worried he didn't understand the bigger picture. "This isn't about grants and legalities of grant management. This is about a set-up, a frame-up, and a cover-up! That's what you need to see!"

Mikovits knew the notebooks would provide the evidence that VIP Dx's XMRV test was worthless, and she believed the Whittemores knew it too. There was also the possibility that the lab data contained in the notebooks would provide clues to help her understand ME/CFS, autism, and other diseases. Who knew what kind of windfall could come from curing those diseases?

* * *

As Judge Adams opened the hearing at 9:03 a.m., things started badly and only got worse. The Whittemores were represented by Ann Hall, whose husband worked for the Washoe County District Attorney's office, which had issued the search warrant for Mikovits's house. To Mikovits, this was yet another corrupt connection between the civil and criminal cases.

In addition to Dennis Jones, Mikovits was represented by criminal attorneys Scott Freeman and Tammy Riggs, who had been retained that Monday after Thanksgiving when she was mysteriously required to appear in Reno for *Science's* mugshot.

Mark Zaloras, who had appeared on Mikovits's behalf on November 22, 2011, at the Reno hearing while she sat in jail in southern California, was standing by on telephone. The judge seemed angry that Zaloras was not physically present. Jones explained they'd been given only one day's notice of the hearing.

Jones then presented his first substantive argument, namely why Mikovits had not provided her requested emails. Jones explained that

Freeman had advised her that she could assert her Fifth Amendment right against self-incrimination in the criminal case. In addition, she didn't have a criminal lawyer at the time and was in jail in southern California at the time of the hearing on the preliminary injunction.

When she was released, she hired Scott Freeman and he had recommended that Mikovits invoke her Fifth Amendment privilege against self-incrimination regarding the emails, which she did. Mikovits believed the stronger argument was the one she'd made in New York to her colleagues, the argument that under federal law she was required to protect the confidentiality of patients in the study, but she had chosen to trust the experience of Freeman.

Jones proceeded to question why Mikovits was arrested on a no bail warrant, but Judge Adams returned to whether Mikovits had complied with the order to produce all of the emails. Jones replied that "counsel believed that the motion to stay would trump the entire civil case, including your prior order."

Adams asked whether Jones understood that the filing of the motion did not stay the court's order to produce the emails. Jones said he did, which was why they had retained a company to copy the emails, but it was taking a significant amount of time. He went onto say that the volume of emails was tremendous, more than 22,000 in all, and that to protect privacy he asked that a Discovery Master—an independent court representative who would review the documents—be appointed to assist in not revealing patient information.

The judge didn't want to offer the services of a Discovery Master to sort through the emails. He simply wanted all 22,000 of them, regardless of the patient information contained within.

Jones protested that they needed one more week to finish copying them, but Judge Adams would not waver, claiming it was inconsistent with the order entered on November 22, 2011. Jones responded by saying it was only inconsistent in date.

Judge Adams replied, "Well, it's inconsistent in terms of process. It's inconsistent in every respect, isn't it?" Mikovits sat next to her attorney, incredulous. She would only fully understand four months later on March 14, 2012, when judge Adams recused himself admitting accepting more than $10,000 from Harvey Whittemore.

Jones said he wasn't sure that the attorney who represented Mikovits in the previous hearing, Mark Zaloras, understood a Discovery Master was available. There was also some concern by Zaloras as to whether the emails simply by being handed over, would become part of the public record.

Judge Adams didn't seem to believe these were reasonable concerns.

Maybe sensing what Mikovits had warned him, Jones pushed the button for the litigator's most incendiary option. "I just have to say that I am very disappointed and upset with the fact that my client has really been denied effective assistance of counsel by a series of events, including what we believe to be undue influence."[3]

In both law and science, the claim of bias or undue influence is a grave charge.

Judge Adams responded by asking who had denied her effective assistance of counsel.

Joes replied, "Well, you would not allow me to appear by telephone even though I was on medical leave following surgery, so her lead counsel was not able to participate."

From that point, the judge seemed to change topics, all with the purpose of not revisiting his earlier decision. "I'm totally unaware of that, but in any event, Mr. Zaloras is also counsel for the defendant. In fact, he's one who has limited but lengthy experience in the criminal law, isn't he?"

Jones replied, "He is."

Judge Adams continued, "And finally, couldn't he have advised the Court and opposing counsel that we couldn't go forward with the Chambers conference because of your unavailability? He never told me that. I certainly wouldn't have proceeded with the Chambers conference discussing broad orders concerning issuance of the preliminary injunction and the specific order on discovery if I had been advised that the defendant wasn't represented by counsel. She was represented by counsel and we had a very extensive Chambers conference concerning the terms of the preliminary injunction order and the terms of the clerk's discovery order."[4]

Jones replied, "I understand that, your Honor. Understand that he did the best he could under the circumstances, having never spoken with his client."

Judge Adams replied, "He didn't tell me that he had never spoken with his client."

Jones tried to explain again. By the time of the November 22, 2011, preliminary hearing in Reno, Dr. Mikovits had been in jail five days and Jones was the only attorney with whom she had discussed the case, one time, by phone, only four days before her jailing.

Jones was well enough to appear by phone, but Judge Adams denied this request. After this denial, the best he could do was get another attorney, Mark Zaloras, to appear on her behalf.

Zaloras couldn't interview Mikovits because she was imprisoned without an arrest warrant from November 18–22 in southern California. What is remarkable about the transcript is that even Judge Adams admitted that Zaloras couldn't comment knowledgably on the Fifth Amendment claim because he didn't know anything about the criminal case.[5]

Mark Zaloras could not interview her to get essential facts and background because she was in jail in southern California.

* * *

Jones continued to press the bias issue.

For those familiar with the severity of the charge of bias against a judge, as well the subsequent revelation that Harvey Whittemore, family members, and business entities had donated $10,400 to the campaign of Judge Adams, it reads as Kabuki Theater.

Jones tried admirably to finesse the accusation of bias, falling just short of outright accusing the judge of corruption: "As I said, your Honor, I'm sure he did the best he could, but he also told me at one point when he called me during the hearing that *the handwriting was on the wall*, your Honor was going to sign a preliminary injunction in his opinion, and the best he could do—" (emphasis added).[6]

Jones's implication must have been clear.

"Just a moment. The hearing was set for the preliminary injunction. That was the time at which the parties were entitled to produce witnesses, evidence and argument on the Motion for Preliminary Examination. Witnesses did appear, even though the defendant didn't. At the request of counsel for both parties, I met with counsel in Chambers, which resulted in the order and the preliminary injunction. If Mr. Zaloras felt that, quote, *the handwriting was on the wall*, end of quote, he was not only inaccurate, because I hadn't heard any evidence on the motion, didn't prejudge it in any way" (emphasis added).

"Secondly, he never told me that he wished to proceed with the hearing, that he wished he'd hear evidence from the plaintiff or present any evidence on the defendant's behalf, or that he felt in any way coerced or pressured or maneuvered into making a decision on behalf of his counsel."[7]

Judge Adams continued. "Again I stress, if he had not literally even talked to his client before the hearing, which I find to be implausible, but if that were the truth, I would expect that would be the number one thing he'd bring to the Court's attention. I would not expect that he would enter into by stipulation not only a preliminary injunction, but a detailed order on the discovery process. That just doesn't make sense to me. Does it make any sense to you?"[8]

Judge Adams refused to accept that substitute counsel had been unable to discuss the particulars of the case with Mikovits while she was in jail in California. Did Judge Adams believe attorney Dennis Neil Jones, the former president of the Ventura County Trial Lawyers Association, was flagrantly lying?

If so, Adams should have thrown Jones in jail for contempt of court and violating the canons of professional ethics.

* * *

To Dennis Jones, there were a number of issues that didn't make sense.

Why hadn't he been allowed to appear by telephone?

And while the initial claim against Mikovits contained no direct evidence, while she was in jail the Whittemore side had dropped the bombshell, a new affidavit supposedly written by Mikovits's research associate, Max Pfost, the day before her arrest, but not part of the November 7th lawsuit.

How were they supposed to prepare to defend against this if they were given multiple documents and no time to decipher the truth? His client was in jail and Jones was recovering from major surgery.

* * *

It seemed apparent that Judge Adams simply did not believe that Mikovits had not talked with Mark Zaloras before the hearing.

> "Can you imagine any judge anywhere in the United States proceeding with such a conference, the lawyers entering these orders we've talked about, and having been told that the lawyer on behalf of the defendant had never met with the defendant, knew nothing about the case? You think there's any judge in the United States who would do that except a judge subject to the due discipline under the code of judicial conduct and, in my view, subject to removal?"[9]

Even if Mark Zaloras *hadn't* let the judge know he'd never spoken to his client, the judge still claimed that Zaloras didn't know what Mikovits had been charged with or even where she was being held. That would have alerted the judge that Mikovits's representation in the case was inadequate.

At one point Adams said to Jones, "And I certainly don't think you're suggesting that all that would have occurred had I been told that defense counsel had no witnesses, had nothing to present, didn't know how to counter the witnesses of the plaintiff, and had never even met the defendant?"[10]

That is exactly what Jones was saying, but for some reason Judge Adams wasn't accepting it. In his summation of the case, Judge Adams claimed, remarkably, that Mikovits was perhaps the greatest villain to ever darken his courtroom, someone who had to be stopped at any cost.

> "The remedy for this conduct is within the discretion of the Court. In my entire tenure on this bench as presiding judge in Department 6, I have never rendered a default against a disobedient party in a civil case pursuant to NRCP 37(b)2(c) . . .
>
> The defendant's willfulness is demonstrated by every fact in this case. She did not comply in any way, shape, or form with the Court's Order, or the terms of the Preliminary injunction. She did not timely advise the Court or adverse party of her unwillingness, her disability to perform it for her."[11]

The Honorable Judge Brent Adams was appointed to his position on July 4, 1989, which meant when he served as the presiding judge in the case against Mikovits he had been on the bench for twenty-two years. According to his own admission, in all that time he had never handed down a default judgment.

A default judgment is the legal equivalent of a nuclear bomb.

It means you have shown such disrespect for the court that you will not be allowed to present a defense. Even those who have just a passing familiarity with the American legal system know that every person is supposed to get their day in court.

Everybody it seems, except for Judy Mikovits when she found herself in a Reno, Nevada, courtroom in front of a certain judge.

A judge who three months later would be revealed to have received a very substantial campaign contribution from Harvey Whittemore. Judge Adam's failure to remove himself from the case, or at least inform the parties and get a waiver from them, would probably have subjected him to an ethics investigation in forty-nine out of fifty states. The only state, which didn't have such an ethics rule, and left such issues purely to the judge's discretion? Nevada.

* * *

The civil action against Mikovits had been filed on November 7, 2011.

Mikovits had been unable to appear at the November 22, 2011 hearing because she'd been in jail in southern California, having been arrested on November 18. Her attorney, Dennis Jones, had just undergone major surgery, and had to get a replacement attorney, Mark Zaloras, who was unfamiliar with the case. Zaloras had shared his impression with Jones that "the handwriting was on the wall" in the case, and Jones even said so directly to Judge Adams.

On December 19, 2011, she was found liable as a result of the default judgment handed down by Judge Adams.

Dr. Mikovits had gone from having a case filed against her, to judgment being taken in less than six weeks, without being able to present a single piece of evidence. How could anybody reasonably claim this was justice?

* * *

When the hearing was over, her criminal attorney, Scott Freeman, approached her. He spoke in a low but urgent voice. "If I were you, I'd get out of town just as fast as I could. Because I and everybody else sitting in this courtroom thought you were going to be in jail. Get out of here before he changes his mind!"[12]

So, Mikovits, her husband, and Jones hurried back to their hotel, grabbed their bags, and got to the airport three hours early. A part of Mikovits wished they had arrested her on the spot. She knew the patient community wouldn't stand for such shenanigans and the effort might cause the whole thing to blow up in the Whittemore's faces.

But then there was David, who was twenty years her senior, and the entire ordeal had hurt him too much already. She didn't know if he could handle her being jailed again. They landed in Los Angeles happy to be free, but knowing that the dark forces were still operating against them.

* * *

As if the courtroom antics were not enough, and as if the mugshot was its own permanent tattoo, on December 23, 2011, *Science* completely retracted the October 2009 article linking XMRV to CFS.

Mikovits wondered why no one was asking about the possibility of common scientific practices creating recombinant viruses plaguing the human race, and causing a myriad of diseases, would never be answered.

* * *

On January 25, 2012, Judge Adams held a damages hearing. The hearing was quickly moved to his chambers. Mikovits was not present but was represented by her attorneys, Dennis Jones and Scott Freeman. The Whittemores were present with their attorneys, niece Carli West Kinne and Ann Hall (whose husband worked for the DA's office).

Since Judge Adams had entered a default judgment against Mikovits, all that was left was to determine the amount of damages Mikovits would pay the Whittemores.

And the Whittemores had their number.

They wanted Dr. Mikovits to pay them *fifteen million dollars in damages.*

* * *

CHAPTER SIXTEEN

The 1st (and Last) International Workshop on XMRV and the World's Most Celebrated Virus Hunter Enters the Fray

When Ian Lipkin chose a career in infectious diseases, he envisioned hunting for pathogens in daring treks around the world. Though disappointed to learn that modern-day virus hunters work largely from the lab, he still wound up as a pioneer. . . . Lipkin has developed groundbreaking techniques that have helped a new generation of disease detectives sleuth out the infectious roots of mystery ills, chronic disease and neuro-psychiatric disorders like autism and OCD.

—*Discover*, May 11, 2012[1]

Dr. Mikovits did not have a clear idea when the planning first began for what was later dubbed the 1st International Workshop on XMRV, scheduled for September 7 and 8, 2010. She was sure that organizing took at least six months and first invitation letters were emailed April 21, 2010.[2]

The decision to have it must have been made sometime in the winter of 2009–2010, probably when the controversy about XMRV in humans was really crystalizing. The confirmation study by Lo and Alter was probably not yet known to the organizers, and thus it was initially set up to focus overwhelmingly on the negative studies that had generated the post-*Science* controversy.

The conference was organized by familiar players through a group in the Netherlands: John Coffin, Stuart Le Grice, Robert Silverman, Jonathan Stoye, a Coffin protégé from the UK, and Charles Boucher from Holland Frank Ruscetti was originally invited to speak, but he protested to Silverman and Le Grice that neither Mikovits or Singh were invited to present, which seemed an unsightly omission since they had done the most important work since Silverman.

Ruscetti, ever the cynic, saw their lack of inclusion as evidence of the Old Boys Club, anti-feminist power infrastructure in science. Annette Whittemore was also upset when she saw the proposed line-up of speakers and wanted to immediately mobilize the patient groups to start a blitzkrieg of protests.

Mikovits vetoed that idea and proposed a different course of action. She contacted anybody she had worked with in the field whom she *knew* had positive data similar to her own, and asked them to submit an abstract.

She had a whole list of giant minds in this category. There was Dr. David Bell, the gentleman doctor who had visited some of his rural patients on horseback and had been treating ME/CFS patients for more than twenty years since encountering a pediatric outbreak in upstate New York. Paul Cheney, the groundbreaker of the Incline Village outbreak, Dr. Julio Blanco and his team from Barcelona, Spain, Dr. Jose Montoya of Stanford University who had great compassion and a fiercely investigative attitude toward ME/CFS. Dr. Michael Snyderman, an oncologist who had tested positive for the antibody to XMRV and chronic lymphocytic leukemia (CLL), and had been treating himself with antiretroviral medications, which—as he documented—had put his cancer into remission.

Mikovits recalled the response of those she contacted being very gracious and supportive. Eventually eleven or twelve abstracts from researchers with positive reports about XMRV were submitted for presentation at the conference. While an abstract lacks the influence of a peer-reviewed paper published in a top journal, it is usually a good indicator of where the field is going and thus an integral part of future funding strategies concerning a given scientific topic. The organizers of the conference had little choice but to respond to this influx of abstracts, although their decisions didn't always seem to make sense.

The only positive abstract submitted that was given a speaking spot was the one submitted by Mikovits entitled "Detection of Infectious XMRV in the Peripheral Blood of Chronic Fatigue Syndrome Patients in the United Kingdom." Mikovits was also invited to chair the fourth roundtable discussion on ME/CFS with John Coffin, as well as participate as a panel member in a question and answer session that would be web-cast around the world. But it was the research Mikovits presented on ME/CFS patients in the United Kingdom that would provoke the greatest firestorm. It even attracted the attention of Dr. Francis Collins, the head of the National Institutes of Health.

After that, a very different "Battle of Britain" began.

* * *

In 2002, Dr. Jonathan Kerr was the Sir Joseph Hotung Clinical Senior Lecturer in Inflammation, as well as the Honorable Consultant in Microbiology at Saint George's University in London, but in 2010 he suddenly left research for private practice. His career downgrade raised the question whether Kerr was like other historical researchers who stumbled upon inconvenient truths and found their careers suddenly jeopardized.

In 2008 Kerr authored what was billed as a "Major Article" in the *Journal of Infectious Disease*, "Gene Expression Subtypes in Patients with CFS/ME." As explained in the article:

> The goal of this study was to determine the precise abnormalities of gene expression in the blood of patients with CFS/ME . . .
>
> Using this approach, we identified differential expression of 88 human genes in patients with CFS/ME. Among these genes, highly represented functions were hematological disease and function, immunological disease and function, cancer, cell death, immune response, and virus infection.[3]

To fully appreciate the importance of this information it's probably necessary to make a brief detour into the field of epigenetics, which had been turning a standard Darwinian understanding of evolution and genetics on its head.

In the typical Darwinian understanding of evolution and genetics, an animal either has a trait (or the recessive trait for the gene), or it doesn't.

What happens to an organism during its lifetime has no effect on the genetic inheritance it passes down to its young. In other words, these traits stay fixed. The classic demonstration of this principal is that rats who have their tails cut off give birth to offspring with tails.

What's being discovered in epigenetics (which means "on top of genes") is that diet, daily stresses, even toxins can turn gene functions on or off like switches, making us more susceptible to disease or infection.

Because of her expertise in epigenetics, Mikovits was immediately intrigued by Kerr's interesting work. Kerr's group found that among the well-defined CFS/ME population, there was abnormal activity in eighty-eight of their genes. It was if their internal switchboard had gone mad.

Genes of people with CFS/ME were behaving in a radically different way than those of healthy individuals. The data had significance not only for ME/CFS, but also for Gulf War illness (GWI), a related controversial disease.

One gene in Kerr's studies, NTE, is the primary site of action of organophosphate (OP) compounds, such as sarin nerve gas (one of the agents some GWI veterans were exposed to during deployment, and organophosphates are also used regularly in civilian life as pesticides), which causes axonal degeneration and paralysis due to inactivation of its serine esterase activity. In the nervous system of adult chickens, OP-modified NTE initiates neurodegeneration.

NTE probably regulates neuron-glial interactions during development and possibly also during adult life. Exposure to OPs may thus trigger CFS/ME and Gulf War Illness.[4] Kerr had crossed a red line by suggesting that some type of chemical exposure, like organophosphates in sarin and in widespread civilian use, might be affecting gene expression, particularly those genes involved in brain function and proper regulation of the immune system.

In such an environment of abnormal gene expression as the result of a chemical exposure, was it beyond imagination that a previously suppressed retrovirus might cause havoc? A chemical exposure, followed by changes in gene expression, could lead to the increased ability of a previously suppressed retrovirus to do harm.

Since the immune system was faulty due to changes in gene expression, it would be hard to say whether it was the immune system, the chemicals, or the retrovirus doing the greatest amount of damage. The picture looked like a "wrong place, wrong time" breakdown caused by multiple and possibly synergistic factors.

* * *

Because of his research and reputation for being precise yet low-key, Kerr was a natural fit for Mikovits. They wrote an application as multiple principal investigators for a federal grant (known as an RO1 grant) to study pathogens and patterns of gene expression in ME/CFS patients.

Submitting their first proposal in June of 2007, they were awarded the grant in September of 2009 after two revisions. It was the first federal grant the WPI received. However, if one of the Principal Investigators in such a study loses his or her academic association, the grant may be recalled, unless another researcher from the same institution can step in or it can be shown that the remaining Investigator can fulfill the research objectives. Mikovits was surprised when Kerr was not given tenure at St. George's University. After that, his name suddenly appeared on a negative XMRV study.

A ME/CFS parent named Chris Cairns, who wrote a blog with the handle the "CFS Patient Advocate" (his daughter has CFS), chronicled the veering course of Kerr over the years in a June 9, 2010, post. While he was not in a position to know exactly what happened to Kerr, Cairns normally attended many of the scientific meetings on ME/CFS and was a jovial colleague to many of the researchers. His article was a window into the puzzling change in Kerr. It also reflected concerns of the patient community that something unusual, and potentially sinister, was happening in the research.

Cairns was formerly professor of fine arts at Haverford college and was honored by the WPI on September 14, 2011, as "Advocate of the Year," two weeks before Mikovits was fired. Kerr begins by noting that the expectation had built through 2009 that "something big" was going to come out of the WPI, and then in October of 2009, the XMRV research was published. Cairns noted that Kerr seemed to disappear after that, then his name was "tacked onto a strangely negative validation study of XMRV." Then came the Invest in ME conference in May of 2010, and the strangeness continued.

> Jonathan Kerr was invited to update his research at the Invest in ME conference in May 2010. He was not in attendance as the conference began and came in halfway through the morning. He gave a very short lecture where he turned away from the microphone towards the slides on the screen and talked in his usual inaudible voice. No one could hear anything he was saying.[5]

Despite his apparently reputation as a soft-spoken mumbler, Kerr was well-respected in both the scientific and patient communities prior to 2010, vie-

wed by the patient community as one of the researchers who honestly searched for answers to their condition. That sharply changed in 2010, as Cairns
continued to describe what seemed like erratic behavior by Kerr at the 2010
Invest in ME conference in London.

A woman asked a question and Kerr turned away from the microphone
as he spoke. His slides were bleached out. At the end of his talks he usually
talked effusively about his research help. This time he showed a slide with
a picture of a group greatly diminished in number, then mentioned his
research was funded by the ME Trust, a group that was suspected by many
in the community as pushing a psychiatric explanation for their disease.
Cairns continued:

> To this observer it was obvious that Dr. Kerr was embarrassed and did not
> want to be at this conference. His entire manner and presentation, or lack of
> presentation, spoke of him just wanting to disappear. He was noticeably not
> interacting as in the past. He seemed like dead weight, and psychologically
> half his size.[6]

For Cairns, who had seen Kerr present a number of times, this was truly
unconventional behavior. The picture Cairns painted was of a scientist
under some kind of psychological siege. Cairns asked how it had all come to
this and suggested his own explanation.

> It had been obvious from the beginning that Dr. Kerr's CFS/ME gene work
> did not win approval from his colleagues in the UK . . . The writing was on
> the wall and things obviously got hotter for him when he collaborated with
> this rogue agency in the US, this WPI place.
>
> And then XMRV hit the fan and that was enough for Dr. Kerr's colleagues.
> They pulled the plug on him. In various countries they put polonium 210 in the
> soup, in the UK they just give the "poison pill" (in a collegial manner.) I imagine
> that Kerr always expected that something bad was going to happen. So he must
> not have been surprised when the late night call came, or someone cornered him
> in the hallway of his tiny research facility—urging him to disassociate himself
> from "those people."
>
> This friendly fellow or gal, a fully supportive, honest soul, must have said
> to him, "Jon, pull yourself together, just take this pill, just take this poison
> pill. Swallow it with a full glass of water".[7]

While Cairns's suspicions about what had happened to Kerr (that is, his
allegation that someone offered a metaphorical "poison pill") could not be

confirmed, the events complicated the federal grant he had been awarded jointly with Mikovits.

Mikovits knew she was on a short timeline to save the grant, and wouldn't let it go without a fight.

* * *

Prior to Kerr's failure to get tenure, they had collected most of the controls they needed but hadn't made the arrangements to get the patient samples. They needed those samples desperately if they were going to have any chance of retaining their grant and finishing the research.

Mikovits called Phlebotomy Services International, a company that hired people to draw blood around the world and knew about grants and institutions, medical protocols, security, and performing work under the Institutional Review Board rules that covered medical research. Mikovits explained to the founders of the company, what she needed and begged for their help.

The couple knew the importance of the project and offered to volunteer their time, but needed somebody to pay their plane fare to England and for a hotel and supplies they'd need for the job. Mikovits put five or six thousand dollars on her own American Express card to get them to England, and a generous anonymous donor came up with the five thousand they needed for supplies.

Mikovits contacted some of the patient groups in the London area to determine if they could assist in shuttling patients to draw locations. Many patients would need a ride, and preferably from somebody who understood how taxing such a venture would be to a person with ME/CFS. The patient groups waged a heroic response and were able to get nearly fifty ME/CFS patients in the very short few days that the phlebotomists were available. A handful of patients were too weak to travel, so one of the phlebotomists made a pilgrimage to their homes to get their samples. Other patients would likely have physical payback so they were heroic in their efforts as well.

After a panicked night when the samples were held up by FedEx, by the next day all of the samples had made it to Frank Ruscetti's lab at the National Cancer Institute and were waiting to be processed. Since questions had been raised that XMRV was simply mouse contamination, Mikovits wanted the samples to go directly to Frank's lab so he could arrange to have them processed in a facility that never handled mouse samples. This should have ended the contamination dispute.

Mikovits believed they had taken all necessary steps to save the RO1 grant and that she could make a credible argument that Kerr's part in the study could be transferred to the WPI.

* * *

Gerwyn Morris was a member of the UK group of patients and it was through an email to Mikovits about the possibility of getting into the study that he struck up a friendship with her.[8] A chemist and microbiologist by training, he joined a couple of ME/CFS forums right before the *Science* paper came out. He had very little difficulty unraveling the design of studies, and so he shared his expertise with the non-scientists on the forums. Morris had gotten sick with ME/CFS around 1971–1972, when he was a pubescent fourteen years old, well over a decade before the first "modern" US outbreak in 1984–1985.

Over the years he went through an oscillating pattern of recovery and illness, a pattern Cheney has noted is more common in pediatric cases, who often have recovery periods, sometimes consuming several years of his life. Morris was a vocal critic of the four negative studies right after the *Science* paper, and in 2013 detailed his objections for this book.

His concerns revolved around the difficulty of the PCR systems used in most of the negative studies, as PCR is a process known to be extremely sensitive to even slight variations in conditions. In an interview for this book, Morris discussed his concerns.

"If anybody believes or reads anything about MuLV viruses, they will know that they don't come alone. A retrovirus never comes in one homologous sequence. It just doesn't. Not even if they replicate by something called clonal expansion, which most retroviruses do, [with] HIV being the exception, not the norm."[9]

"So if you actually are going to target one particular sequence in your PCR assay, because the way they designed their assays was to calibrate or optimize them, so that they could actually detect a low copy number of viral plasmids, VP-62, and in sparkling brand new human DNA, that is for a start, never, ever going to be anything like getting a virus out of biological sample and the various corruptions of DNA that can occur."[10]

Morris also thought the scientists were setting their sights on a too-small bull's-eye, the VP-62 clone of Silverman, rather than the broader sequences found by Mikovits and Lo and Alter. "You can't say for a fact that this PCR assay which they replicated ad nauseum in all of the studies, could even detect a very slight variation on the VP-62 clone. So it's fairly obvious that they weren't actually genuinely trying to find a retrovirus in my view. I think they had [already] decided for one reason or another that VP-62 was a contaminant, and they would only bother to look for examples of the contaminant."[11]

"But by using the PCR testing they did, they were capable of detecting only VP-62. And if one is totally honest about it, the way that retroviruses integrate into the CpG islands, they're very high, density of guanine and cytosine, cross-linked, which means they're very difficult to pull apart and re-anneal. And you have to have very special cycling conditions to be able to pick up anything in those regions. So, even if they were looking for VP-62, the virus, in my view, and in my background in biochemistry, they would not be able to detect it. But the basic question is, whether you can be certain from the scientific level of certainty whether the virus you're looking for is in that sample, and that question was never answered."[12]

To put it in simpler terms scientists have shown that false negatives as well as false positives can occur in PCR assays.

He was also clear about what he considered Mikovits's strengths and weaknesses "She is one of the most honest, caring individuals I've ever had the pleasure of meeting. Her weakness, I think, is naïveté. I don't think she can contemplate that other scientists are not quite as honest as her. And I don't think she can contemplate how rich and powerful interests can manipulate the population as a whole. And I don't think she can buy into the fact that she's been manipulated as well."[13]

Morris also voiced strong opinions about whether animal retroviruses were circulating in the human population. "I think that there are gamma-retroviruses out there. Vaccines were passed through animal tissue, like smallpox was passaged through mouse brains, before they knew anything about what they were doing. They were creating replication recombinant retroviruses with gay abandon in the early vector research until they realized what they were doing."[14]

"If there aren't some gamma-retroviruses circulating in the human population there bloody well ought to be. What I think the *Science* results were: I think Judy might have been picking up a HERV-W, which is a gammaretroviral HERV [human endogenous retrovirus], otherwise known as the multiple sclerosis retrovirus because it was talked about as a cause of MS for the better part of fifteen years. It's been around, [but] whether it's a primary cause or its activation is a secondary cause of pathology . . . I don't think anybody knows. I suspect that oxidatively modified envelope protein [a reaction in the cell to foreign proteins] could have triggered Judy's antibody test."[15]

Gerwyn couldn't give a definitive identification of the current virus plaguing patients, but it was clear something of a viral nature was going on.

"Judy's results cannot be explained by contamination. The 'contamination' occurring in her lab was just Fantasy Island. A lot of people would like to make the facts fit their theory, but they don't. The multiple lines of evidence in Judy and Frank's study [demonstrates] there's no way that's contamination."[16]

* * *

But Mikovits's lobbying to get positive data into the 1st International Workshop on XMRV still paid off. More than ten abstracts confirmed basic parts of the retroviral theory advanced by Mikovits.

Maureen Hanson of Cornell University submitted one entitled "XMRV in Chronic Fatigue Syndrome: A Pilot Study" in which she took 30 adult subjects, 10 with severe CFS, 10 who had "recovered" from the disorder, and 10 controls and tested their blood. Markedly, even a significant number of ME/CFS patients identified as "recovered" showed evidence of XMRV. Hanson stated clearly in her abstract, "Our results corroborate those of Lombardi et al."

Dr. B. P. Danielson from Baylor College of Medicine contributed "XMRV Infection of Prostate Cancer Patients from the Southern United States and Analysis of Possible Correlates of Infection." The abstract agreed with findings that XMRV did not seem to be correlated with the genetic defect in the RNase L pathway suggested by Robert Silverman of the Cleveland Clinic. Tissue testing—which might find latent virus not as easily cleared from blood—seemed a key determinant in testing to the team from Baylor.

From Emory University came more detailed evidence of the sequences of XMRV integrated into prostate cancer tissue, in "Variant XMRVs in Clinical Prostate Cancer." This abstract concluded that, "Clinical prostate cancer samples contain integrated XMRV pro-viral sequences, which can be amplified in nested PCR approaches with approximately 1000 more bases than can be attributed to XMRV alone.[17]

Silverman and Das Gupta of the Cleveland Clinic weighed in with the positive finding of the "Presence of XMRV RNA in Urine of Prostate Cancer Patients". RNA is the messenger molecule of DNA and can be used to detect active infection. If a test were developed utilizing urine instead of blood or tissue, detection in men at risk of prostate cancer would be hugely simplified.

Another abstract by Silverman explored XMRV and gene function, "XMRV Infection Induces Host Genes that Regulate Inflammation and Cellular Physiology," and found that:

> "The chemokine IL8 is one of the most highly induced genes in response to XMRV infection of prostate cancer cell line DU 145. XMRV induction of the 30 host genes identified in this study suggests a profound effect of the virus on fundamental cellular physiology and inflammation.[18]

The Mikovits lab had also found that the chemokine IL8 was the inflammatory marker most strongly correlated with XMRV infection, so they had located the same signature. The data that XMRV affected genes were implicated in cancer was not surprising but sobering.

The researchers seemed to be slowly mapping the mechanism by which XMRV led to prostate cancer.

* * *

Paul Cheney must have had the gratifying feeling of coming to the end of a long journey when he submitted his abstract. Many credited him for originally suggesting that a retrovirus underlies ME/CFS, although he modestly claimed that the media firestorm around AIDS would have put the seed in someone's mind eventually.

Cheney's abstract, "XMRV Detection in a National Practice Specializing in Chronic Fatigue Syndrome (CFS)," began with a historical recap, then his conclusions corroborated Mikovits's work and he asked the question of whether family members of ME/CFS patients were also at risk. He concluded:

> "XMRV was detected in 74.5% of 47 consecutive cases of well-characterized CFS from 24 states and two countries. There is a high positivity rate (50%) of XMRV in non-CFS exposure controls within family members.[19] Patients reported significant illness in first order family members."[20]

Troubling results also came from a collaboration between Emory University, the Cleveland Clinic, and Abbot Laboratories in the abstract "XMRV

Induces a Chronic Replicative Infection in Rhesus Macaques Tissue But Not in Blood."

This research found the virus seemed to disappear relatively quickly from the blood, but remained in tissue where it could replicate to detectable levels upon immune stimulation.[21] The immune stimulant they used was an injection of bolus peptides that mimicked a vaccination.

The virus in other words raised an under-the-radar immune response, rapidly left the blood, and settled into tissues. It could be a stealth invader.

Nobody knew whether these individuals with no knowledge of their XMRV infection were walking time-bombs, waiting for an immune crisis that allowed the retrovirus to emerge from hiding to cause significant damage.

Harvey Alter and Shyh-Ching Lo of the FDA presented an abstract of their paper previously published in the *Proceedings of the National Academy of Sciences* that had confirmed Mikovits's work and caused such consternation the previous summer that its publication was delayed for several months. They reported:

> "We found MLV-like virus *gag* gene sequences in 32 of 37 patients (86.5%) compared with only 3 of 44 (6.8%) healthy volunteer blood donors . . . No evidence of contamination with mouse DNA was detected in the PCR assay system or any of the clinical samples. Seven of 8 *gag*-positive patients were again positive in a sample obtained nearly 15 years later."[22]

Mikovits and her team found that 67 percent of her well-characterized ME/CFS population showed evidence of XMRV infection. Alter and Lo found evidence that 86.5 percent of their well-characterized ME/CFS population showed evidence of infection by a broader group of murine leukemia viruses.

Mikovits and her team found that 3.7 percent of a healthy control population showed evidence of XMRV infection. Alter and Lo's research showed that 6.8 percent of a healthy control population showed evidence of infection by this wider group of murine leukemia viruses.

Statistically broken down, this meant that between eleven and twenty-one million individuals in the United States were potentially infected by this group of related murine leukemia viruses.

The emerging picture was of a simmering plague that could take years or decades to show its explosive potential.

* * *

Along with Mikovits's work, the most controversial was an abstract coauthored by Frank Ruscetti and Mikovits's research associates, Max Pfost and Katy Hagen, "Detection of Infectious XMRV in the Peripheral Blood of Children." The study contained 66 subjects: 37 parents and 29 children. 17 of the children were diagnosed with autism, a pair of three-year-old twins had Niemann-Pick Type C, a neurodegenerative disorder, and 10 healthy siblings.

They wrote:

> "an understanding of XMRV infection rate in children may be particularly helpful, given that 1 in 100 children in the US are diagnosed with neuroimmune disorders, including Autism Spectrum Disorder (ASD) and that CFS and childhood neuroimmune disorders share common clinical features including immune dysregulation, increased expression of pro-inflammatory cytokines and chemokines, and chronic active microbial infections.
>
> XMRV was detected in 55% of 66 cases of familial groups from 11 states. Sequencing of PCR products of *env* and *gag* confirmed XMRV. The age range of the infected children was 2–18. 17 of the children (including the identical twins) were positive for XMRV (58%) and 20 of the 37 parents (54%) were positive for XMRV. 14 of 17 autistic children were positive for XMRV (82%)."[23]

This was earth-shattering news for autism parents, because along with the macaque study it also supported the theory that these children might have been harboring an undetected retrovirus in their immune cells that could be activated through an immunization.

The immune challenge of a vaccination could have offset the body's delicate suppression of the retrovirus, bringing it out of hiding. For other children, a simple fever could have begun the immune cascade that led to autism.

Mikovits knew that among the autism communities there were many who traced the onset of their child's problems to a vaccination, while others felt their child had exhibited signs from birth. An immune challenge to the mother during pregnancy could have also stimulated the retrovirus to rampage. Mikovits knew this was controversial territory.

But her job was not to shy away from or even indulge controversy, but rather to seek scientific truth.

* * *

In a harrowing twist, Mikovits almost didn't get to present at the 1st International Workshop on XMRV because her laptop broke down. This would have been a tragedy since Francis Collins—head of the NIH, and former head of the Human Genome Project, which was also centered in Cold Spring Harbor (and possibly the most ambitious government science project since the Manhattan project that built the first atomic bomb)—planned to sit in on her talk.

She was staying at the home of her twin sister and planning to celebrate the birthday of her beloved niece who was born the day she turned in her PhD thesis, September 6, 1991. But relaxing family plans over Labor Day weekend took a backseat to the panic that set in when her laptop crashed and had swallowed her presentation. She emailed Pfost back at the lab saying that she needed him to reassemble it. When she later read his emails, it was clear he had burnt midnight oil and coffee until at least 4:00 a.m. reassembling it. She had never met a more dedicated research associate or student than Pfost. What he sent to her was perfect.

The meeting took place at the NIH main building in Bethesda, Maryland, in the Masur Auditorium. Mikovits was slated to be the last speaker before lunch on the second day, followed by a short question and answer session.

That session opened with an invited lecture on ME/CFS by Frank Ruscetti, a talk by Brigitte Huber of Tufts University regarding her inability to detect XMRV in patient samples, Maureen Hanson of Cornell University on her success in detecting XMRV in ME/CFS patients, Lo of the FDA on his positive study with Harvey Alter of the NIH, and finally, Mikovits on the UK patients.

Francis Collins attended the opening, listened for a little while, then left. On the second day, shortly before Mikovits spoke, he reappeared, trailed by his staff, and sat quietly. Most people knew Collins for his work as director of the Human Genome Project, but others saw a less conventional side, at least for a scientist. He'd detoured from expectations by writing a book called, *The Language of God,* in which he detailed his belief in a supreme being and a unique destiny for every living person, and a companion book *The Language of Genes,* which tried to unify science and religion by postulating that an almighty creator did not conflict with his abilities as an objective scientist.

With perhaps the most powerful man in the US public health system sitting in front of her, Mikovits took the stage and began her talk about the patients from the United Kingdom.

Her presentation followed point by point of her submitted abstract. She detailed how they'd found XMRV in a subset of ME/CFS patients in England and they needed to develop more information about replication and pathogenesis. There was also a great need to develop tools for screening and treatments.[24]

Despite the panic of the previous day, Mikovits felt serene and confident in the middle of her talk. She had run through the important parts of her research in a steady, methodical manner. She described how her team had carefully and strategically weeded out any potential obstacles. They'd been concerned about the possibility of contamination so after the samples were drawn they were shipped directly to the NCI and processed in lab that did no mouse or retrovirus work.

Since their team believed patient selection criteria was likely to be the most important variable in detecting XMRV or a closely related retrovirus they'd used the world's most rigorous definition for the disorder. She ran through the five different methods they'd used to detect evidence for the virus and noted that it appeared that for best results they needed to utilize some form of molecular or biological amplification.

They'd developed a new set of primers to detect the envelope of the virus and in following along with Alter and Lo's work, they were finding a wider variety of genetic diversity in the virus, including xenotropic and polytropic sequences. This meant that certain strains of the virus were clearly able to infect different species.

All in all, the study seemed to be an elegant counter-point to the negative study by Jonathan Stoye and her former collaborator, Jonathan Kerr. One of the pieces of data she added to her presentation that had not been in the abstract was that approximately 5 percent of the controls were positive for XMRV, a number which was comfortably between their original research and Alter and Lo's confirmatory study.

Collins raised a question after her talk. He wanted to know where she'd gotten the control samples. She told him the samples had come from the London Blood Bank and were procured by her collaborator, the suddenly elusive Kerr. Collins thanked her and left the workshop a few moments later. Based on his later behavior, he must have been impressed by what he'd heard.

The question and answer session, chaired by Coffin and Mikovits, with Jonathan Stoye, a protégé of Coffin, serving as a kind of master of ceremonies. Eventually, Mikovits was queried on her opinion about using antiretrovirals. A commentary co-authored by Andy Mason that appeared along with the Alter/Lo confirmatory study in the *Proceedings of the National Academy of Sciences* had suggested that it was time to seriously consider a trial usage of existing and approved antiretroviral medications for CFS patients, so it seemed timely.[25]

Mikovits simply quoted the last line of Mason's commentary in part: "At this juncture, studies to establish proof of principle are justified to determine whether safe antiviral regimens can impact ME/CFS." If an opinion expressed in the *Proceedings of the National Academy of Sciences* wasn't good enough, then what was?

* * *

A day or two after the meeting, Mikovits received word that Collins had mandated a multi-center study to be directed by Ian Lipkin of Columbia University.

Unlike the planning for the 1st International Workshop on XMRV, the WPI and Mikovits were chosen for the multi-center study. Despite the difficulties, Mikovits and her collaborators had experienced the last few years, it finally looked like the ME/CFS community (and possibly other impacted groups) were finally going to get the scientific investigation they deserved after decades of limbo.

The best news was that she learned the RO1 grant, which she'd been in danger of losing since Kerr had lost his academic affiliation, now it would be awarded to Mikovits of the WPI. Mikovits and Ruscetti were encouraged by the positive data from Hanson, Lo/Alter, and all of the prostate research groups that, in often-sexist science, never seemed to generate heated attacks.

It was difficult to believe that these divergent groups coming to the same conclusions were all victims of exotic biology and contamination falling from the sky like some kind of blood rain.

* * *

From the beginning of his academic career, Ian Lipkin was unconventional. He was in the first class of men at Sarah Lawrence College where he studied anthropology and shamanism. Rather than pursue cultural anthropology, he wanted to become a medical anthropologist and bring back traditional medicines to the West.

After medical school, he took a residency at the University of California, San Francisco (UCSF) just before the AIDS epidemic began. When AIDS (known then as GRID, or Gay-Related Immune Disease) blitzkrieged the Bay Area, he was one of the few physicians willing to see patients who were infected and had one of the neurological disorders associated with the disease. The medical community would eventually realize that HIV was turning the body against itself, a hallmark of an autoimmune disease.

One of the early treatments for HIV was a process called plasmapheresis, in which blood is removed from a patient and put into a centrifuge, with white cells and platelets in one area, red blood cells in another, and plasma in yet another. The plasma was known to contain antibodies, that in HIV/AIDS caused part of the problem. The plasma was replaced with albumin so that the blood volume was maintained, then the blood was returned to the individual. Many patients experienced significant relief from this treatment, though they weren't cured.

Lipkin thought one of his patients, a ski instructor from Vail, Colorado with HIV/AIDS, might benefit from it, but many physicians weren't interested in plasmapheresis. In a May 2012 article from *Discover*, Lipkin recalled those days, showing the kind of concern for patients that also made so many patients adore Mikovits. Lipkin had heard about a Russian guy at Pacific Medical Center with a centrifuge. Lipkin recalled:

> He was willing to help me, provided two things: number one, I would take responsibility for inserting the needles in my patient's arm. And two, he would be paid in cash—$600 up front for every treatment. So I put my guy in my Ford Fiesta, and I drove him over to this medical center. He could barely walk, so I walked him in. He lay down on this hospital bed, and I put in two large-bore needles, one in each arm.[26]

Lipkin then described how the man modestly improved with this treatment, was able to go skiing again, but eventually progressed to AIDS and died. It was after that experience that Lipkin decided to study infectious diseases.

Although there were times when Lipkin would anger Mikovits and she would swing from gratitude to wanting to curse him out, it was always clear that the compassion of that younger Lipkin was still vitally alive in him as he got older. He did seem like he wanted to help.

* * *

When Mikovits first met Dr. Ian Lipkin in his office at Columbia University in New York City on November 4, 2010, she wore a business suit, wanting to project a professional image to the group that would form the backbone of the multi-center study to determine whether XMRV was associated with ME/CFS.[27]

By contrast, Lipkin wore jeans, a button-down shirt, and sneakers, looking more like a Silicon Valley wunderkind than a senior scientist.

In that meeting in a small conference room were Lipkin, Switzer from the CDC, Harvey Alter from the National Institutes of Health, Suzanne Vernon and Nancy Klimas—who were invited to the discussion on behalf of the patients and doctors—Dr. Simone Glynn who was part of the already assembled Blood Working Group, and Judy Mikovits. Dr. Cathy Laughlin, of the NIH/NIAID, which funded the study, was there as well. The main issue they discussed at the meeting was which physicians would provide samples.

This was when things started to go awry.

The investigators were told that each of the principal investigators (Alter, Mikovits, and Switzer) could veto a decision by the group as they were the stakeholders, but this protocol quickly broke down in practice.

Lipkin, who sat at the front of the small conference table, was curt, to the point, always professional, but made it clear that he had taken this assignment as a favor to Francis Collins and Harold Varmus (the head of the NCI), who wanted the study to be scientifically and ethically unassailable. Of the physicians to select study participants, it was decided that Dan Peterson's patients would be included but that none of the original cohort that formed the basis of the original *Science* paper would be used.

Incredibly to Mikovits, the top 300 patients who she had worked with in all the WPI studies would not be included. Mikovits vehemently disagreed with the inclusion of Peterson for the simple reason that his sickest patients, those with the most dysregulated immune systems and all of the footprints upon which she had built her hypothesis, would not be included. Her veto was denied.

Next Klimas suggested Dr. Lucinda Bateman who Mikovits also vetoed on the grounds that Bateman was a fibromyalgia specialist and not an ME/CFS specialist, hence increasing the danger of "idiopathic" fatigue patients or patients without the same disease thrown into the mix. Bateman was the doctor Peterson sent his patients to when they didn't match his profile of an ME/CFS patient with natural killer cell and RNase L abnormalities who seemed prone to develop lymphoma. Bateman had participated in patient selection on a negative study already, so clearly the patients she treated were not those likely to have evidence of XMRV infection.

Everybody agreed that David Bell, the physician who handled the pediatric outbreak of ME/CFS in Lyndonville, New York would be a good choice, but he had retired. Mikovits made two suggestions: the obvious choice of Paul Cheney, and Dr. Joseph Brewer, a respected doctor in the Midwest who specialized in ME/CFS and HIV/AIDS.

Both were vetoed by Klimas and Vernon, under grounds that they were "controversial," as Cheney had tried to help his patients with stem cell therapy and Brewer had suggested doing a trial of antiretrovirals.

This angered Mikovits since she knew several leading physicians were prescribing the same treatments and afraid to talk about it too loudly. Moreover, Brewer and Cheney were experienced at diagnosing ME/CFS patients with evidence of XMRV infections. The group also decided to cut from the study any patients with known thyroiditis.

But Mikovits had seen an abundance of thyroiditis and thyroid cancer in ME/CFS patients and families with evidence of XMRV infection. She had, in fact, proposed the thyroid as a possible tissue reservoir for the virus. Mikovits felt that excluding this population was unwarranted. When Mikovits flew home from the first meeting with Lipkin, she thought the study was going to be a complete disaster.

From the Denver airport while on a layover she had a long conversation with Annette Whittemore where she poured out her frustrations. Unbeknownst to her, while she was safely tucked away at the meeting, Annette had announced online that VIP Dx was coming out with a serological test for XMRV shortly, using Mikovits's name but never showing her any data or asking permission to use her name on the company website.

Of course patients were excited and couldn't wait to buy in, since it would be concrete proof for their doctors, but it was the Whittemore's haste that would again cause Mikovits serious trouble.

* * *

The second time Mikovits met Lipkin she came away with a more favorable impression than the first meeting where she'd merely felt that Klimas and Vernon had run roughshod over him and he had not fended them off and managed the meeting. Both Lipkin and Mikovits had been asked to give a talk at the New York Academy of Sciences on March 25, 2011.

Opened in 1817, the New York Academy of Sciences had counted among its members President Thomas Jefferson, Thomas Edison, Charles Darwin, and Albert Einstein. It currently boasts twenty-seven Nobel Laureates. Dr.

Lipkin was also a member. The event where they were invited to speak was a series of lectures on the ominous topic, "Pathogens in the Blood Supply."

Lipkin's talk was "Microbe Hunting" and Mikovits's was "Strategies for Detection of XMRV in Blood." Prior to the event, Lipkin had not discussed his presentation with Mikovits, but she had a general idea of the type of talk he was likely to give.

As she listened to his presentation, it seemed to her that he had specifically crafted the perfect lead-in for her. He recounted his discovery of West Nile Virus, how it had been a difficult virus to detect and caused a controversy among public health officials who had failed to take quick action, allowing the virus to spread much farther than it should have. He again reiterated his idea that most of the pathogens that were easily detectable had already been found: the future belonged to scientists who used the latest technological tools to follow the "footprints" of these elusive viruses and bacteria to their lairs, the true tracker-hunters of virology.

Mikovits felt a shiver of excitement, knowing she was on such a prestigious stage with the world's most famous virus hunter setting up her presentation. Lipkin finished up, turned to her and said, "Go get 'em, tiger!"

* * *

Afterward, Lipkin approached her and said his team wanted to take both her and Annette to dinner. They had a pleasant meal and Mikovits saw another side to Lipkin. He talked about his role as a consultant for the movie, *Contagion*, directed by Steven Soderbergh. The movie was a fictionalized version of the SARS outbreak in China, in which Lipkin led the investigation.

Mikovits talked about Frank Ruscetti and their common difficulties with Gallo. She told Lipkin she was never going to accept an order she believed was unethical or immoral, no matter who gave it.

Lipkin mentioned he wanted to retire in five years to pursue other interests outside of science. Joining them that evening was Lipkin's collaborator, Dr. Mady Horning, who was widely expected to take over the Center for Infection and Immunity when Lipkin retired. Hornig had a child with autism and had to excuse herself early to relieve her babysitter.

By the time the dinner was over, Mikovits had warmed up to the charming Lipkin. She had seen some evidence of his reputation for arrogance at their first meeting, how he was a "take-charge" kind of guy, but he reminded Mikovits of her father in many ways. Once his shell was cracked, Lipkin's good heart and compassion were evident.

The two of them might never become the best of friends, but Lipkin was a person she could respect, work with, and with whom she might even enjoy a lively dinner discussion.

* * *

As the attacks on XMRV increased in the months after that night, and the criticisms expanded from detection of retroviral sequences being nothing more than mouse contamination, to include a recombinant virus that had been created in the lab in the early 1990s with no natural history of infection, Lipkin rose to her defense and showed his "take-charge" persona also meant he could hold the line for a scientist treated unfairly.

When Mikovits got back from the scene at the 2011 Belgium retroviral conference, where scientists had shouted, "Stop spending money on these people!" there was an email from Lipkin telling her, "Judy, ignore the noise, and simply focus on the science."

Lipkin was, in fact so concerned that he offered to come out to the WPI to show his support and confidence in the work. Since it was already mid-June then and Annette and her family always went away for summer vacation in the month of July, they had to scramble to find a suitable time frame. They settled on Lipkin flying out to the WPI from New York on June 24, and staying for three days. During that time he toured the labs, conducted a physician's roundtable, and gave an open address in the WPI auditorium that was open to the public and thus expected to attract local ME/CFS patients. But Mikovits was miffed when he conducted just a cursory tour of her research lab, with a dismissive, "If you've seen one research lab, you've seen them all," and then choosing to spend a great deal of time at the clinical lab, VIP Dx.

She later found out that Lipkin had been very concerned by failings in the Lombardi-run clinical lab, and even wrote a twelve-page letter to the Whittemores about these problems. His cursory interest in her research lab turned out to be a confidence in the quality of her work.

Harvey and Annette seemed to make sure that one of them always shadowed Lipkin. Mikovits was familiar with the Whittemore charm offensive and expected as much. The night before the physician's roundtable and Lipkin's public talk at the WPI, the Whittemores hosted a dinner for him, inviting Mikovits, Ruscetti, as well as Jamie Deckoff-Jones, the former emergency room doctor who had contracted ME/CFS and about whom the *New York Times* reported: "Some ME/CFS patients are already trying

HIV medications prescribed 'off label.' One patient, Jamie Deckoff-Jones, a physician in Santa Fe, N.M., has been keeping a popular blog about her improving health while taking antiretrovirals prescribed by her doctor. 'I think the sickest patients have the right to try the drugs,' she commented."[28]

Deckoff-Jones had recently been named director of clinical services for the Whittemore Peterson Institute when negotiations had stalled out with Paul Cheney.

Harvey gestured the server to pour expensive bottles of wine for Lipkin, who fancied himself a connoisseur, while others sipped their own drinks. The libations loosened everybody's tongues, and Lipkin started expressing opinions that offended Mikovits.

Lipkin declared it just short of criminal that some physicians were prescribing antiretrovirals for ME/CFS. Mikovits glanced sideways toward Deckoff-Jones, who wouldn't have been able to sit up for a whole social dinner in the year before she had started on antiretroviral therapies: she was living proof why Lipkin's words were dangerous.

Mikovits also thought of Michael Snyderman, effectively treating both his CFS and cancer with antiretrovirals. Lipkin went on about why people like Brewer with his antiretrovirals and Cheney with his stem cells weren't serving the patient population. Mikovits soon reached her limit. She threw her napkin in the chair and stormed out, refusing to return to the table. Lipkin seemed baffled as to how he had offended her.

"Well, you just insulted pretty much everything she's done in her career," declared Frank Ruscetti, "including her association with me."

Lipkin backpedaled and seemed genuinely contrite, promising to apologize the next day. Unlike Judy, Frank believed Lipkin's influence was so huge that they had to swallow any anger.

The next day Lipkin arrived a half-hour late for the Physicians Round Table, cutting it short for Lipkin's public presentation in the large auditorium at the WPI.

Mikovits felt she'd heard most of the speech before—the need to follow the "immune footprints" of the virus, and so on—yet smiled when Lipkin told the assembled group about the technology he possessed at Columbia and challenged, "Give me a million dollars and I'll find this virus!"

Many local ME/CFS patients showed up for his talk. A few were being pushed in wheelchairs by caregivers with ready oxygen tanks. Afterward, many patients approached Lipkin in the lobby, shook his hand, and gushed how appreciative they were that he was trying to help. Although Mikovits was incensed about the conversation the night before, she witnessed a shift in his expression. After the group left, Lipkin approached her. "CFS patients aren't supposed to be in wheelchairs," he said, obviously shaken by how disabled they were.

"The ones we see are," she answered somberly.

They talked for a few more minutes and she genuinely thanked him for coming. Harvey whisked Lipkin away to the airport, where the world's most celebrated virus hunter hopped his plane to the Big Apple.

* * *

CHAPTER SEVENTEEN

The Name Game and the "Immaculate Recombination"

How many have we created, John? How many XMRVs are out there?

—Judy Mikovits[1]

There's an old joke about a man who goes over to his neighbor's house to borrow a lawnmower. "I'm sorry," the neighbor replies as he stands in his driveway near the mouth of his open garage, "but I lent it to my cousin who lives on the other side of town."

But the man sees the lawnmower in plain sight, just on the other side of his neighbor's bike propped on a garage wall. "But it's right there," argues the man.

"What does the reason matter?" asks the neighbor. "I'm never going to give it to you anyway."

In her research on XMRV, Mikovits would often feel like the man in the old joke asking to borrow his neighbor's lawnmower. The reason for her lack of support never seemed to matter. No one in authority would consider the notion that a retrovirus was associated with ME/CFS, regardless of the evidence she and others presented. She suspected the scientific hierarchy had already made the decision that retroviruses were not involved in ME/CFS, and they would not bend.

* * *

Mikovits remembered all too well the stories of Frank Ruscetti from 1970s when he was working as a postdoc in the Gallo lab. Coffin and most of the other famous scientists in the field emphatically told him not to bother because there were no such things as human retroviruses.

Ruscetti and Bernie Poiesz were presenting their research at the 1980 Cold Springs Harbor conference that indeed they'd isolated HTLV-I, an infectious human retrovirus from patients with Adult T Cell Leukemia (ATL). It was so heretical that Gallo, who always attended the CSH meeting in those days and had promised to come, never showed up, leaving Bernie and Frank to face any dissonant music themselves.

David Baltimore and Howard Temin, who both received the Nobel Prize for the discovery of reverse transcriptase, and other giants of the field usually sat in the front row. Coffin had worked in the Temin lab as a postdoc and published with him in 1971.[2]

More than half a century had been spent trying to unlock the secrets of the first known animal retrovirus.

Ruscetti and Poiesz's presentation of their isolation of the first human retrovirus had been slated for the Sunday morning of the 1980 CSH gathering. At that time, any session about human retroviruses was always held on Sunday morning, after all the important people had already left and very few of the influential leaders of the field stuck around.

The majority of the remaining audience greeted their findings with silence, while the anti-human retrovirus critics simply walked out. It reminded Ruscetti of the tense walkout for Robert Koch when he give a lengthy dissertation of his discovery on the TB-causing bacteria. All the great German scientists like Rudolf Virchow who had been critics just exited without a word.

After the discovery of HTLV-I, Coffin hadn't been much of a factor in Ruscetti's life for the last three and a half decades except when they criss-crossed at conferences.

Mikovits really didn't have much of an opinion about Coffin until he showed up as one of the organizers of the hastily called July 22, 2009, meeting at the NIH, prior to the publication of the *Science* article. Perhaps Coffin was briefly worried that he hadn't pursued relevant data on other retroviral sequences witnessed thirty years ago. Maybe that was when he decided to write the accompanying editorial in *Science*, entitled "A New Virus for Old Diseases."[3]

Brigitte Huber, who had a laboratory near Coffin at Tufts, first suggested that XMRV was mere mouse contamination at the XMRV Workshop at the Cleveland Clinic in November of 2009. She said that in some very preliminary testing she had not been able to find evidence of the XMRV retrovirus sequence as defined by Silverman and DeRisi. Her explanation for the failure was that the labs involved—the Cleveland Clinic, the National Cancer Institute, and the Whittemore Peterson Institute—must have inadvertently had small pieces of mouse droppings, blood, or hair, that had somehow found their way into the testing materials. Ruscetti thought that Huber was a stalking horse for Coffin.

Mikovits and Ruscetti wittily dubbed it the "Mouse Crap Theory" of XMRV. The situation had started to spiral out of control and somebody needed to put a stop to it.

The Cold Springs Harbor Retrovirology Conference of May 2010 was something of a counterstrike when Alter and Lo reported the confirmation of the *Science* paper as well as their unpublished findings of a larger family of mouse (murine) leukemia viruses present in these patients.

However, publication of their confirmation study had been cagily delayed, while the negative study by the CDC by Switzer zipped through to publication like it was on a jetstream, the eventual publication of the Lo/Alter paper in August of 2010 finally seemed to debunk and mark the death-knell of the "Mouse Crap Theory," turning it back to compost.

Francis Collins's decision that the data presented at the First International Conference on XMRV now required a multi-center study to reconcile the positive and negative studies under the direction of Lipkin further undercut the "Mouse Crap" theory of Coffin and his cronies.

Coffin attended the November 4, 2010, meeting with Lipkin to set up the multi-center study. He sat on the other side of the conference table from Mikovits, disengaged from her, although the end of the meeting presented an awkward comedic opportunity.[4]

Right then, Harvey Alter asked if anybody wanted to share a ride to the airport. Mikovits and Coffin said yes so they all squeezed into the back of a cab. Because she was smaller than them, Mikovits was flanked in the back seat by Coffin on her left and Alter on her right.

Alter was friendly, asking how a cancer and AIDS researcher had become involved in ME/CFS. She told him the story and he listened with complete attention. By contrast, Coffin on her left huddled up against the left window, ignoring their presence.[5]

* * *

Coffin's impending publications would support the mouse contamination theory. Despite the Lo/Alter findings, the drumbeat of contamination accusations continued, bolstered by the journal *Retrovirology*, edited by Kuan-Teh Jeang, which published five articles on December 22, 2010, reporting the presence of mouse DNA into human materials and PCR reagent kits, as well as presence of MLV sequences in some widely used reagents. Others reported that sequences giving rise to clones of Silverman XMRV were identical to those in the prostate cell line 22RvI.

As a reviewer of one the five papers that he thought needed more work to warrant publication, Ruscetti was annoyed when it was published anyway, but Jeang told him that together the five papers supported one another, thus they surely didn't need the same level of vetting the Mikovits paper had gone through with *Science*.

Shortly after those papers came out in the winter of 2010–2011, Ruscetti was unexpectedly called into a meeting in his supervisor's office, the head of Center for Cancer Research (CCR), with Coffin and six of his colleagues.

At that time, Coffin and Dr. Vinay Pathak presented their evidence that Silverman's XMRV was actually a laboratory-generated virus as a result of a recombination between what they called preXMRV-1 and Pre XMRV-2.[6] This recombination had such a low chance of occurring that XMRV in any specimens that contained evidence of XMRV must have been contaminated through routine lab procedures like tissue culture and concluded there was no evidence that people or any animal species were naturally infected by XMRV.

This meant to them that all the positive data had either to be a result of contamination with mouse products or a laboratory-made viral artifact generated by recombination. Coffin told Frank Ruscetti that "*Science* started this and *Science* will end it."

Ruscetti thought to himself then, "What had *Science* started?"

Their paper had gone through standard review process and then a few other hoops and had not been Coffin part of "starting it" as author of the news and views accompanying the paper? As far as Ruscetti was concerned, if recombination theory was true, it was the first time a retrovirus infecting human cells and had no natural history of infecting people was created in the lab, who could have known beforehand?

Nobody would have then "started it"—it would have been a self-generating accident resulting in exotic biology for which no scientist is responsible for knowing.

The State of the Knowledge Conference in April of 2011 was the next opportunity for Mikovits and Coffin to spend considerable time together, and the patient and scientific communities seemed to have chosen their champion.

Mikovits felt there was no room for personality squabbles and preferred to focus on healthy scientific debate. If somebody could come up with a good argument backed up with data that didn't completely rewrite the rules of retrovirology, she was ready to listen to any refutation.

But as the "Mouse Crap Theory" of contamination continued to crumble, assisted by a German study that found that about 10 percent of immune-compromised patients in that country had XMRV sequences in nasal swabs, the argument of doubters again started to swing toward the XMRV recombination theory.

* * *

In addition to the "Mouse Crap Theory," doubters who took the tack of that joke's projecting husband came up with a second theory, which could best be described as "Holy Mouse Crap!—We Created a New Retrovirus!"

The theory was that XMRV was a lab-created retrovirus (a "recombinant virus"), the type of deformed sci-fi monster that would strike mortal panic into the general public, but added caveats: *We don't know if this recombinant retrovirus has ever infected people. If it has infected people, we think the normal human immune system would keep it from being a problem. Remain calm.* The guardians of public health will get to the bottom of all this!

From the beginning of their investigation the Mikovits Ruscetti labs had never been able to find Silverman's XMRV. The only evidence of that sequence in the paper was done in Silverman lab. It was only when she'd told her research associate Max Pfost to relax the PCR conditions did they detect sequences related to XMRV-Silverman. They had, in other words, been looking in the general library, but needed to open the secret door behind a bookcase to get to the hidden editions.

At one point during their face-off at the State of the Knowledge workshop in April of 2011, Coffin made a statement that puzzled Mikovits even years later. As she went over the various pieces of information supporting an XMRV infection, including the electron micrographs, the immune response from multiple proteins, and strain differences when sequencing larger parts of the envelope of the virus, Coffin said in a very heated voice, "I'm not saying it isn't another retrovirus, I'm just saying it's not *that* retrovirus!"[7]

Was this all just some kind of name game?

What did it matter to the millions who suffered from chronic ME/CFS, the millions with prostate cancer, and possibly the millions of children with autism and other neurological problems if there were variations in the retroviral sequences? Virus strains could have variations in the genetic sequences. From the beginning they'd been suspicious as the full-length sequence of XMRV in patients was only done in Silverman's lab.

DeRisi and Silverman hadn't named the virus based on the classification of the sequence. They called it xenotropic because it wasn't found in a mouse, meaning it had leapt species. It had been found in human prostate tissue. Nobody knew then that the sequences were not actually *in* the tissue but in the RNA made (cloned) from the tissue. It was a murine leukemia related-virus because it looked like a mouse retrovirus. DeRisi had laid it all out during his *TED talk* in 2006 when he talked about his ViroChip technology.[8]

Mikovits couldn't help but wonder if all of the shifty semantics with Silverman and DeRisi were an attempt to control intellectual property that could be worth substantial fame and money after a giant discovery.

Was Coffin interested in *un-discovering* XMRV so he could later *re-discover* it?

At one point during the roundtable discussion while Coffin was talking about contamination, Mikovits had finally had enough. She said, "These people aren't contaminated! They're infected! They make antibodies! An immune response is not a contaminant! I don't really care if you call it contamination or infection, but these people need a treatment!"

Antibodies are not cell phones picking up signals from tissues in a faraway lab. Why wasn't anyone listening to her explanation about the antibodies? Suffering patients needed help.

* * *

Mikovits and Coffin both presented their points with searing passion. In the Old Boys' network, Mikovits had to go by different rules: a woman could not be too fiery, and passion for patients was immediately viewed as overly soft (toward the "phobic, hysterical, maladaptive" patients in her case) and unscientific.

When it came time to do a wrap-up, Harvey Alter was troubled.[9] He took Mikovits into a backroom and said, "Judy, what do I do? He's so vehement that this is recombination or contamination, or something. What do you make of all this?"

In the time she'd come to know Harvey Alter, she found him to be a decent, mild-mannered man who lived for the integrity of the scientific process. Like Alter, she was convinced there had been no evidence of contamination in a single experiment. She had been fully transparent with all data since day one. Mikovits and Ruscetti did not cherry pick the data.

Mikovits had insisted on leaving conflicting data in figure 1 even when Silverman wanted it removed. Ruscetti stood by Mikovits. They had done all of the proper controls. The confounding data, which showed one sample 1118 negative for gag or env sequences but positive in another test for Env and Gag proteins remained in figure 1. She walked through the science again. If it was contamination, how could you explain the antibody response? What about the sequence variation and the proteins detected by a monoclonal antibody that recognized all family members ever tested? They'd been able to take two slightly different viruses out of the same person, separated by a period of years (thus indicating it was not a one-time lab event).

The proteins of the virus had come straight from un-manipulated tissue and the electron micrographs of the virus budding from the cells were difficult to dismiss. How could that have been contamination?

Eventually, Alter sighed and said, "Well, I've got the unpleasant task of summing this up." She shrugged. As the sports expression went, she'd left it all on the field. Alter had the ball now.

* * *

Alter gave an excellent summation of the two competing theories. Mikovits knew Alter had been momentarily troubled but his anxiety had melted into pure grace. He was well-informed, funny, and humble in the way a scientist should be in the face of things he does not fully understand.

Alter began, "When two women claim to be the mother of the same child, King Solomon made a judgment to cut the baby in half and resolve the dilemma, and it was resolved when the real mother gave up her share to save the baby's life. Well, I'm no Solomon, but I'm stuck here between the diametrically opposed viewpoints of Judy Mikovits and John Coffin, both of whose science I greatly respect, and I don't truthfully know who to cut or what to cut, except that I'd like to cut out of here.[10]" The audience laughed at Alter's endearing discomfort. They had their own jitters.

Were they looking at a new pandemic like AIDS, or had this all been some terrible misunderstanding or misinterpretation of data? Alter said he'd try to review the two presentations and see where they led. He talked about how gag and env sequences for XMRV and closely related viruses in dramatically higher numbers in ME/CFS patients than in controls and had also detected antibodies in the majority of the patients and there was no evidence of contamination.[11]

Alter was summarizing beautifully. The *gag* and *env* sequences were translated into viral proteins unique to this pathogen and she was finding sequences of greater diversity, which was an indirect argument against contamination that scientists should have understood.

Contaminants usually don't show wide genetic diversity, but are more "one note." Evolution generally proceeds at a predictable rate and thus the finding of sequence variation was another indicator that this virus had been present in humans for some time.

The finding of antibodies in the patients with these *gag* and *env* sequences was key since antibodies are only generated in a human body as the immune system responds to a perceived invader. Antibodies could not be generated in a test tube.

Mikovits had been able to detect XMRV sequences and proteins in samples collected at different times from the same patient, thus lowering the possibility of a mistaken result. Her team had also cultured virus from blood cells and unmanipulated plasma, and tested again and again for contamination and found no evidence of it. They had tried to make it fail-safe.

Alter continued, now discussing the Coffin data.

What Alter was describing at this point was the beginning of the shift away from the "Mouse Crap Theory" of XMRV to the "Holy Mouse Crap!— We Created a Retrovirus! But Don't Worry, It's Not Dangerous!" theory that Coffin would now put forth as the main objection to the work of Mikovits and Ruscetti.

Coffin was continuing with the "Mouse Crap Theory" of contamination by claiming he'd found evidence of mouse mitochondrial DNA in his assays, but was shifting the argument by claiming a recombinant mouse-human retrovirus was possibly infecting 22RVi, a prostate cancer cell line. This cell-line was widely used and it was reasonable to suspect this was the reason for the results reported in the positive studies, although Alter mentioned it had not been in the Mikovits lab. Alter continued with Coffin's argument that XMRV had been created in a lab:

"He further shows that XMRV might be a hybrid, a recombinant of two pre-XMRV viruses, one of them, which is replication competent, the other of which is not, and if you put these two together it almost perfectly matches XMRV.[12]

A "pathogenic exogenous" virus is one which can infect the organism and spread to other organisms of the same species and cause disease. But eventually over time as Darwin's ruthless "survival of the fittest" works its way through a species, those who survive at the highest rates are the people who have rendered the virus harmless. They are many ways to do this.

The immune system can learn how to destroy the virus. Some can vary the receptor that a specific virus uses. As they'd discovered with the "elite controllers" in HIV/AIDS, some lacked a receptor, which would allow the HIV retrovirus to gain access to the cells and cause disease.

The most common defense (thousands of retroviruses have been rendered nonpathogenic this way) is to silence the virus and eventually cause enough changes that it is no longer a competent virus A virus, which had become "endogenized" meant it was present in the host genome and could be passed to offspring but had lost the ability to harm that organism.

Specifically, Coffin was suggesting that the passage of prostate cancer tissue through a certain species of laboratory mice, (the "nude mice") had generated a new hybrid replication-competent (able to reproduce) retrovirus that matched very well with the XMRV sequences reported by Silverman. Alter continued with his summation of Coffin's argument:

"Thus, Coffin claimed that XMRV evolved when XMRV-negative prostate cancer xenografts were passaged in these nude mice and then became infected over time.

That XMRV is a recombinant of two pre-XMRV strains, only one of which is replication competent, that XMRV was created in a lab long after ME/CFS was recognized as a clinical entity, and that XMRV is a contaminant that does not warrant further study. This was a second kind of contamination."[13]

These relatively brief restatements by Alter described the central battlefield where Mikovits and Coffin would fight it out over the next six months before their confrontation at the Ottawa meeting in September 2011.

In spite of the fury of their arguments, they did agree on important components.

Mikovits had no difficulty believing that the XMRV sequences as found by Silverman in his prostate cancer sample tissues had been generated by the passage of prostate cancer tissue through nude mice as described by Coffin.

Coffin was alleging that there was a Frankenstein-like recombinant retrovirus related to mouse xenotropic virus that had been created in a lab and readily infected human cells, but that scientists shouldn't worry about this freak monster.

The unanswered question to Mikovits was that if one recombinant retrovirus had been created through the passage of human tissue through mice tissue, how many others had been created in the last fifty years of human /mouse xenotransplantation?

Could something very similar have been created by an earlier passage of human tissue through mice in another lab?

Why couldn't that question be asked?

Why was Coffin arguing that XMRV was a contaminant that didn't need to be studied any further? Wasn't that an abandonment of scientific curiosity, a patronizing parent telling his scientific "children" there was no monster under the bed?

Alter continued with his summation:

> "So, where do we go with these two very divergent pieces of data? I do want to give you this cautionary note, the fact that contamination can occur, and nobody doubts that, does not mean that it has occurred in any given laboratory. There is as yet no direct evidence for contamination in either the Mikovits lab or the Lo laboratory.
>
> So, what are we going to do? I think, rather that cutting the baby in half, the resolution may come—first of all we have to resolve this issue, is it contaminant or is it real, and I think that it will be addressed by two coded panels that are currently in preparation."[14]

The answer was that there would be two investigations, one by the National Heart, Lung and Blood Insitute. (this would be known as the Blood Working Group). and a second investigation funded by the National Institute for Allergy and Infectious Diseases (headed by Anthony Fauci), which would come to be known as the Lipkin multi-center study.[15]

Many in the ME/CFS community were skeptical about Fauci, an opinion described by Cort Johnson's *Phoenix Rising* website, when he wrote that Fauci:

"has been no friend to ME/CFS—in fact, he's been considered a major problem for years. He appointed Stephen Strauss to head the CFS team in the 1980s and moved for ME/CFS to get kicked out of the Institute ten years ago, an event which lead to years of research declines as the program slowly disintegrated in the backwater that is the little Office for Research on Women's Health.[16]

Mikovits hoped Fauci wouldn't have too much influence over the investigation. The selection of patients who'd never been tested for XMRV concerned her. If there was contamination it would have taken place in the lab as the samples were being processed.

To do a replication study, the logical step was to go back to those patients who had previously tested positive. There was a possibility their samples could have been contaminated, but the patients would not have been contaminated by a needle stick, so newly-drawn samples processed in a different lab would not show contamination. Once positive by this redundant method, they were clearly infected. To Mikovits it seemed there was an awful lot of effort going into reinventing the wheel. The very process itself, called into question the group's intention. In the months and years to come she'd wonder if their intention hadn't been to actually invent a square and proclaim a wheel had never existed.

Alter finished up his summation:

> "So I think that's where we stand. These panels are going to come late this spring, early summer. They'll take quite a while to do, because they are so complex, but hopefully we'll have some answers this year."

The audience members applauded warmly. Alter ended by noting that it was late on a Friday afternoon, but he wanted to give Mikovits the opportunity to make concluding remarks. She did have a few things she wanted to remind the assembled group.

> " . . . We took the systems biology approach, collected samples from well-defined patients who had the infectious characteristics that Tony Komaroff talked about, that were in the Lo study, that Dan Peterson had done, these patients had data on them for decades.
>
> We took samples over three years. So it wasn't that we found the sample in every sample indicated—like the macaque study where it quickly went into the tissue, like the mouse study that came out this week that said that the antibody responses were weak and transient. We don't know everything about this virus, but HIV does not cause AIDS. The CDC definition is HIV and one of 25 co-pathogens, so the Lyme, the EBV, the enteroviruses . . .

> We've developed a cytokine signature that is distinct from Nancy's (Klimas) cytokine signature and from Ben Nadleson so this is a marker to follow on clinical trial improvement."[17]

If one used HIV as a template for the damage this retrovirus caused it fit like a glove. The CDC's own definition of AIDS was infection with the HIV retrovirus and one of twenty-five other co-pathogens, including Lyme, Epstein-Barr virus, and a host of others.

It made sense then that antivirals, antibiotics, or similar medications, which went after some of these co-pathogens, would have the effect of lessening the severity of the disease (just as they did with AIDS). They were developing inflammatory cytokine signatures, which would allow them to figure out which medications might be most effective for each individual patient.

In the view of Mikovits the future beckoned with hope for treatment of these patients if the researchers would but open their eyes, remember their training, and see the evidence in front of them. She continued:

> Whatever their disease is, they are infected and sick, and I know John Chia has patients that are coinfected [with enteroviruses] and they don't treat the same way. So we can get together with the physicians who have coinfected patients, even Lyme doctors who we're working with across the country, and start doing something now, take it out of CFS. It's not about CFS, it's about a retrovirus we don't understand very well. As Frank Ruscetti said in a meeting a month or so ago, if this were HIV, it would be 1983. That's all.[18]

Mikovits's urgency made perfect sense, as did her statement that the apparent retrovirus plus coinfections they were seeing put them in a time capsule, trapped like AIDS patients in 1983, when the scientific grasp of the illness was initially like the blind man and the elephant parable, with researchers only seeing a tail but not an ear (or one coinfection but not another): patients had been suffering for decades, and viable, safe, and effective treatments might be already approved.

But even though AIDS had thrust the term "fast tracking" into the scientific lexicon to refer to the urgency of getting medications to needed patients, and even though the FDA soon declared ME/CFS to be a "serious or life-threatening" condition, dealing with affected ME/CFS patients was hugely controversial in retroviral research.

Some physicians would argue that antiretrovirals, even though they'd been prescribed for the better part of two decades and ME/CFS still lacked FDA-approved drugs of its own, presented unacceptable risks, and should not be used until more definitive information was available. This was the position Ian Lipkin had taken.

But she had seen how devastating the condition could be, and Mikovits also thought it was perfectly reasonable for people to attempt to use antiretrovirals under an astute doctor's care. The same drugs were given to pregnant women with HIV and their children to insure the mothers didn't give their kids AIDS.

She felt a trial of antiretrovirals—drugs that had received rigorous trials and funding, with proven use around the globe—was perfectly reasonable for patients with a "serious and life-threatening" condition. Mikovits told the assembled group that even if they didn't think anti-retrovirals were medically sound, they could consider interventions like immune-modulators with a better safety profile. Mikovits thought the research should be accelerated.

Mikovits felt that Coffin seemed intent on burying it.

* * *

The unpublished research that John Coffin had referenced at the May 2011 State of the Knowledge Conference at the NIH finally came out in *Science* on June 1, 2011. The title was "Recombinant Origins of the Retrovirus XMRV"[19] and while Mikovits may have questioned the research, especially since they were focusing so much on Silverman's VP62 clone of XMRV, his conclusions particularly alarmed her.

> "The retrovirus XMRV (xenotropic murine leukemia virus–related virus) has been detected in human prostate tumors and in blood samples from patients with CFS, but these findings have not been replicated. We hypothesized that an understanding of when and how XMRV first arose might help explain the discrepant results . . .
>
>   . . . We conclude that XMRV was not present in the original CWR22 tumor but was generated by recombination of two proviruses during tumor passaging in mice."[20]

Coffin was suggesting XMRV might have been created by the passage of human prostate cancer tissue through mice. They were able to find XMRV in the prostate cancer cell lines and it was their belief that XMRV was created by the "recombination of two proviruses during tumor passaging in mice." The authors emphasized that the probably of this exact recombination was so low that it would never occur again.

Mikovits couldn't help but think that her colleagues had all lost their collective mind. Didn't they see the foretold danger in their words? Did this research actually make them feel safe containment had taken place? For Mikovits this could be a possible catastrophe.

Coffin's work, at face value, showed that a replication-competent retrovirus had been inadvertently created by passing human tissue through mice. A month later a publication from Adi Gazdar's team showed that murine leukemia viruses from many different laboratory lines could become airborne, infecting human cells grown in the same lab. No retrovirus had ever demonstrated the ability to efficiently spread through the air, so this was alarming. The virus didn't generate much of an immune response and disappeared quickly from the blood and moved into tissues.

With all of this information, Coffin was still suggesting this was a novel, singular event, something that—while unprecedented in scientific history—deserved absolutely no further investigation.

Mikovits believed Coffin was effectively claiming that this event was something akin to an "Immaculate Recombination," an event so unprecedented and rare in history that it would be a waste of future research dollars. But Alter and Lo's confirmatory study had found something very close to what Silverman had reported as XMRV, not Silverman's sequences. Mikovits's conclusions from the data were the opposite of Coffin's.

A potentially airborne, recombinant retrovirus had to be studied.

It was unlikely to be "immaculate" or "one-time only." More likely was that it had happened before, maybe not XMRV but closely related viruses. Scientists had been passaging human tissue through mice for more than seventy years. Was this truly the first time retroviruses had recombined? What said that recombination events had to lead to the exact same replication-competent virus? Scientists lived in fear that the next pandemic would occur when a virus might jump from one species to another among animals living on a farm.

For decades, they might have been accelerating that process by passing tissue from humans directly through animals. Had they fully understood the ramifications of what they were doing?

This was why, during the Ottawa Conference of August 2011, Mikovits asked the important question "How many have we created, John? How many XMRVs are out there?"[21] But it didn't seem anyone was interested in the question.

CHAPTER EIGHTEEN

Whittemore Justice

Whittemore's former partners at Wingfield Nevada Group—Tom Seeno and Albert Seeno Jr.—have sued Whittemore, saying he used his smooth-talking, sleight-of-hand tactics to steal millions from the Wingfield company. Whittemore shot back with his own lawsuit claiming the Seenos threatened to kill him and his family if he didn't follow their orders and sign over his assets.[1]

—Martha Bellisle, "Wit and Work made Lobbyist, Harvey Whittemore, 'An Institution.'" *Reno Gazette Journal*, February 12, 2012

A federal grand jury indicted former Nevada power broker Harvey Whittemore today on charges of funneling $138,000 to an elected official's campaign through family members and employees and then lying to the FBI about it, the US Attorney's office said.[2]

—David McGrath Schwartz, "Federal Grand Jury Indicts Former Nevada Power Broker Harvey Whittemore," *Las Vegas Sun*, June 6, 2012

Dr. Mikovits was at the Physicians Round Table in Tampa, Florida in late January 2012 when she heard about the forty-four million dollar Seeno lawsuit against the Whittemores.[3] Since the suit was public record, she hurried to download it and poured over the forty-two page complaint that named Harvey Whittemore, Annette Whittemore, and the Lakeshore House Limited Partnership, which took its name from their historic home in lush Glenbrook.[4]

The mystery of the previous year when she had seen such a change in the Whittemores came into stark relief. She thought of the time in late March 2011, when she had begun to see an unsavory side of the Whittemores, but this was a whole new scale. She thought she had been in a world of wealth and privilege with sincere dedication to unraveling neuro-immune diseases. The accusations in the lawsuit painted a picture of a dark and dangerous world:

> This is a case involving the misappropriation, breach of fiduciary duties and embezzlement of tens of millions of dollars. Acting as manager of Wingfield, Whittemore has admitted and confessed to engaging in over 20 different financial transactions designed to deplete Wingfield of its assets for the sole purpose of enhancing and promoting Whittemore's financial condition and to further his standing in the political community of Nevada.[5]

The Seenos alleged ten causes of action against the Whittemores, including breach of fiduciary duties, breach of contract, civil conspiracy, and unjust enrichment. In early 2004, the Whittemores and Seenos forged a business partnership and the Whittemores sold fifty percent of their interest in various companies to the Seenos, including Argus Media, Wild West Sound Company, Redlabs USA (since 2007 DBA VIP Dx), Dr. Pepper/7-Up Bottling Company of the West, and two companies associated with the massive Coyote Springs development-in-planning.

All of these entities became part of the newly formed Wingfield Nevada Group Holding Company that was created on or around January 1, 2005. Harvey Whittemore was a manager of the Holding Company and, according to the Seeno lawsuit, "From February of 2007 through 2009, Whitemore exercised complete control over the financial books and records of Wingfield."[6] Around September of 2010 the Seenos began to notice financial discrepancies:

> In or around September 2010, the Seenos confronted Whittemore with their suspicions. However, they quickly learned their suspicions, while well-

founded, had not even scratched the surface of Whittemore's fraud, deception, and malfeasance. Confronted with some of the evidence uncovered regarding his potential fraudulent conduct, breach of fiduciary duties, and misdeeds, Whittemore confessed, admitted, and disclosed a multitude of acts that revealed years of theft, conversion, asset misappropriation, and breach of fiduciary duties to Wingfield.[7]

According to the lawsuit, Harvey Whittemore himself had prepared a written confession of his misdeeds on September 16, 2010, and gave it to the Seeno representatives.[8] Mikovits could not get over the irony of the situation. Harvey had tried to get her to sign a confession (in the form of an apology) in order to get her out of jail, and wrap up the civil case. She had refused. Now the Seenos were claiming he had written his own confession regarding the theft of millions. As Mikovits read about the various monies allegedly embezzled by Harvey from Wingfield, the story got more incredible. To her surprise, she stumbled upon her own name in the Seeno lawsuit:

> For example, Defendant Annette Whittemore, who lacked management authority under Wingfield, mandated to the Wingfield accounting staff that 75% of the salary of Dr. Judy Mikovitz (sp.), a research scientist and employee of WPI, would be charged to and paid by Wingfield.[9]

She had no idea she was indirectly salaried by the Wingfield company. In 2006 she had gone with Annette Whittemore to appear before the Nevada state legislature to raise funds for the new institute. They were granted six hundred thousand dollars for each of the first two years and four hundred thousand dollars for the two years after that, for a total of two million dollars in funds that included salary for Mikovits and the start-up staff. Her understanding was that her salary came from these funds and the Whittemore family foundation. But if the Whittemores had not used the state money to pay for Institute staff, and instead took salaries out of Wingfield, where had it gone?

She called her attorneys, thinking this revelation meant things could turn in her favor. Her hearing on damages was set for a few days away. Maybe it would now never happen. The Seeno lawsuit was seemed to be unmasking Whittemore lies, but even in the volley of lawsuits her attorneys were still dismissive. "These charges have nothing to do with your case," they told her.

* * *

A few days later, the Whittemores filed a countersuit against the Seenos, claiming that the Seenos owned them sixty million dollars.[10] This amount of damages claimed by the Whittemores would later balloon to $1.8 billion dollars.[11]

Mikovits delved into reading Whittemore's litigation as well, and it clarified some of the inexplicable moments she had observed in the last year of her work for the Whittemores. She recalled in April of 2011 when Annette Whittemore had missed a meeting by a couple of hours. Annette typically appeared at the WPI in crisp business suits, her hair nicely-styled, and her makeup perfect, but this time she seemed shaken, was plain-faced without even mascara, and pulled Mikovits aside.[12]

Distraught, Annette told Mikovits that she had been at the house when an armored truck pulled up and she was forced by a menacing thug to empty their safe of fifty thousand dollars and expensive jewelry.

Harvey paced the halls near the café, talking on two cell-phones, as if frantically trying to solve a serious problem. The story seemed suspicious to Mikovits and made her wonder if something more sinister was happening. When the Whittemores filed their sixty-six million dollar countersuit against the Seenos on February 1, 2012, Mikovits read an account of this event:

> On one occasion, a large, very ominous and burly man named "Ray" demanded that Mr. Whittemore open a safe in the house. They assured Mr. Whittemore that they only wanted to see what was in it and that they would take nothing from the safe. Mr. Whittemore got the combination from Mrs. Whittemore, as it was her safe, and opened it for inspection.
>
> The large, burly man then dumped everything in the safe into a bag and took it with him. Taken from the safe was a portion of Mrs. Whittemore's jewelry and cash, leaving behind only Mrs. Whittemore's passport. They also took Mr. Whittemore's watches and jewelry. The Seenos demanded, and the Whittemores transferred, most all of their family automobiles and most other assets that they had to the Seenos, except for certain retirement accounts and insurance policies.[13]

Mikovits recalled another peculiar incident with Harvey.[14] Her condominium at the River Walk had suffered mold damage and at the Whittemore's urging she had moved to a condo they owned at the Palladio.

As the situation with the Seenos deteriorated the Whittemores had moved out of their houses at Lake Shore and Boulder Glen, eventually settling in another condominium at the Palladio, a few floors above Mikovits. They had become neighbors.

In September 2011, the month in which she would later be fired, Judy saw Harvey pull up in a ten-year-old SUV and park on the street in front of the Palladio. Mikovits was dressed in her workout clothes returning from the YMCA across the street as Harvey stood next to the vehicle, glancing from right to left in a seemingly paranoid way to see if anyone was watching. The SUV was filled with Whittemore possessions as they were apparently finishing up their move into the Palladio. Mikovits approached him and said, pleasantly, "Hi, Harvey!"

She startled him, but when he saw her he seemed to instinctively relax, as if he had been expecting someone or something else.

Mikovits knew about some vague financial hot water, as Annette had at one point muttered something about how they had to "give everything back," Harvey seemed humbled right then, even deflated. "They took our cars," he said, inhaling the crisp, pre-autumn air. "They took our cars because we're bankrupt."

Mikovits could only imagine how difficult the confession was for Harvey, who valued his ability to make a showman's entrance and impress everybody. Harvey's eyes darted up and down the street, as if he hoped nobody else would discover his secret shame.

"Well, at least you have something to drive," said Mikovits brightly. "You don't need *stuff*." Mikovits had not known what to say as the world of the uber-rich baffled her. Were the Whittemores actually destitute? If so, how could the WPI still function? What about their properties? Surely their vast fortune couldn't just evaporate? But of course it didn't—she never saw the old SUV again and, their Cadillac SUV and Range Rover quietly reappeared.

This just convinced Mikovits that they were on their way to solving their problems, and Harvey's woes had only been a minor blip.

* * *

As described by the Whittemore lawsuit, the anger of the Seeno family exploded into volcanic rage:

> Albert Seeno, Jr. threatened to go to the FBI with information that he alleged he had regarding these alleged improprieties; that he would personally bring down every member of the political "machine" in Nevada including references to US Senators; that in the course of this threat asked whether Mr.

> Whittemore believed in God; whether Mr. Whittemore went to church; that
> Mr. Whittemore should gather his flock on Sunday and pray.
>
> Albert Seeno, Jr. threatened that when he was through with Mr. Whitt-
> emore he would be disbarred and behind bars; that they would spend Mr.
> Whittemore into the ground and there was no way he had the finances to
> defend himself; that he would bring down any business or friend associated
> with Mr. Whittemore; and that at Mr. Whittemore's funeral there would not
> be one person in attendance.[15]

Mikovits could not help but notice that the actions supposedly taken by the
Seenos mimicked those Harvey and Annette had done to her.

She couldn't help but wonder if her September 2011 public disavowal of
the XMRV test, which had potentially been worth millions, had thrown the
Whittemores into a panic. It might have been the only significant asset they
had to appease the Seenos, since the real estate market was in such trouble
because of the recession.

All of these allegations by the Whittemores painted a significantly
different picture than the one soon presented in the pages of the *Reno
Gazette Journal* later that month. Recounting the history of the Seeno family
business, the article of February 25, 2011, reported how the grandfather of
Albert Seeno, Jr., Gaetano Seeno had helped rebuild San Francisco after the
1906 earthquake and had started Seeno Homes in 1938, and how a friendship
with Reno businessman Bill Paganetti led them to open the Peppermill
Hotel Casino in the late 1970s. Of the Seenos, Paganetti said:

> Paganetti called Whittemore's allegations "beyond fiction." He said he had
> been in business with the Seenos for 33 years and has never had an argument
> or cross word with either brother. "These people are the salt of the Earth.
> These people are good businessmen. I could live 100 lifetimes and not have
> another partnership like this."[16]

While the Seeno image was fluffed a bit in the article, it still pretty much
matched their track record. In the course of running various business entities with a value estimated at more than four billion dollars, the list of alleged violations by the Seenos comprised less than a handful.

In the *Reno Gazzette Journal,* Albert Seeno Jr. pointed out that as a
casino owner he was continually under scrutiny of state regulators and that
Nevada gaming laws required he avoid all activities which discredit the state:

> Gaming officials have access to and can take control of all records,
> including phone and travel at any time, he said. Seeno said he has complied

and cooperated with all regulators and regulations. "I'm not aware of any disciplinary actions involving the Seenos," said Mark Lipparelli, chairman of the Gaming Control Board.[17]

The Seenos'. picture of Harvey Whittemore was, on the other hand, consistent with what Mikovits had observed since her discovery of the problems with the Lombardi and XMRV serology in the blood working group in the months before she was fired.

The Seeno lawsuit painted Harvey Whittemore as a devious, manipulative man-of-cutthroat-ambition who would stop at little to keep his position. His connections also meant he could pull some pretty large strings to defend his high status, even when his connections didn't know they were being played.

* * *

Mikovits believed that the Whittemores had set things up so she would be in a one-down position. The joining of the criminal and civil cases created a situation in which she could not defend against either case.

Invoking her Fifth Amendment rights in the civil case was a strategy proposed by her lawyer, Scott Freeman, as a result of the two cases being so closely intertwined. Even if the judge eventually found that she could not invoke her Fifth Amendment right against self-incrimination, it was beyond belief that he had summarily taken away her right to a jury trial and entered a default judgment, something Judge Adams had never done in twenty-two years on the bench.

A common cliché about the legal system is that the wheels of justice moved slowly, but in her case it seemed like a runaway train. Less than six weeks had passed between the filing of the civil suit on November 7, 2011, to the December 19 hearing where Judge Adams had entered a default judgment against her.

In retrospect, Mikovits believed the December 19, 2011, hearing before Judge Adams was probably the high-water mark of the Whittemore's campaign against her. The lawsuit by the Seenos suggested a pattern of dishonest dealings by the Whittemores, and more fallout seemed to be on the way.

Mikovits could never have imagined that the next act in this drama would involve the most powerful man in Congress, the majority leader of the United States Senate, Senator Harry Reid of Nevada.

* * *

On February 8, 2012, roughly two dozen FBI agents served subpoenas on business associates and friends of the Whittemores in about thirty locations in southern and northern Nevada.[18] The *Las Vegas Review Journal* ran a story a few days later about the chain of events leading up to the FBI action.

> Contributions made on one date—March 31, 2007—to the re-election campaign of Sen. Harry Reid, the Democratic majority leader, have attracted the interest of FBI agents, sources said.
>
> On that day, Reid's campaign received $115,000 from about two dozen Whittemore employees and their family members, most of whom each contributed the maximum allowed $4,600, according to federal campaign reports.[19]

As the *Las Vegas Review Journal* noted in an article published on February 20, 2012, regarding the gathering storm that centered on Harvey:

> "The big splash FBI agents made February 8 when they simultaneously served subpoenas across the state in their campaign contribution investigation of power broker Harvey Whittemore was designed to interview witnesses before they had a chance to get their stories straight."[20]

When Mikovits first read about the FBI actions she recoiled in horror. She realized she had inadvertently been part of such a conduit scheme designed to funnel money to Harry Reid. Only it had not been in 2007 when Harry Reid was raising money for other candidates, but in February of 2010 when Reid himself was up for re-election and facing a formidable challenger.[21]

A Harry Reid fundraising event took place at the Whittemore home.

It was a keystone soiree since it was slated to be Harry Reid's last election and he faced stiff competition from his Republican opponent, Sharron Angle, who had Republican Party support and was also a Tea Party favorite. Like all Whittemore events, this one was catered to the hilt.

After getting a drink, Mikovits mingled and greeted Harry Reid, wishing him well in the re-election. Harvey came up a few minutes later and said he needed to talk to Mikovits. In the front of the house, Harvey and Annette shared an office, with two grand mahogany desks facing each other holding computers. Harvey gathered a few other people together along with Mikovits and shepherded them into his office, one at a time, where Annette joined them.

Harvey rhapsodized about the importance of the WPI, emphasizing how it wouldn't exist without Harry Reid's help, and impressing the point that everyone should contribute to his re-election. There was no doubt, he said, that Reid's Republican opponent would feel very differently about the WPI. Mikovits said she would be happy to contribute, but did not have that kind of money. She noted she hadn't been paid her bonus in a couple of years. If she got her bonus, she could probably make the maximum donation of $9,400 with her husband but David complained. They needed the money for their own families but Harvey made it clear, if Reid was not re-elected she would have no job.

She thought it had all been legal. But it was not.

The *Las Vegas Review Journal* article captured the attributes of Harvey and the case:

> In 2007, Whittemore, who considered US Sen. Harry Reid, D-Nev., among his closest friends, was orchestrating the development of Coyote Springs, a master-planned community in southern Nevada. With the help of Reid and other members of Nevada's congressional delegation, Whittemore sought to overcome several governmental hurdles because of county water issues and federal land issues.[22]

To Mikovits, it seemed as if the FBI believed it had assembled most of the puzzle. She wished they could see the even-greater picture, including what had happened at the WPI.

The Seeno case and the FBI investigation were not isolated events to Mikovits but seemed typical of how Harvey Whittemore behaved when squeezed in a corner. The scale of what Harvey had attempted was breathtaking.

But there it all was in black and white in the pages of the *Las Vegas Review*. The article noted that the contributions made on March 31, 2007, to the Reid campaign had attracted the attention of FBI agents.

> On that day, the Senate majority leader's campaign received at least $133,400 from 29 Whittemore associates, including his family members and his employees and their spouses, most of whom contributed the maximum allowed $4,600, according to federal campaign reports.[23]

The article reported that the number had gone up since the initial reports, and there were those who saw this as one of the largest campaign contribution cases in recent history.

> Gross, former head of enforcement for the Federal Elections Commission in Washington, said he had never seen that amount of money bundled all at

once in a federal race. "I can't recall another situation where that much came in on a (single) day," he said.[24]

Mikovits told Freeman and Riggs she wanted to go the FBI and tell them everything she knew about Harvey's dealings.

Her criminal attorneys told her she was crazy to consider this as she was still under indictment, released to their recognizance, and if she insisted on going forward she could find another attorney. Their opinion was that if the FBI didn't come looking for her, she shouldn't go looking for them. Mikovits felt much differently. She felt a moral obligation to tell the truth, regardless of the cost.

It wasn't until a few months later when she talked to her bankruptcy attorney David Follin, whom she then learned was also an experienced criminal attorney, that she realized she had a legal professional willing to back her and she could wager the risk of her speaking to the FBI.

In late July of 2012, Follin arranged a conversation for her with the FBI. The FBI had a program with the unofficial title "Queen for a Day." A potential witness could come in and talk about a pending investigation and the FBI promised it would not use any of the information gathered in the meeting against the informant. Mikovits had to pay her attorney more than a thousand dollars for the three hours the FBI agent spent in the attorney's office talking to the informant. She detailed everything to the FBI agent, who in turn, expressed concern that her testimony might be too complicated.

"What's complicated?" she responded, incredulously. "He did the same thing in 2010 that he did in 2007." In November of 2012 she received word that the FBI did not believe it would need her testimony, although that might change. She had wanted the opportunity to stand up in a court of law and tell the truth about Harvey, but there was nothing else she could do.

She may have made a mistake by giving money to Reid in 2010 at Harvey's urging, but she wasn't the only one, and she had come clean about it to the authorities.

* * *

Just when it seemed that things couldn't get any stranger, they did.

In March of 2012, while talking with Jones in the midst of an unrelated hearing, Judge Adams disclosed he had received a total of $10,400 in campaign contributions from Harvey Whittemore, his family, and his company.[25] This was the same judge who had issued the default judgment against Mikovits in the civil case, for which the WPI was looking for damages from Mikovits of fifteen million dollars.

If the civil action against Mikovits had taken place in California where Mikovits was living with her husband, the judge would have had to consult the California Code of Judicial Ethics, which runs fifty-three pages.[26]

By comparison, the Nevada Code of Judicial Conduct is six pages long.[27]

The rule in California is that a judge must disclose to the other side any contribution made to his campaign of more than $100 by a party or lawyer who appears before him.[28] California law also holds that a judge must disqualify himself if one of the parties before him has given more than $1,500 to his campaign.[29]

Nevada has no rules on the amount of money a judge may receive before he is required to disclose that information to the other side, or disqualify himself from the case. While judges in Nevada are required to "avoid both impropriety and the appearance of impropriety,"[30] the judge himself makes a call about these ethical gray zones.

In essence, the judge gets to judge his own behavior.

If Judge Adams had been on the bench in California and had failed to reveal the $10,400 in contributions, he would have been subject to an investigation for violation of both California law and the California Code of Judicial Ethics.

Punishment could have ranged from a fine to being stripped of his judicial robes.

* * *

With the Whittemores now under swarming legal and media scrutiny, Mikovits noticed that her own attorneys were treating her more favorably.[31]

It was suddenly as if they fully believed her and were listening with sharper ears. She had been puzzled by the sometimes rote behavior of her lawyers. She knew she had been under stress (as anyone would be), in disbelief at the unfolding events, and on most occasions in their presence she had been angry and distraught. But all of this seemed a reasonable reaction to something so overwhelming and traumatizing.

In fact, she thought she behaved a good deal like the cancer patients she dealt with in support groups who were facing the very real possibility they might die. She had just never taken it personally when they became angry and upset. She had also never yelled at any of the patients. She felt her criminal attorneys seemed to yell at her quite frequently when they should have been levelheaded. On March 7, 2012, a settlement offer was made by the Whittemore's attorney, Ann Hall, and conveyed by email. The offer listed several things which had been discussed before: return of the notebooks, an apology from Mikovits, paying more than $450,000 in legal fees, the two sides would refrain from publicly criticizing each other, and the WPI would tell the district attorney that they considered the matter closed.[32]

For Mikovits, this offer bent the facts and was as ludicrous as both the civil and the criminal cases, and she conveyed her feelings clearly to her attorneys. On March 8, 2012, she wrote back to her attorneys:

> All original notebooks are in custody. I did absolutely NOTHING wrong and will NOT apologize for being wrongfully terminated in an attempt to prevent Whittemore crimes and will not pay a penny for their frivolous lawsuit.[33]

Her professional and financial life was at stake. An apology was an admission she had committed wrongdoing, and she strongly believed she had not done so. If she gave a scripted, bogus apology she doubted she would ever work in science again. More importantly, how could she face the patients if she sold them out to protect herself?

The truth was the truth.

For his part, Dennis Neil Jones was trying to settle the matter in a way which did not devastate her financially, although he appreciated $450,00 was financial devastation to someone who rarely had more than a few thousand dollars in any bank account. On March 9, 2012, he wrote to Mikovits:

> I understand how frustrating the civil and the criminal cases are for you, but please don't lash out at me. Im trying to help you. I am not the judge. I do not make the legal rulings in the case . . . I am also obligated to provide my analysis regarding any settlement offer, so that you can make an informed decision. You say that you will never apologize and you will never pay WPI anything. I get that. I just want to make absolutely sure that you understand the probable implications of your position. The judge doesn't care about the new civil case against the Whittemores. He doesn't care whether politicians are refunding the Whittemores' political contributions. Those new developments will not cause the judge to throw the case out of court.[34]

No wonder everybody was screaming at each other. It was like a medical and criminal action thriller on the big screen, too unreal to comprehend.

But with the news media reports about Harvey, at least one of her attorneys seemed to understand that he needed to broaden his grasp of what was happening. Her criminal attorney, Scott Freeman, told Mikovits she needed to fly to Reno and he would have a "Dr. Judy Mikovits Day." She could talk to him as much as she needed until she had nothing else left to say. *He* had copies of all the notebooks and other documents in his office.

So on March 12, 2012, Mikovits flew to Reno to spend the day with Scott Freeman and Tammy Riggs to prove her innocence.

The first item they tackled was the notebooks, which Mikovits had not seen since she had been fired on September 29, 2011. In their damages claim, the Whittemores were asking for the salary they had paid Mikovits for the previous five years and the cost of the entire research program: including Peterson's salary, WPI parties and travel costs paid by patient organizations.

Using her calendar and the copies of the notebooks, she had been able to correlate every day she was at the lab with the work she had written down in a notebook. The notebooks did not leave her lab/office ever, so whenever she had taken a trip she had recorded the information separately, then transcribed it when she returned.

In Katy Hagen's notebook, she was able to go to the page marked March 23, 2009, and find where Lombardi had directed Hagen to culture approximately thirty patient samples together with the LNCaP cells producing XMRV in Mikovits's research lab rather than the clinical lab, curiously while Mikovits had been out of town. As she explained to Freeman and Riggs was a direct violation of all sound scientific protocol and from what they had learned subsequently about the ability of XMRV to spread through the air, it was a serious problem, and the reason Mikovits had been so methodical about dividing up the lab work to prevent contamination. The entire staff had been told this and Hagen had taped the instructions in her notebook. Ruscetti and Mikovits were furious at this breach of their instructions.

While Mikovits believed that Silverman's VP-62 clone and 22RVI prostate cancer line had never been in her labs, which meant that she had cultured an XMRV-like retrovirus from at least one patient, it did raise the issue of contamination in the research lab.

Lombardi had denied that he had ever cultured an XMRV sample in the clinical lab or research lab and Mikovits believed him until she found the entry in Hagen's notebook on July 8th, 2011, as things started to deteriorate. *Technically,* he had told the truth. He had in fact directed the lab technician Katy Hagen to do the culture.

When Mikovits had uncovered this fact and the next day disclosed it to Harvey and Annette Whittemore. Annette blamed it on Mikovits, saying in an email that Mikovits "should have known what her employees were doing."[35]

A scientist needed to have absolute trust in the integrity of her collaborators or else the entire structure fell apart. On March 12, 2012, Mikovits noted that the only missing item from her desk drawers was the loose four pages Lombardi had copied from his notebook when she had confronted him over Silverman's assertion that Lombardi had sent him samples in March 2009.

However, Lombardi was never authorized to send samples to anyone.

What the hell was going on?

She had demanded he document to her what he had been doing since he had been transferred to the clinical lab after the debacle at the Cleveland Clinic in November 2009, a time period of almost two years. So then he had produced for her just four xeroxed pages documenting that he had indeed done no work on the NIH grants while still requesting salary support. She had put those pages in Hagen's notebook marking the page and told both Carli West Kinne and Harvey and Annette Whittemore that he was not to be paid for FY 2010–2011.

Now they were missing from the materials in Freeman's office.

Another issue that had confounded the attorneys, vital to the criminal case, was the alleged contract she had signed on February 22, 2009, with the WPI in which she allegedly promised that all copies of her data and intellectual property were the sole property of the WPI and would be returned upon her termination. The only glitch was she had never signed the document, nor even seen it, until a month after she was fired when it was mailed to her home.

Mikovits explained to Freeman why she believed it was a forgery. For one thing, she had started three years earlier, on November 15, 2006, and this was certainly not the email contract agreement she had worked under since that date. Even more bizarre, the date on the document had been changed to October 2008, in handwriting that was not Mikovits's.

She had signed the collaboration agreement with the National Cancer Institute and the Cleveland Clinic on January 22, 2009. That document made it clear that any intellectual property generated by the collaboration was owned equally by the three institutions. She would not then have signed a second contract the next month contradicting the January one.

The pagination on the signature page was also different than the rest of the contract and the font was different. Mikovits wondered if it had been taken from an unrelated document she *had* signed on February 22, 2009, that was not the cut-and-pasted "contract" they now had. But the signature of the new date clearly did not belong to her. She made her Os and *Ts* in a very distinctive manner that was apparently, by the forger, inimitable.

Freeman listened carefully and looked at examples of her signatures and date, along with the signature and new date on the so-called "contract" page. Freeman did so with a magnifying glass and then asked Tammy Riggs to come over and peer closer. They concurred it did not look like her handwriting.

Mikovits felt they must have added in the October 2008 date when they realized the February 2009 date was after they had generated preliminary data on XMRV and the inflammatory cytokine signature. In fact, all of the intellectual property belonging to Mikovits had been generated before February 2009. When Scott and Tammy had magnified the seeming forgery and saw it with their own eyes, Mikovits could almost feel the climate shift.

Mikovits went on to show she could account for every notebook entry and every date when she was not in the lab for the entire five and a half years!

Clearly, there were no missing notebooks!

At the end of the day, Scott Freeman was beaming. "How come you never lie?" he asked her. Mikovits couldn't get over the question. Scientists were supposed to tell the truth, just as was supposed to happen in the legal system.

Why was actual honesty such a rare thing?

* * *

The following day, March 13, 2012, Scott Freeman and Tammy Riggs met with Harvey and Annette Whittemore, Lombardi, and their attorney, Ann Hall. Mikovits was not there, but when her attorneys called later that day they sounded elated.[36]

They had reviewed the notebooks to show that Mikovits had worked diligently for five years and everything was recorded and present, so the Whittemore team's claim that she owed them her salary for that time or that original notebooks be returned was baseless. They also showed that every notebook was accounted for. They directly pointed out the page on which Lombardi directed Katy Hagen to culture XMRV in Mikovits's research lab.

Freeman and Riggs went on to say how they were able to discuss how Lombardi had not been entitled to the cell line, which they had used to justify firing Mikovits and said they felt comfortable they could prove that in a court of law. Then they discussed the supposed contract and the difficulties they thought the Whittemores would have in proving that Mikovits had actually signed it.

They described Annette Whittemore and Lombardi as "white as ghosts," and Hall still attempting to maintain there were missing notebooks.

About two weeks later Mikovits got a courtesy call from Scott Freeman. He wanted to tell her personally that he had been appointed to the bench and would now be a judge. He would commence that position immediately so could not continue representing her, but his partner would do so.

* * *

Two days later, on March 14, 2012, Mikovits was in San Francisco meeting with a bio-tech company when her civil attorney, Dennis Jones, and her husband, David, called.[37] Jones had been meeting with the Whittemores to discuss the damages in the civil suit.

Jones said there was a new offer on the table and suggested she take it.

If Mikovits signed a letter of apology, the civil suit would be settled and they wouldn't even ask for legal fees. The criminal charges would be dropped immediately. Her signature on a letter of apology and promise never to speak of these events again would end everything. David, battle-weary and wanting to move forward, suggested she sign it as well.

Mikovits wasn't tempted. She explained to her husband and her lawyer that they did not understand that to sign an apology that was a lie and never discuss the crimes committed against her and the patients would end everything. It would effectively end her life.

Dennis Jones objected, but Mikovits remained firm.

The offer was withdrawn.

* * *

The Seeno family lawsuit against the Whittemores, as well as the Whittemore counter-suit would eventually be settled in February of 2013, but the truth or falsity of the accusations in both cases would remain hidden from view.

At that time, James Pisanelli, the Seeno family's Las Vegas lawyer, released a statement which read: "The parties have reached a comprehensive settlement of all pending litigation. The financial terms of the settlement are confidential and the Seeno/WNG parties offer no comment as to the litigation or any aspect of the agreement at this time, except that all litigation by and between the parties will be dismissed."³⁸

The outcome was disappointing to Mikovits. She'd wanted the facts to be revealed to all in a court of law, not silenced forever, as they attempted to extort her into silence.

* * *

Jones continued to press her case.

On April 25, 2012, he filed a motion for the new judge to set aside the default judgment. The motion was everything Mikovits could have asked for, although Jones was not optimistic about its chances.

The new judge, Janet Berry, who would rule on the motion still had to work with Judge Adams, and Jones felt no judge would rule against another sitting judge. Still, it felt good to see her side of the story presented in the motion:

Judge Adams recently confirmed the existence of bias, because:

1) he accepted susbstantial campaign contributions from Harvey Whittemore—co-founder of the Whittemore Peterson Institute and President of its several affiliated companies, from Whittemore's family members and from a company with which Whittemore is or was closely affiliated,

2) Mr. Whittemore is the target of an FBI probe and a Grand Jury Investigation into illegal campaign contributions,

3) The criminal investigation of Mr. Whittemore, as well as a fraud suit filed recently by Thomas and Albert Seeno, alleging that Mr. Whittemore converted and improperly disbursed funds from the same corporation that contributed $5,000 to Judge Adams's re-election campaign, has garnered intense media scrutiny,

4) two Congressmen recently called upon everyone who accepted campaign contributions from Mr. Whittemore, including Judge Adams, to donate those contributions to charity; and

5) Judge Adams is unable to donate the money he received from Mr. Whittemore and his affiliates to charity.

None of the foregoing facts were disclosed to Mikovits or her counsel until the day before Judge Adams recused himself (March 16, 2012), even though Judge Adams knew about the tainted donations the very first day the case was assigned to him.

Judge Adams also admitted to imposing against Dr. Mikovits the most severe sanction he has ever imposed in a civil case—the striking of her answer—because she attempted to assert her constitutionally protected right against self-incrimination in response to a preliminary injunction issued by Judge Adams. Significantly, at the time of the hearing on Plaintiff's motion for preliminary injunction, Dr. Mikovits was in custody in California, on a bail imposed by the Washoe County District Attorney, presumably at Mr. Whittemore's request.[39]

Mikovits was pleased with how her attorney had set up the facts, but also how it impacted the fundamental justice of her case. Jones went through the enormous political influence of Harvey Whittemore and the allegation he'd orchestrated Mikovits's criminal arrest, just as she was trying to respond to the preliminary injunction in the civil case.

Then, Judge Adams struck Dr. Mikovits's answer, denied her motion to stay this action pending resolution of the criminal case and denied all other requested relief—thereby handing Plaintiff a win by default.

This case and the companion criminal case stink of undue influence and bias. Judge Adams' belated recusal does not restore the scales of justice to an equal balance, because he has already irreparably damaged her case by ordering her answer stricken, entering her default, granting a permanent injunction and ordering that she cannot take any discovery or even introduce evidence at the damages hearing.[40]

Jones also argued that bringing in a new judge just to hear damages did not restore the parties to an even playing field.

The civil case should be restarted from the beginning.

The motion had presented all the arguments Mikovits wanted the court to hear, but also what her supporters in the patient and scientific communities needed to see exposed. Jones pulled no punches in laying out exactly what he viewed as the facts. Mikovits was a nationally renowned scientist and a Principal Investigator on several government grants for which she held the ultimate authority for funds, materials, and protocols. Jones continued:

Dr. Mikovits relationship with WPI deteriorated after she discovered a series of improper, unethical and possibly illegal acts by the Whittemores and Vincent Lombardi, the lab director at VIP Dx. These acts included, but were not limited to,

1) the Whittemores arranging to pay one half of Dr. Lombardi's salary out of WPI funds obtained via federal grants, even though Dr. Lombardi did no work at WPI but was instead employed full time at VIP Dx;

2) ordering a research assistant to misuse cells allocated for non-profit research to harvest supplies for the for-profit VIP Dx clinic, and then covering up the misappropriation for two years;

3) selling blood tests through VIP Dx, even after those tests had been invalidated as unreliable, thereby committing Medicare fraud; and

4) failing to validate serological assays.[41]

As Jones had predicted, the new judge was unwilling to go against another sitting member of the bench. The motion was denied. Judge Berry, set the date of the damages hearing for September 6, 2012.

That hearing would never be held.

* * *

While the Whittemores were prevailing in the civil case against Mikovits, they were not faring so well with the FBI and the Grand Jury.

On June 6, 2012, Harvey Whittemore was indicted on federal charges of election violations. An article in the *Las Vegas Sun* a few days later captured the contradictions of this near-mythical figure.

Harvey Whittemore once bounded through Nevada's Legislature as the embodiment of influence—a lobbyist for casinos and other powerful interests, a lawyer, a gregarious personality and sharp mind. He was also a lucrative campaign contributor for elected officials, which helped open doors.[42]

The article described Whittemore as somebody who pushed boundaries, like the installation of a private pier at his home on Lake Tahoe. Mikovits recalled the many times she'd stood on that very pier as Harvey pulled up in one of his speedboats, encouraging everybody to jump in and speed across the lake to a restaurant.

And there were the darker aspects to his character, such as the tall glass walls that the Legislature had installed in the Senate and Assembly chambers to keep those in the public gallery separated from the lawmakers. They quickly became known as "Harvey" walls because of Whittemore's well-known penchant for reaching into the chambers before votes on legislation.[43]

But all of that was of little importance, as Whittemore entered the courtroom "in a dark suit and shackles around his ankles" to offer his plea to the four federal charges against him for crimes againt the democratic process.[44]

The article described the courtroom hearing as surreal, as the room was filled with reporters, friends, family, and lawyers. The reporter noted he couldn't get many people to go on the record to be quoted, but the connections between Harvey and the powerful in Nevada ran all the way to the judge, William Cobb, who'd known Whittemore for close to forty years and admitted to having a close relationship with Whittemore.[45] The article recorded that Judge Cobb even admitted feeling a little bit ridiculous when he asked Whittemore to state his full name for the record.[46]

* * *

When Mikovits's criminal attorney Scott Freeman had phoned to let her know he could no longer serve as her attorney on March 29, 2012, he had assured her that the criminal charges were on their way to being dismissed in the next week.

But that did not happen. Each hearing scheduled in March, April, and May, was postponed with no reason given.

The waiting was agony for Mikovits, but her advisors and friends told her it was simply a matter of time before things started turning in her favor.

The district attorney for Washoe County filed for dismissal of the criminal charges against Mikovits on June 11, 2012, only four days after Harvey was indicted. Her supporters cheered this new direction, but Mikovits could not help but believe the charges had not been dropped sooner in order to prevent her from bringing evidence of another conduit scam against Whittemore. This scam would directly involve and benefit the WPI.

Jon Cohen, the writer for *Science*, even popped up with a quick email a few days later, asking for an update of her current work.[47]

Mikovits could not believe his casualness, as if she had just emerged from some quotidian debate. Was he completely oblivious about what had happened?

She wrote him back and tried to keep any anger tempered, saying simply, "The publication by you and *Science* of my mug shot without my permission and with the refusal of the Ventura judge and my attorney has caused irreparable damage.[48]

Cohen responded by writing back, claiming that the mug shot brought sympathy and goodwill her way, and if there was any damage caused to her reputation, it was the fault of the Whittemore Peterson Institute and the Washoe County District Attorney's office.[49]

Mikovits ripped into him in a subsequent email for what she considered to be their tabloid-like sensationalism of her case:

> "*Science* promoted my defamation. That picture provided no sympathy only disgust and fear. Many of my colleagues found its publication so reprehensible as to vow never to read or submit to *Science* again. You may not have created the story but you sensationalized it without merit or balance to the destruction of my 30+ year career. Others lied to you or covered their mistakes. You chose to not report the Silverman mistake, which was paramount as a mistake, but instead to destroy me when I was conveniently charged with a fictitious felony and could not defend myself. You should be ashamed of yourself. At least I have my integrity."[50]

Cohen continued on in his defense, writing back that he was proud of the coverage by *Science*, and that there was never any attempt to harm her reputation. He said he was contacting her because of what he imagined was very good news, the dropping of the criminal charges, which had always struck him as bizarre and extraordinary.[51]

Eventually, Mikovits relented and gave a short interview to Cohen.

What else was she going to do? Retreat into silence? How would that help her repair the damage? He then published a brief article in the *Science Insider*, hardly the type of thing to restore her reputation, but perhaps better than nothing. At least Cohen seemed to make an effort at remediating the harm. The article was entitled, "Criminal Charges Dropped Against Chronic Fatigue Syndrome Researcher Judy Mikovits."

Cohen's article noted that the charges had been dismissed without prejudice, which meant that a complaint could still be filed in the future, then noted the bizarre twists in the case. The judge in the civil case who ruled against her, removed himself from the case when it was revealed he'd received significant campaign contributions from Harvey Whittemore, who was himself now charged with making illegal campaign contributions. Cohen continued.

> Assistant District Attorney John Helzer, who filed the dismissal, says Whittemore's legal troubles factored into his decision. "There's a lot going on with the federal government and different levels that wasn't occurring when

we first became involved with prosecuting this case," says Helzer. "And we have witness issues that have arisen."[52]

The witness issues that had arisen, were of course related to Max Pfost, Mikovits's former research associate, who had disavowed his earlier affidavit.

* * *

"Let me give you some advice," her criminal attorney, Tammy Riggs, told Mikovits, minutes after informing her the criminal charges were dismissed.[53] "I want you to understand what it means when the district attorney dismisses charges against you 'without prejudice.' It means he can come back at any time within the statute of limitations and say they have new evidence and refile the charges. I also want to explain slander. "

"But that's ridiculous," Mikovits had interrupted. "They don't have any evidence and it's not slander if it's the truth. I have emails and witnesses to back up every word I have ever said."

"I don't know where Harvey is getting the money but he has it and he will bury you the next time. Go live your life, because you are lucky you are not in jail in Nevada as you would never get out," Riggs said.

"What life do I have? Since I was twelve years old, I have dedicated my life to the chronically ill. I am a research scientist!" But, Mikovits had a sinking feeling that Riggs was right. The idea that she had eluded Harvey's clutches but still was not out of danger, was shared by her civil attorney, Dennis Neil Jones.

Mikovits was ready to make the damages hearing scheduled on September 6, 2012—the civil trial she had been denied. She compiled a list of ninety-seven witnesses, patients, and scientists, including Ian Lipkin, Frank Ruscetti, and even John Coffin, who could testify that she had done her job and had not only done no damage to the WPI, but had been a benefit to the institute.

"We're not going," Jones said.

Mikovits became very agitated and excited, saying that this needed to be done.

"Look at you!" said Jones. "You're upset and you're yelling."

"You'd yell, too! I have put up with this farce long enough! I'm going to defend myself! I'm going to show the truth! I have hundreds of emails! I have all of these witnesses and they will come and I will pay for it!"

"You don't have the money," Jones said.

Mikovits replied, "I'll sell my houses. They're coming and I will—"

Jones swiftly cut her off. "You're being irrational. Do you think that if you go into that courtroom and you tell the truth, and you're calm and collected, you expose the fraud, you expose Harvey and Annette Whittemore, and all the reporters are there—do you *honestly think* you will walk out of that courthouse? Do you think that the DA won't be standing there with handcuffs and leg shackles to take you back to jail for new evidence that he just discovered?"

"That's ridiculous. There's no new evidence!" replied Mikovits.

In a cold, but not unfriendly voice he said, "They didn't have any the first time, did they?"

She got that nauseated feeling again. She knew that if she was arrested again, this time in Reno, that they would not have the money to pay a bail bondsman and she might be locked up for a very long time. The previous time she had been in jail it had nearly destroyed David. What would another false imprisonment do to her husband? For several weeks Jones had been trying to convince her to file bankruptcy. "You want me to file the bankruptcy, don't you?"she asked him.

He started going over the points he had already made to her several times. "Taking a chapter seven bankruptcy stays everything. There's no liability to the default judgment against you. You'll be able to get a job. It freezes everything. There is nothing in the courts. There's nothing that says you did anything. You can go on with your life."

It seemed like a lawyer's trick, but there was no other viable option. She had kept herself from being convicted as a criminal, but even if uncuffed, her hands were still tied.

But she was not free, either, not with the threat of fifteen million dollars in damages being assessed against her by a biased judicial system. Taking the bankruptcy was the only way to stop the madness.

But it was still not the justice she wanted.

It was not even close.

The plain truth was she could not go back to her life. She was a research scientist and she had been unjustly disenfranchised. Nobody would touch her. Now like the patients, she had to face the fact that she could never again do those things she loved the most.

* * *

On May 29, 2013, following a two-week jury trial, Harvey Whittemore was convicted on one count of making excessive campaign contributions, one count of making contributions in the name of others, and one count of causing a materially false statement to be made to the Federal Elections Commission.[54]

In a press release issued by the Federal Bureau of Investigation, Acting US Assistant Attorney General Mythili Raman said, "Today, a jury convicted Mr. Whittemore of using dozens of straw donors to evade contribution limits so he could make good on a campaign fundraising promise. The cornerstone of our campaign finance laws is contribution limits and transparency, and Mr. Whittemore's crime was designed to undermine both. Today's verdict demonstrates our resolve to aggressively pursue those who use illegal tricks to corrupt our democratic process."[55]

The conviction had the potential to send Harvey Whittemore to jail for five years and pay a $250,000 fine on each of the three felonies of which he was convicted.

* * *

The government's sentencing memorandum summarized the case against Harvey Whittemore and the three charges of which he had been convicted.

"After promising Senator Harry Reid, a candidate for reelection to the US Senate, in February 2007 that he would raise a total of $150,000 in campaign contributions, Whittemore sought to make good on his promise by using employees of his real estate development company, his family members, and many of their spouses as "straw" or "conduit donors . . . On a single day, he caused one of his employees to transmit $138,000 in contributions, the vast majority of which he had unlawfully funded."[56] The government asked for Whittemore to be sentenced to four years and three months in prison, three years of supervised release, and a fine of $133,400, the amount of Whittemore's illegal campaign contributions.

The sentencing hearing was held on September 30, 2013, before District Court Judge Larry Hicks, starting at 1:33 p.m. and not ending until nearly six hours later at 7:14 p.m., an uncommonly long time for such a hearing.

To say it was a hard-fought hearing on both sides would be an understatement.

Whittemore's attorney, Dominic Gentile discussed Whittemore's well-documented charitable activities and good character as grounds for a reduction in sentence.[57]

Eventually, Harvey Whittemore took the stand to accept responsibility for what he had done, but also to offer an explanation. "I didn't do this for greed. I didn't do this for power. I didn't do it for any other reason other than a friend asked me to raise money. I didn't want anything. I didn't need anything. I didn't need to be ingratiated. I didn't need a pat on the head. A person asked me to help him. I did it in a way that was wrong. I humbly ask that you look upon this life's work and my conduct and find that it is worthy of a variance below the guidelines."[58]

This drew a swift rebuke from Assistant United States Attorney Steve Myhre who said, "This is a very serious offense. And we've heard sort of an almost schizophrenic argument from the defense: On the one hand, it's serious, but, oh, really not that serious: I accept responsibility but not really. Mr. Whittemore said that he believed his conduct was lawful, that he was naïve and perhaps arrogant to believe that his conduct was lawful, but, oh, my goodness, I gave myself bad advice, and I accept that the jury found me guilty . . . it's a far cry from any expression of remorse . . . He's guilty of knowfully and willfully violating the law. And the law he violated strikes at the very heart of our election process, the very cornerstone of our government."[59]

Myhre went onto note that at one point during the trial, Harvey Whittemore's attorney represented that Whittemore was worth $300 million dollars[60], but then proceeded to what he considered to be the heart of the matter. "This was about a campaign contribution. Three years before the election. A campaign contribution pledge that he made at the very time that he's sitting on 30,000 acres in southern Nevada that will forever seal his fortune, seal his wealth . . . And that pledge, that campaign pledge, he went to the extent of committing a felony offense so that he could maintain his status, his level of influence, his level of accessibility to the senator."[61]

Judge Hicks had a few things to say about Harvey Whittemore and the case. "Now, before me I have this man, Harvey Whittemore, who I have personally known over the years and known the family, or known of the family, not close personal friends, but at least to the level of people you say hello to when you pass in the hallway and you appreciate seeing them . . . I have a man here who he and his wife have, I mean, in the way they've treated their family, the way they've treated their friends, are reflected in the letters that have been received by the court . . . And I can tell you that those letters are absolutely incredible. This man is a fine family man. This man has done great things in this community. And these are not things that are to be treated lightly."[62]

During trial the issue of Harvey Whittemore's character had come up, and the letters he had received. Assistant United States Attorney, Steve

Myhre compared the Whittemore saga to the Wizard of Oz. Myhre argued before the court. "And you recall in that story how Dorothy and her three imaginary friends are confronted with a big screen of this larger-than-life person, and there's all sorts of smoke and mirror and loud noises and booming voices and these images that just scare the daylights out of those people.

"But it's not until Dorothy's little dog Toto runs over and pulls a curtain back that you see that behind all the loud noises and the smoke and the mirrors and the lights is someone manipulating some controls, manipulating the scene, if you will.

"And you recall the classic response that the wizard manipulating the scene, what he says when the curtain's pulled back. He says, 'Ignore the man behind the curtain.'"[63]

Myhre then linked the Wizard of Oz narrative to those who had written letters on Harvey's behalf. "Those witnesses have not seen or heard the testimony that came out of the witness stand from all of the people who have testified in this case. Those witnesses have not seen all the documents that you have and that you will see during your deliberations. They see one thing. They see the façade on the wall with the smoke and the mirrors. They don't see the man behind the curtain with the levers funneling money into the Reid campaign. They don't see that. You, however, have that benefit."[64]

There was a good deal more discussion, then Judge Hicks moved towards a conclusion of the sentencing hearing. While it seemed that Judge Hicks had heard a a great deal of the government's argument, it did not fully convince him. "You take this serious of a law that goes to our campaign financial disclosures and our reliability and how much faith we can place in our electoral process and in our democracy itself, and take examples of all of these conduit contributions, in one day family, friends, business associates, every one of those is a felony offense . . . the evidence shows black and white where the money came from, Mr. Whittemore's personal account; who it went to, his employees, his relatives, and their spouses, all for the purpose of making a phony contribution to Harry Reid's campaign. The evidence was unassailable . . . No one is above the law. And there simply must be just punishment for criminal conduct involving intentional felony offenses such as these, which many would argue that the court should not depart from the sentencing guidelines that are the beginning point for the court to consider sentence in this case. ***However, I'm doing that.***"[65] (bold and italics added.)

Instead of the fifty-one months in jail the U.S Attorney's office had argued for in the sentencing memorandum, Judge Hicks sentenced Harvey

Whittemore to twenty-four months in jail, imposed a hundred thousand dollar fine, two years of supervised release, and a hundred hours of community service.

After exhausting various appeals, on June 5, 2014, Harvey Whittemore was ordered to surrender to federal prison authorities on August 6, 2014, to begin serving his two-year prison sentence.[66]

It had taken the federal authorities a little over two and a half years from their raid on Harvey Whittemore and his family and business associates to prosecute him, handle the various appeals, and for Harvey to spend his first day in jail.

By contrast, it only took six weeks for Judy Mikovits to go from having a civil case filed against her to having a judge who took significant campaign contributions from the the Whittemore family and other associates to impose a default judgment on her, all without her having a single day in court to present her case. (And during those six weeks, she had spent five days in jail with no warrant or probable cause of any crime being committed by Mikovits.) As much as she might have wanted to, it was difficult for Mikovits to believe that anything close to justice had been done.

CHAPTER NINETEEN

The Silverman Mistake

John Coffin, had his own theory; XMRV had been accidentally created in a lab at Case Western University in Cleveland between 1993 and 1996. Coffin described how lab workers there had transplanted human prostate tumor cells into an immune-deficient lab mouse, a common procedure for procuring a colony of cells, or a human cell line for further study.[1]

—Hillary Johnson, "Hunting the Shadow Virus,"
Discover, March 2013

When Joe DeRisi created his Virochip in 2003 and claimed it could identify every known virus, the scientific community hailed it as a major breakthrough in viral detection, eventually awarding DeRisi the $250,000 Heinz Award for technological innovation in 2008.[2] Scientific advance depends on the development of new technology. However, the tools often fall short of being able to evaluate complex natural or biological phenomena.

DeRisi's Virochip contained 72 base pairs of DNA from every known virus. The average replication-competent retrovirus contains approximately 8,000 base pairs of DNA (DNA viruses like EBV and HHV-6 are much larger).

Can a test, which contains less than 1 percent of the DNA of a virus accurately identify a virus?

The answer is—it depends.

All scientists are taught that any procedure is only as good as the quality of materials used. This is why it was so stunning to learn in 2012 the prostate tissue Silverman used in the original study did not contain XMRV sequences, but that the RNA made from this tissue did contain XMRV sequences.[3] Six years later, they report an obvious source of contamination.

Had they used appropriate controls, they likely would have detected their source of contamination in the beginning.

* * *

While it's true that some viruses evolve and mutate quickly, it's also true that many regions of their DNA do not. When a "conserved" region of DNA is discovered it provides scientists with an effective means of identifying the virus. However, this approach is fraught with the potential for error. The identification of a "conserved" region of DNA, one that does not change over millennia, could indicate the existence of a broader family of viruses.

What about changes in the remaining DNA of the initial virus?

When Bob Silverman and Joe DeRisi put the prostate cancer tissue samples on the Virochip micro-array, it lit up with what looked suspiciously like a mouse leukemia virus (MLV).[4] There were a few base pairs of difference but that was to be expected since the virus was found in human tissue. Since it was not exactly a mouse leukemia virus they named it murine leukemia virus-related virus.

And since it had been found in human tissue rather than mouse they named it "xenotropic" which means it was replicating in the cells of another species, namely human beings.

That is how the virus came to be called XMRV, or xenotropic murine leukemia virus-related virus. All they had at this time was 72 base pairs out of an estimated 8,000 in a typical virus. If they wanted to make real progress, they had to get more genetic information about the virus, and the traditional way of getting this information was to extract and isolate the virus from an actual human, the same way that human retroviruses HTLV-1 and HIV-1, and almost all animal retroviruses had been discovered.

By 2006, science had come a long way since Montagnier had isolated HIV, and some, like Lipkin and DeRisi, believed the shortcuts developed by the explosion of molecular technology to the well-known methods of viral isolation were valid and could be solely relied upon.

Using a needle biopsy of three different prostate cancer tissue samples, Silverman and his collaborator at the Cleveland Clinic, Jaydip Das Gupta, extracted a sample from VP-62, one of the prostate cancer tissues. With the help of enzymes available in the laboratory, they took the 72 base pairs identified by the Virochip, and just like one might see on a television pathology show, they used PCR (polymerase chain reaction) to amplify the genetic sample so that enough existed for appropriate testing.

They call it "genome walking," and the idea is to literally extend the 72 known base pairs out into other regions of the genome. This technique may produce errors because of "low specificity and efficiency, short walking distance, and complexity of the methodology."[5]

But from this kind of clonal research all the way up to Dolly the Sheep,[6] all scientific cloning is a work in progress.

Silverman and Das Gupta couldn't get enough of the 8,000 base pairs to create an actual infectious molecular clone. Mikovits would not learn until 2011 that they were not able to obtain enough genetic materials from the VP-62 sample to create a clone. They would then go to other samples and start again. This yielded additional genetic sequences, but still not enough for their full-length molecular clone.

They sequenced additional material from another sample to get their final sequences of the 8,000 base pairs in order to complete their infectious molecular clone. Then, with their sequences from three different patient samples, they attempted to piece together the sequence of what XMRV might look like in an actual human being.

It was as if Silverman had created something of a "Frankenstein" virus from three different patients. What if there was significant variation between the viruses in these three patients? Or what if they had "stitched" the virus together incorrectly into a hastily-made "crazy quilt"? They might not have duplicated what existed in nature, but created something entirely unique.

For Mikovits and those of her generation who had been trained to isolate a virus from a single patient sample, this was, at best, work that needed to be carefully confirmed by an additional demonstration that this artificially created virus had a natural history of infection in humans or any species—something Silverman could not demonstrate. Without that step, his work remained incomplete.

Since 2009, Mikovits had proposed that Silverman's infectious molecular clone could vary slightly from what actually existed in human beings, since all viruses evolve and change and Silverman's "crazy quilt" was not a natural creation. She contended that this could cause problems with viral detection. Mikovits had argued that point in a roundtable discussion at the Invest in ME Conference in May of 2010.

Mikovits claimed that they had to "use a natural isolate or all bets are off as far as negative studies are concerned."[7] Others disagreed and contended it made no difference. The timing, however, was a critical factor. If Silverman had revealed to Mikovits or Ruscetti that the information that their infectious molecular clone had been created from three different samples rather than one—in 2009 rather than belatedly in the summer of 2011—Mikovits's team would have had additional ammunition for their concerns, and likely would have prompted them to ask for additional controls.

While the process of viral isolation from a single patient sample was much more time-consuming than what Silverman and Das Gupta had done, Mikovits would continue to believe it was the superior method.

Mikovits couldn't help but view Silverman's two-year delay in telling her this information as a sharp diversion from standard scientific protocol. The scientific community expects that a scientist will reveal both the strengths and flaws of their work so that other researchers, particularly collaborators, might accurately consider it. A scientist wasn't supposed to conceal the potential flaws of their work like some TV commercial pitch-man.

This failure would make it difficult for Mikovits to trust the work of Silverman and Das Gupta again. If they had revealed this information at the beginning of the investigation, the molecular clone could have served many useful purposes, such as being an analog for the virus until a complete sequence was generated from an actual human derived isolate.

And while Mikovits considered their infectious molecular clone to be just an approximation for the virus from the beginning, for the majority of the field it was the only viral sequence they'd consider allegedly linked to ME/CFS or other diseases, until the publication of the Lo/Alter paper in September 2010 that found variant mouse leukemia viruses (MLVs).

* * *

Once Silverman and Das Gupta had engineered their full-length molecular clone in a plasmid construct, it was transfected into cells to generate large volumes of virus for study. To explain further, plasmids are best understood as molecular factories for generating viruses and cellular proteins. A plasmid is not a virus.

A plasmid is a small, circular, double-stranded DNA molecule, which can be cut with certain enzymes, new DNA inserted, then reattached behind a promoter which turns on the production of the genes in the DNA. The plasmid is then inserted into cells through a process called transfection. Transfection involves altering the cell to open up pores in the cellular membrane so that the plasmids can enter into them. This is commonly performed using either electrical current or detergents.

Once inside the right cells such as LNCaP, the plasmid will churn out virus.

A scientist working with plasmids needs to have a way to identify them, and luckily there are a few ways to make such identification. A plasmid can have certain genes inserted, like those for antibiotic resistance, or even the genes that make a jellyfish glow, which can then be used to mark those cells that contain the plasmid (in the case of jellyfish genes, through bioluminescence that actually illuminates the plasmid). Another way of identifying plasmid contamination is to look for the promoter sequences not contained in human cells.

There was much evidence beyond the early electron micrograph that XMRV was a *natural* infection and neither plasmid contamination nor contamination with 22rv1 lab contaminant had anything to do with the evidence of a related XMRV that Mikovits and her team isolated from the ME/CFS patient samples.

First, Silverman had collaborated with a specialist in viral integration at the University of California, Los Angeles, to confirm that prostate tissue samples showed the retrovirus had integrated into human DNA.[8]

Second, there was a team from Emory University that had also shown the virus in human tissue, although they were only able to detect sequences of the virus, not the entire virus.[9]

Third, Ila Singh had detected viral proteins in prostate cancer tissue. All of these studies had been published between 2006 and the publication of their work in 2009.

Whenever a question would arise regarding the sequence of the virus, Mikovits would contact Silverman to ask whether he was certain of the data.

Silverman would remind her of the published literature he had generated at the Cleveland Clinic, the work done by the Emory team, and others.

By early March of 2009, because of the difficulty in obtaining full length sequences, Mikovits and Ruscetti would raise the possibility the virus isolated from ME/CFS patients was not Silverman's XMRV. Silverman told them everything was done correctly in his lab at the Cleveland Clinic and they believed him. After all, he had "discovered" XMRV.

The envelope of XMRV Silverman was closely related to spleen-focus forming virus (SFFV) envelope, the virus envelope upon which they had based their antibody test. The antibody test detecting reactivity to SFFV Env could detect not only antibodies to Silverman XMRV, but also to related virus envelopes. Mikovits was confident their test was detecting a related XMRV as they had validated it using appropriate competition studies.

* * *

Plasmid contamination was another issue altogether and Ruscetti was astounded when Silverman first raised the plasmid contamination issue in a cryptic, roundabout manner with Frank Ruscetti in June 4–8, 2011, at the 15th International Conference on Human Retrovirology, HTLV, and Related Viruses in Leuven, Belgium.

The conference was an opportunity to discuss recent data on the pathophysiology of human retroviruses with American as well as international researchers. One of the sessions was chaired by Robert Silverman and dealt exclusively with XMRV.[10]

As chair of the session, Silverman invited himself to give a talk on XMRV that he titled "Human Infection or Lab Artifact: Will the Real XMRV Please Stand Up?"[11] His presentation left little doubt that as of June 2011 Silverman considered XMRV to be a human infection. A part of the abstract is reproduced below:

> Among the confounding factors are the potential for lab contamination with similar or identical viruses or viral sequences originating in mice. In some studies, relatively contamination-resistant methods (e.g. IHC, FISH, and antibody detection) suggest that either XMRV or a similar type of virus is present in some patients. Evidence for and against genuine infections of humans with this intriguing virus (and/or related viruses) will be discussed."

Following Silverman was Shyh-Ching Lo, the collaborator with Harvey Alter on the XMRV confirmation paper.

After Shyh-Ching Lo's talk, during what was supposed to be the question and answer session, Mikovits saw Kuan-Teh Jeang, the editor of the journal *Retrovirology,* stand up with some other researchers and start shouting, "Stop this waste of money! Stop this research now! We should be spending these resources on people with real viruses and real diseases!"[12]

As the chair of the session Silverman tried to calm the crowd, but they didn't appear to be mollified. Mikovits had never seen any behavior like it in the course of her thirty years in science, even though these gatherings had grown increasingly dramatic.

Although the outburst had come after Lo's talk, it was clearly aimed at Frank Ruscetti and Judy Mikovits, who were seen as the investigators who started this research. Counter-attackers were trying to shout them down with as much subtlety as an unruly mob. Wasn't this supposed to be science?

The title of Frank Ruscetti's presentation was "Development of XMRV Producing B Cell Lines from Lymphomas from Patients with Chronic Fatigue Syndrome."[13] Ruscetti presented the details behind the virus infection of B cells and the increased rate of certain types of cancers among the patients with ME/CFS, the issue (B cells) that had first drawn Mikovits to study the disease. The meeting abstract gave a condensed account of Ruscetti's presentation at the Leuven Conference. Ruscetti noted that the incidence of non-Hodkin's lymphoma in the general US population was 0.02 percent, but was nearly 5 percent of the ME/CFS patients. This translated into a long-time ME/CFS patient being 250 times more likely to come down with non-Hodgkin's lymphoma than a typical healthy person, an astronomical figure. He then observed that:

> Additionally, development of cancer coincides with an outgrowth of gamma delta T cells with specific clonal T-cell receptor gamma. We hypothesized that infection with XMRV and/or other viruses can trigger a dysregulated immune response which favors the development of B-cell lymphoma.[14]

Accompanying the non-Hodgkin's lymphoma was an abnormal population of gamma delta T cells, an important component of the immune system.

These data led Ruscetti and Mikovits to suggest that infection with XMRV either alone or in combination with other coinfections (such as those commonly found in ME/CFS patients, just as with HIV/AIDS patients) might trigger an abnormal immune response that was favorable to the development of this type of cancer. Ruscetti detailed the numbers of ME/CFS patients with these rare types of cancer, their lab results, then concluded:

> Therefore XMRV infection may accelerate the development of B cell malignancies by either chronic stimulation of the immune system and/or by direct infection of the B-cell lineage. Since viral load in peripheral blood is

low, these data suggest that B cells in tissues such as spleen and lymph nodes could be an in vivo reservoir for XMRV.[15]

Ruscetti's presentation covered several key points. The 11 ME/CFS patients with lymphomas or associated disorders all tested positive for XMRV, and 9 of the 11 were positive for the T cell rearrangements, which seemed to be a key indicator of a poorly functioning immune system.

Four B cell lines from patients began spontaneously producing XMRV virus.

One of the arguments against XMRV as a disease-causing pathogen was that even if the virus was able to infect humans, the human immune system would be able to tamp it down and control it. Specifically, the researchers argued that APOBEC3G, an antiviral enzyme, would fight off any such retrovirus.

Ruscetti and Mikovits found that the virus was replicating in these patients even though APOBEC3G was present.

The researchers who opposed further investigation of XMRV and neuroimmune disorders presented various objections, but all these lacked an explanation as to why people with ME/CFS had a rate of non-Hodgkin's lymphoma 250 times greater than the regular population. A retroviral hypothesis fit the presentation of ME/CFS and retroviruses had twice before been associated with ME/CFS.

Although Mikovits was aware of how heated scientific controversies can become, it was in Leuven she saw the first evidence of the angry mob of scientists that Coffin would say a few months later in the journal *Science* would "burn her at the stake."

* * *

Mikovits could not forget the behavior of the chief editor of *Retrovirology*, Kuan-Teh Jeang.

Why was the anger directed only at XMRV in ME/CFS and not in prostate cancer?

Was it because Mikovits had suggested this retrovirus might be linked to other diseases besides ME/CFS, like autism?

Was it because Mikovits was raising the possibility that there might be other as yet unidentified retroviruses associated with neurological conditions, such as autism and even cancers, which were rising at dramatic rates?

Was that so horrible a possibility that a scientist testing viable hypotheses deserved to be burned at the stake?

And though Mikovits was worn down by the controversy, first over suspected mouse DNA contamination, then Coffin's hybrid virus, and soon enough over the Silverman plasmid contamination, she knew there were many dramatic and pressing questions raised, and that she had to persevere.

It was during one of the breaks between the sessions that Silverman approached Ruscetti and said to him that Das Gupta really felt bad about what happened.

Ruscetti was confused.

What did Das Gupta have to feel bad about?

Silverman told Ruscetti that the some of the positive samples from WPI sequenced in his lab were contaminated by sequences present in the Silverman XMRV plasmid. If plasmid contamination was suspected, the controls he had told them he had done in the beginning should have shown plasmid contamination two years ago.

Silverman had just given a talk in which he strongly supported the idea that XMRV was an actual human infection. Was he now suddenly bailing on ME/CFS and XMRV? There had been multiple lines of evidence of a system-wide immune dysfunction, and a retrovirus was still a viable hypothesis.

Silverman had been at their side, defending their research against the negative studies, and even refusing calls to retract their publication in *Science*, but all that was about to change.

* * *

In an attempt to distance himself from the ME/CFS and XMRV controversy, Silverman was planning to give his collaborators little time to determine whether his XMRV plasmid theory of contamination was valid.

In an email Silverman sent at 5:59 p.m. on July 6, 2011, to his coauthors, he attached the data showing contamination in his lab and asked them to let him know in the next two days if they had any comments or whether they believed different actions should be taken.[16]

All of the coauthors were both shocked and disturbed by this email. He was now giving them a two-day window to respond to his discovery of plasmid contamination before presenting it to *Science*, and exposing what could be a disastrous scenario, before they could offer any defense. This was not standard protocol. Frank Ruscetti composed a reply to Silverman and sent it at 8:01 a.m. on the following morning, July 7, 2011.

Dear Bob and co-authors:

I find this study incomplete. In the interest of getting the scientific truth out, before it is submitted, one should do the complete science and figure out the source of the contamination. It would seem appropriate that the Silverman lab do an equal number of XMRV positive DNA samples made from prostate cancer samples at roughly the same time as the CFS samples for cmv and neo primers. I also think the Silverman lab should test the DNA made from the same patient samples (same time and WPI processing) made at the NCI from Mike Dean. Our labs have some of the same DNA but those from the Dean lab would be a less contentious set of DNA to use.

Let me know your thoughts.

Regards,
Frank.[17]

But Silverman did not seem interested in discovering the source of contamination, only in putting the XMRV in ME/CFS investigation behind him. He wrote back later that day, saying that while the source of the contamination interested them, it would not change the fact that figure 1 of the *Science* article was wrong. He also stated that they might never know the source of the contamination, an assertion that Ruscetti and Mikovits found to display an astonishing lack of scientific curiousity.[18]

How could Silverman not be concerned with finding the source of the contamination and how it might have affected their research? What infuriated Ruscetti and Mikovits more than anything was the material he was planning to submit to *Science*.

The letter implied that the contamination had taken place at Mikovits's lab at the WPI. On the one hand he was saying in his email to them that they might never figure out the source of the contamination, he was clearly saying where he believed it had taken place in what he was sending to *Science*.

Ruscetti wrote him back a strongly worded email on July 7, leaving little doubt as to his opinion.

Dear Bob:

I find your answer disingenuous to say the least. *The source of contamination is of interest to us, but having that answer will not change the fact that Fig. 1 is erroneous.* (italics added) Depending on the source of contamination, it will determine whether all the figures or just fig. 1 is discredited. All statements in this field about contamination pinpoint the problem, science demands that we do so also. *We may never figure out the source of the contamination, but we need to make public that the interpretation of figure 1 is just plain wrong.* (italics added) The reality is since the publication of the paper in 2009 nobody has believed fig 1 is correct so why the rush to publish before doing the experiments we suggest. Second we all suspect (you may already know thus the comment you said to me in Belgium that Joy feels awful about it) the source of contamination is your lab.

These facts should be either experimentally tested or written in your letter or we will send to *Science* in a letter from the other authors these facts . . .[19]

Ruscetti then listed a number of ways that they might test the samples which had been relied upon in the Science article, and brought up what he believed were accusations that contamination had taken place at either the Whittemore Peterson Institute, or the National Cancer Institute. Ruscetti finished off the email:

"You protest too much. All this will be seen as an attempt to say that the contamination could not have occurred in the Silverman lab."

"The CFS and healthy control PBMC DNA samples were treated identically. Then why did you insist that the samples not be blinded? Why did you coerce Lombardi into culturing more patient samples without our permission or knowledge? These were the identical PBMC DNA samples in the original tubes that were used to produce figure 1 of Lombardi et al., *Science* (2009)."

I plea with you to do the experiments I suggested. What difference is another month? If your LAB is as pure as Caesar's wife as you suggest, then you have nothing to worry about.

Regards,
Frank.[20]

Ruscetti was laying out what seemed to him be a logical, well-reasoned argument for a careful process, not just for the scientific community at large, but for the patients who had waited decades for answers. Silverman relented and delayed submission. Ruscetti got DNA from the 29 of the 30 samples including the 11 samples published in figure 1, and split them for plasmid detection in WPI and an independent lab.

None of the samples showed evidence of plasmid contamination.

Ruscetti told this to Silverman in late July 2011. Neither Mikovits nor Ruscetti ever heard from Silverman again. Silverman submitted his partial retraction to *Science* in August.

* * *

Mikovits and Ruscetti belatedly came to see what they believed was the bigger picture.

They concluded that Silverman had been having lengthy discussions with John Coffin, who was trying to convince Silverman that his viral discovery had been an artifact, an account which would be backed up by Hillary Johnson's article in *Discover* magazine in March of 2013.[21] At Coffin's urging, Silverman had investigated and found plasmid contamination with XMRV in *some* of his samples and wanted to retract his contribution to the original *Science* paper.

Ruscetti suggested they test other sets of samples for XMRV plasmid contamination, using the identical samples from the NCI and the WPI. Silverman responded with what seemed like evasions, noting that while the source was of interest to him, it still wouldn't change the fact that figure 1 was inaccurate.

There was little arguing with this point. However, Silverman continued to say, "We may never figure out the source of the contamination," which, while superficially true, begs the question of how does one know until you performed the experiment?

Error analysis in science is supposed to be just as rigorous as the preceding experiments.

* * *

Mikovits and her collaborators still fully believed in their evidence regarding an association between an XMRV-like retrovirus and ME/CFS, despite what Silverman had suggested.

They had been rigorous in sorting out the issue already.

There were the positive results from the Spleen-Focus Forming Virus antibody test, their own gag sequences, partial env sequences, gag and env protein, protein detection in unmanipulated plasma, and even the electron micrograph pictures of the virus budding from the cellular membrane of patient cells. Still, they worked to determine whether there was any validity to Silverman's claim of XMRV plasmid contamination.

On or about July 8, 2011, Dr. Mikovits called into her office Hagen, Pfost, and Lombardi, who worked on project X, and asked them all to bring their notebooks.[22]

Lombardi brought in four Xeroxed pages from his notebook. "This is it?" Mikovits asked. "This is all the work you've ever done in the research lab?" She was aghast at his paltry contribution, standing there next to the other associates with their rigorously detailed work. Lombardi answered in the affirmative, saying that the rest of his notebooks were proprietary information, which belonged to the for-profit clinical company, VIP Dx.

Mikovits made a mental note to revise his level of effort in the ROI annual progress report she had recently filed with Carli West Kinne. Mikovits took the pages from Lombardi and put them aside. Her only concern was trying to determine if any possibility of contamination existed in their earlier work.

In her review of Katy Hagen's notebook, Mikovits found a peculiar entry for March 23, 2009. Katy kept a meticulous lab book, but her handwriting was often difficult to read.

Mikovits could make out that on March 23, 2009, Lombardi had directed Hagen to culture the identical 30 samples that WPI had sent to Silverman twice before so the Silverman lab could do single round PCR, which he wanted for Figure 1 in the manuscript.

Apparently after Mikovits told Silverman they were not going to send the samples again because neither lab could get positive results from single round PCR, Silverman had asked Lombardi to send them. Mikovits asked Katy if one of the entries said XMRV positive (+) or XMRV negative (-). Katy looked at her entry in the lab book. "That's XMRV positive."

Mikovits felt like she might have a stroke. Culturing infected cells with patient cells would *not* result in plasmid contamination, but *could* have resulted in infection of those cells through aerosolization as Gazdar's lab had just discovered. She recalled how on January 24, 2009, a few days after signing the collaboration agreement with Ruscetti and Silverman she'd emailed the HIV decontamination protocols to Katy as soon as she could. Katy even taped the protocols in her lab book.

The critical thing to do for the integrity of the work was to prevent against the possibility of contamination.

Mikovits immediately called a meeting with Harvey to explain what she'd uncovered, including the four pages of work in the research lab for which Lombardi had been paid for 50 percent effort. Harvey asked why she thought Lombardi had violated the protocols.

In an email to Harvey, she cited Lombardi's arrogance and refusal to take direction from a woman. Asked why she thought he'd hidden it for two years? He was a coward! Mikovits was furious.

Annette's reaction to the latest Silverman controversy was immediately protective of their fiduciary interests. The Whittemores were making a presentation to a major diagnostic company, which had given WPI $150,00 per month since October 2010 for the intellectual property rights to the serology test.

Annette sent Judy an email directing her to make sure that Silverman plasmid contamination theory did not get out until they could make their presentation to the company July 24th, 2011, so that the next financial installments could come in. Cultured virus is not plasmid contamination, so now they had two sources of potential contamination to track down.

* * *

An idea Mikovits had been considering from the very start of the research was that the disease-causing potential of XMRV might be related to the body's reaction to proteins on the envelope of the virus.

There existed all sorts of variations on this idea, with one notion that some were defective viruses expressing some proteins, but not what is known as a replication competent virus. For a virus to be infectious and transmissible, it must be replication competent.

If a virion consisted mainly of an envelope, with defective genetic sequences inside, most scientists will consider the organism to be of little concern to human health. Mikovits's postdoctoral work had shown defective HTLV-1 virions could still express pathogenic proteins.

Another group of scientists, Sandy Ruscetti among them, have spent their careers trying to determine whether proteins in the envelope of the virus might be causing an abnormal immune system response and inducing autoimmune disorders.

The process is similar to an allergic reaction.

A peanut is not inherently dangerous. Most people can eat a peanut butter and jelly sandwich with a smile and a satisfied belly. But in some the body mistakenly identifies peanuts as dangerous, floods the system with antibodies, and causes lethal reactions, including anaphylaxis and death.

Some believed it was likely that in some cases the envelope of a virus might be causing an autoimmune reaction in which the body was attacking certain types of cells, leading to diseases like cancer and even neurological conditions.

In a sense, this was a solution that would allow *both* camps to be right.

Coffin and others could proclaim that XMRV Silverman, or whatever closely related retrovirus might eventually be linked to ME/CFS and other conditions, was not an infectious exogenous virus as the term is *generally* understood.

Mikovits and Ruscetti could say in turn that an XMRV-like retrovirus was linked to conditions like ME/CFS and possibly autism, but the mechanism of action for damage was not a classic infection per se, but the body's autoimmune response to proteins of a poorly infectious defective virus.

This approach could open up many intriguing avenues of investigation. If the envelope of the virus acted as an autoantigen, the question would turn to the immune response of the individual. Just as many people could eat peanuts without suffering harm, perhaps the majority of the population could harbor this virus without any ill effects.

In the case of autism, maybe infants could be tested for the presence of the virus, and then if they tested positive, a different vaccination schedule would be provided, just as was done for HIV-positive children. Perhaps by simply monitoring of the immune functions of a child, one could track whether the vaccination was likely to cause an autoimmune response.

This was not such a radical idea, just often happening on scientific "islands" outside of the Mikovits team.

Even study into the long-term progression of treated AIDS—that is, those patients on antiretroviral therapy to control the HIV virus—found that due to the now-longer lifespans of treated AIDS patients, inflammation and autoimmune-like reactions were major players in ongoing symptomatology and even accelerated aging. As reported by AIDS journalist Liz Highleyman:

> "Untreated HIV infection causes inflammation and, despite ART, it does not normalize," Steven Deeks, professor of medicine at the University of California-San Francisco (UCSF), explained at a post-CROI workshop sponsored by Project Inform. "This leads to all sorts of 'badness.' Some 20 presentations showed this same phenomenon linking HIV, inflammatory biomarkers, age-related symptoms, and immunosenescence [accelerated aging of the immune system]."[23]

So even in more accepted retrovirology, as Highleyman continued, the end result of classic autoimmunity and the inflammation from a persistent or controlled infection could lead to similar end results:

> This fine-tuned system [of the immune system clearing problems and then turning itself down] can go awry, however, when the immune system is faced with a threat it cannot overcome. This occurs, for example, during persistent

infection. Other causes of chronic inflammation include autoimmune conditions (in which the immune system attacks the body's own tissues), obesity, chronic stress, and exposure to toxins such as tobacco smoke.[24]

But first, before diverging into questions about envelope material and allergic or inflammatory responses, everyone wanted to know whether Silverman's XMRV plasmid had contaminated the data in the 2009 *Science* paper linking XMRV with ME/CFS.

It was an important question and nobody wanted the answer more than Mikovits.

* * *

Mikovits, Ruscetti, and Pfost would spend the summer of 2011 trying to answer that question. Unfortunately, she had to direct most of her attention to the Blood Working Group.

The WPI and NCI focused the month of July on completing the BWG panel due July 31 and Judy Mikovits asked Frank Ruscetti to sit on the WPI staff meetings. There had developed a great deal of tension for Mikovits as Harvey Whittemore insisted on running Mikovits's lab meetings. The Whittemores were unhappy about the negative data from Mikovits and Ruscetti when the BSRI decoded the samples.

Lombardi would never say anything and would never bring any raw data to evaluate until last meeting before the deadline (August 1). When they saw the data, Mikovits and Ruscetti were stunned beyond belief.

In a written recollection Ruscetti stated, "The data presented by Vinnie for us to send to the BWG as a serology test was the worst data I have ever seen presented as a finished product. Immunoblots with no molecular weight markers, most have background, which makes identifying the bands difficult if not impossible. The positive bands were marked with arrows, which was good because without the arrows they were indistinguishable. The positive bands were only guesses. Every scientist to whom I showed it only laughed."[25]

Furthermore, despite the fact that VIP Dx was charging patients for the culture method and reporting positives to patients, Lombardi had failed to produce any usable culture data. It brought in focus ever more clearly the problems with the clinical lab.

Annette Whittemore responded to these concerns by writng in an email to Mikovits copying Ruscetti, "This complete and utter failure of the BWG was no one's fault but yours and blaming the clinical lab for your poor decision making is unacceptable."[26] In a separate email written that same day, Annette thanked Ruscetti for his help, but did not blame him for making the same decision. Ruscetti goes onto say that Annette then "ordered Judy to submit the clinical data to the BWG," which Mikovits refused to do.[27]

When the head of the Blood Working Group, Dr. Michael Busch came to the WPI and toured the clinical lab a few weeks later, Ruscetti says he sent an email which stated, "I was under impressed by Vinnie to say the least and what I saw and understood in terms of validation and QC [quality control] of the assays in the clinical lab was disturbing."[28]

* * *

On September 23, 2011, Mikovits presented her findings at the International Association for ME/CFS, which was held in Ottawa, Canada. The session was chaired by Dr. Jose Montoya of Stanford University who also presented his work on the "Role of the Immune Response in CFS."[29]

Mikovits then presented "The Case FOR Human Gamma Retroviruses (XMRV/HGRV) in ME/CFS," while John Coffin presented the case "AGAINST XMRV."[30]

As the issue of XMRV was becoming problematic, Mikovits was now referring to the possible pathogen by a more generic name, a "human gamma-retrovirus," since data still suggested she was seeing evidence of a gamma retrovirus in patients. She started her presentation by showing an electron micrograph of a human gamma-retrovirus budding from the cell of a ME/CFS patient.[31] The showing of a similar electron micrograph by Luc Montagnier had convinced the world that HIV retrovirus existed. She even had the same microscopist that Montagnier had used taking the picture.

Despite being depicted as some kind of renegade scientist, her work was both "by the book" of classic science, and "in the footsteps" of a giant HIV predecessor, because she wanted to get it right.

The next slide showed the direct isolation of XMRV/HGRV protein from the plasma of ME/CFS patients. This was the test in which they used the antibodies almost like magnets to draw out the proteins from the plasma. And those proteins could only be created in an actual living organism. If they weren't in the blood at the time it was drawn, no contamination could explain that level of protein.

Next she showed that these retroviruses could infect another cell line, a key test of whether a virus might be pathogenic.

Taking Silverman's latest theory head-on she had a schematic of the circular DNA of the XMRV-containing VP-62 plasmid, showing the various markers that would allow them to detect it in any of their samples.

Mikovits was fond of saying that virology "is not rocket science," and often thought the field was more akin to the type of knowledge a good car mechanic had. If one suspects contamination with a certain agent—in this case, Silverman's XMRV/VP-62 plasmid—one knows it has certain genetic markers that will allow a scientist to identify it as a manmade creation. After knowing those characteristics, one can test for them. If they are present one has contamination. If they aren't, there's no contamination.

It's like taking your car to a mechanic and testing to see if the brake pads are still good.

Mikovits showed that in figure 1 of the original *Science* paper that was prepared by Silverman of the Cleveland Clinic, *some samples did contain the VP-62 plasmid*. However, the original DNA samples, which were housed at the WPI were tested by Max Pfost and Frank Ruscetti *were free of the VP-62 plasmid*.

An independent reanalysis of the same samples also showed the samples to be free of the plasmid, or other mouse contamination, the previous theory Coffin had held against XMRV.

There was no "crap"—mouse or otherwise—in their argument. The next slide showed the cell-free transmission of XMRV to LNCaP cells.

The following slide showed results from a single patient, demonstrating that they appeared to have two separate strains of a human gamma-retrovirus, a result often found with other human retroviruses. The next few slides contained information about the results of the Blood Working Group, which Mikovits had been working with since shortly after the original *Science* paper was published.

The report was to be published during the Ottawa conference, as well as the article "False Positive" in *Science*. All of this negative press was slated to be published on the same day as her talk at the Ottawa conference. She wondered ruefully which story would get the greater coverage, the article in *Science*, the results from the Blood Working Group, or her explanation of the latest findings?

She'd never been much of a believer in the idea that, public perception on certain vital issues was manipulated by powerful forces, but in this instance she felt increasingly that only one side of the debate was being presented.

The Blood Working Group had been created with a single goal in mind, to determine whether XMRV posed a threat to the blood supply. It had become something else, a stalking horse for the association of XMRV with ME/CFS.

Specifically, the Blood Working Group had been using Silverman's VP-62 clone as the control for finding the virus in individuals. As far as anybody could tell, Silverman's VP-62 clone was different enough in its sequence to the sequences Lo and Alter were detecting (after their paper was published they became members of the blood working group.)

And while the Blood Working Group had been finding evidence of XMRV-like gag and pol sequences in some patients, they tightened the standards for calling a sample positive by saying there had to be both gag and pol sequences in the sample in order to call it positive.

Silverman's infectious molecular clone was useful for doing animal studies to determine what might happen to the immune system of the infected animals. The Emory study showed that viral and proviral signals from Silverman XMRV disappeared from the blood within a month, but remained in the tissues where they could be reactivated by immune stimulation, such as happens when one gets a vaccination. In most of the Emory animals, the testing showed that the antibody responses were generally low in magnitude and short in duration.[32]

Would this explain the lack of positivity in the blood? When extrapolated to the human population it made detection problematic.

* * *

Mikovits next went on to the test, which she believed to be the most reliable in detecting this still elusive pathogen, or whatever degraded virus might be causing an immune reaction. She fell back again on the assay used to detect antibodies to the envelope of the Spleen-Focus Forming Virus, which had proven the most reliable evidence of XMRV/HGRV detection.

In her next slide she showed the antibodies they had detected in human samples reacting to multiple proteins of XMRV, including P12, P15 (Ma/E), Rec Env, P30, and gp70.

She also showed how the gold standard of molecular virology, the Western Blot method, could also detect the gp70 and P30 proteins in this population. Mikovits was careful to note on the slide that the "Ability to recognize XMRV proteins does not mean that XMRV was the immunogen. It could be any HGRV (human gamma-retrovirus) or a cross-reactive protein."[33]

Still, the data was showing that *something* was generating these aberrant proteins. Her answer to the problems of detection was the use of next-generation sequencing and she showed the results of some of that research she'd conducted in collaboration with the company Hemispherx Biopharma. The next generation sequencing had detected XMRV/HGRV in seven of eight ME/CFS patients and in only two out of seventeen controls.

Her conclusion was that the next generation sequencing offered a solution to the problems of the detection of these viral proteins. For future plans regarding next generation sequencing she suggested increasing the sequencing coverage for each DNA sample, using appropriate age, sex, and geographical matched non-ME/CFS controls, and looking at more than the seventy-five base pairs already identified. This would enable them to investigate viral-human chimeras, which would show the integration of the virus into human DNA.

Mikovits showed the research Frank Ruscetti had presented in Leuven, Belgium—specifically how three B-cell lines from patients who tested positive for XMRV had spontaneously immortalized, a precursor of cancer.

The take-away message for a patient with ME/CFS was basically a very disheartening, "Life sucks and then you die." In her summary Mikovits made the following points:

- We have shown that we can detect HGRV/XMRV footprints in the blood by serology and nucleic acid analysis without any evidence of contamination.
- Some CFS plasma contains HGRV/XMRV proteins and antibodies that recognize XMRV/HGRV viral antigens.
- XMRV producing Hematopoietic cell lines (B and NK-like) were developed from CFS patients.
- XMRV/HGRV-infected individuals exhibit cytokine profiles characteristic of inflammatory processes.
- Sequence data indicate there are different strains of XMRV/HGRVs that can infect humans.

In the conclusions section she wrote:
- The pathogenic potential of HGRVs in ME/CFS deserves further exploration.[34]

* * *

During the question and answer session, she'd been asked about the XMRV test being sold by the clinical lab founded by the Whittemores, VIP Dx.[35] The VIP Dx test had been used by the Blood Working Group. The VIP Dx test had used Silverman's XMRV VP-62 infectious molecular clone as the control.

She answered the question the only way she knew how, by telling what she thought.

Mikovits said that given the results of the Blood Working Group, the VIP Dx test could no longer in good conscience be used.[36] Prior to the Ottawa conference Mikovits knew that more than three thousand people had been tested at VIP Dx, and that the cost for the various tests ranged from approximately $500 to $700 dollars, making it somewhere between a million and a half and a little over two million dollars that the test had thus-far generated.

Mikovits might have been overstepping her boundaries in answering the question. She was head of the research lab, not the clinical lab. In the end, all a scientist has is his or her integrity, and she didn't want anybody to ever believe she would shade the truth, regardless of whom it might offend. They had asked her opinion and she had given it, without regard for the consequences.

That was the duty of a scientist.

*　*　*

And now began what Mikovits considered to be the wholesale abandonment of the patients.

The medical community did not want to look at the question of whether mouse retroviruses (probably of the type which had contaminated research labs for decades) had somehow jumped into the human population. Once again, scientists turned their backs on ME/CFS patients, like a herd of skittish deer.

The release of the findings from the Blood Working Group on the day of Mikovits's talk at the Ottawa Conference was part of the strategy. It was curious how the mission of the Blood Working Group had morphed from the question of whether XMRV posed a danger to the blood supply, to whether XMRV was associated with ME/CFS.

Also released on the day of Mikovits's talk (September 23, 2011) was an article in *Science* entitled "False Positive" by Jon Cohen and Martin Enserink.[37]

The article noted the original findings published two years earlier in *Science* linking a gamma retrovirus to CFS, and that questions had been raised. And leading the charge was the man she had tried her entire professional life to avoid. In the article, Gallo is quoted as saying,

> Once claims of etiology were made, I just gasped for breath. My own experience argued to me that it's best to stay away from this one.[38]

For somebody attempting to figure out the investigative approach of the public health authorities, the real question did not seem to be whether scientists had found a promising lead into what was causing ME/CFS, but whether they were defending XMRV as if the retrovirus was the wealthy defendant in a high-profile murder trial.

Discussing the Lo and Alter study, the *Science* article by Cohen and Enserink mentioned how many previous efforts at discovering what lay behind ME/CFS had failed to pan out, suggesting that this was just one more such regrettable failure.[39]

And of course they had to include Coffin's remark comparing Mikovits to Joan of Arc and that "the scientists will burn her at the stake, but her faithful following will have her canonized."[40]

Cohen and Enserink gave the final words in this section of the article to Mikovits, noting that she viewed the Lo/Alter paper as confirming her work and that from the beginning:

> she viewed XMRV as one of many gammaretroviruses, which includes the MLVs involved with CFS. In the Lombardi study, some patients tested negative in PCR tests for XMRV and yet produced MLV-related proteins, she claimed, but they decided to count them as negatives. She has another serious regret about the paper. "I'd not put XMRV into the title," Mikovits says. "We never considered it would be a single sequence[41]

The number from the healthy controls in the Mikovits study meant that more than ten million Americans were infected with a mouse leukemia related retrovirus.

Then along came Lo and Alter who found another mouse leukemia retrovirus present in 86.5 percent of ME/CFS patients and 6.8 percent of healthy controls.

The 6.8 percent number in the Lo/Alter study translated to more than twenty-one million citizens in the United States infected with a mouse-related leukemia retrovirus. With somewhere between ten and twenty-one million Americans infected with a mouse related leukemia retrovirus whether or not linked to a debilitating condition like ME/CFS, and possibly other neuroimmune diseases like autism, it didn't appear that anybody was raising an alarm.

Why?

Has it become bad form in science to ask inconvenient questions?

* * *

A few days later, after she left the lab on September 29, 2011, Mikovits learned that Lombardi, who worked at the clinical lab and not in her research lab, was claiming that a cell line purchased with WPI grant money belonged to him.

Mikovits knew it was also the end of the government fiscal year and Lombardi had not done any work on the grant and had stated as much to Carli West Kinne and Annette and Harvey in mid-July. If he had not done any work during the reporting period it would be a federal crime to claim part of his salary for the grant.

In the past few months Mikovits had scientifically defended the XMRV ME/CFS research against claims of mouse contamination, Coffin's hybrid virus, which could only have been created after the appearance of ME/CFS, Silverman's infectious molecular clone, VP62, which had contaminated samples, plasmid contamination, and then she'd told her employers that they couldn't submit their clinical lab's data to the Blood Working Group because it was substandard, and that they had to stop selling their potentially lucrative XMRV blood test.

It's probably not surprising given this chain of events and the stress everybody was under that, when Mikovits answered her cell phone as she walked home after work that day, that it was Annette Whittemore, telling Mikovits she had been fired for insolence. It seems it's rude to refuse a direct order to commit a federal crime, thought Mikovits.

Mikovits and Annette Whittemore had experienced their share of disagreements in the past and had been able to resolve them. Mikovits expected they would be able to sort out all their problems in the next few days.

She was very wrong.

CHAPTER TWENTY

The Lipkin Study

*She has come across as a scientist who really believes in the
importance of truth.*[1]

—Dr. Ian Lipkin, speaking about his collaboration
with Dr. Judy Mikovits

Mikovits had, in the XMRV skirmishes, developed more respect for the
Dr. Ian Lipkin, "World's Most Celebrated Virus Hunter," according to
Discover magazine.[2]

But Mikovits found that Lipkin could run hot and cold.

He was capable of startling generosity, but could also be brusque and
spout negative opinions with such vehemence that Mikovits felt she had to
get up and leave. Despite his place at the pinnacle of the medical research
community, Mikovits thought Lipkin must have often been frustrated by
how little he could affect the course of events when they fought the tide
of the prevailing medical establishment. But he also had a savvy way of
politicking his way through the hierarchy.

He would have several opportunities during the course of their
collaboration to distance himself from Mikovits without even a formal
break, but he didn't take them. He could have just pushed the study into
obscurity like the illness ME/CFS had done to patients—through an
unfavorable chance comment to a reporter or a choice to remain silent when
things got heated—that might have signaled a lack of support.

Yet in every early instance when the fate of the endeavor hung in the balance, he acted like the quintessential humanist, and rose up on her behalf.

When Lipkin spoke to the press and said things she herself might have said, despite his esteemed position, it was as if all the scientists around him had suddenly gone deaf, yet he wasn't attacked the way she was.

For example, when Lipkin spoke about autism as he did in an interview with *Discover* he could've been speaking from the main stage of AutismOne, which Mikovits and Montagnier had been criticized for attending. Asked if autism could be another version of a pediatric autoimmune neuropsychiatric disorder (PANDAS), he replied:

> It's possible, in some people. There is probably a group of people, who have a genetic component to autism, and for them, there may not be much of a trigger or any trigger at all required. Another group is genetically predisposed, and if they encounter some factor or factors, individually or in combination, it could result in the onset or the aggravation of a neurodevelopmental disorder; by factors I include everything from heavy metals to infection.[3]

When Lipkin switched the channel to a program the medical community and scientific press liked, they tuned back in, and lavished him with praise. Despite the role Lipkin tried to play in this growing Greek tragedy, it seemed like someone else had written their roles.

If Mikovits was to be Joan of Arc, burned at the stake by her fellow scientists, perhaps Lipkin would at least stop the scientists from lighting the fire.

* * *

The trip to New York from southern California on December 15 of 2011, to finalize plans for the Lipkin study was complicated by a bizarre request from Lipkin.[4] Lipkin told Mikovits she couldn't fly through Washington Dulles Airport on her way to New York. It all seemed in line with the bizarre edict from Anthony Fauci and Harold Varmus that the National Cancer Institute would be her home lab for the Lipkin study, but she couldn't set foot on the grounds or else she would be escorted away by security.

"What do you mean I can't fly through Dulles?" Dr. Mikovits asked.

Lipkin was very kind and patient and the situation rankled him as well. "We all know you didn't do anything, Judy. Let's just try and get your reputation back."

So she steered clear of the nation's capital as requested.

* * *

Complying with the Fauci/Varmus edict made it difficult for Mikovits to work on the study. In doing his part Ruscetti needed to take pictures of the cells he was working on, then email them to Mikovits. She would look at them on her computer as she sat in a coffeehouse in southern California talking to Frank about what she was seeing. Even with all of these obstructions the work proceeded relatively smoothly, and by May they were ready to unblind the samples.

She had expected the association of XMRV with ME/CFS to go down significantly based on the new exclusion of patients with:

- serologic evidence of infection with human immune-deficiency virus (HIV),
- hepatitis B virus,
- hepatitis C virus,
- Treponema pallidium,
- B burgdorferi (the Lyme disease spirochete),
- medical or psychiatric illness that might be associated with fatigue,
- abnormal serum characteristics, and
- thyroid functions.

Instead of the association of XMRV to be around 67 percent as she'd found in her study, or the 87.5 percent Lo and Alter had found for a wider range of mouse leukemia viruses, the way the population had been whittled down through exclusions rendered it indistinguishable from a normal cross-section of the population.

Had they excluded all of those patients who might have the virus?

Still, the spleen-focus forming virus (SFFV) antibody test had come through. Six percent of the controls showed evidence of antibodies to something close to a mouse leukemia virus and 6 percent of the patients with ME/CFS showed the same. An antibody to SFFV env was in 6 percent of the population, a value confirmed in controls of all the positive studies. But it didn't show an association with this population of so-called "chronic fatigue syndrome" patients.

* * *

This was the end of a very bad year for Mikovits: the partial retraction of the *Science* article; her firing and jail time; full retraction by *Science* in December 2011; Lo/Alter retraction in Jan 2012; and now the Lipkin results.

After the shock had worn off, she did a post-mortem on what had happened in the Lipkin study. She wondered what the point had been. Were they at all interested in finding out what was happening in ME/CFS and other neurological disorders?

Paul Cheney (one of the two original investigators with Dan Peterson in Lake Tahoe), was also baffled by many of the exclusions in the Lipkin study. The exclusion of patients with thyroid problems especially troubled him. "Thyroiditis is fairly common in this illness and there was a recent report, I believe out of the UK, which said that 82 percent of people complaining of chronic fatigue had evidence of thyroiditis. Another paper showed that one of the tissue foci for HHV-6A is the thyroid. Jim Jones, way back when in describing the first cases of CFS, said 85 percent of them had auto-antibodies against the thyroid. So by definition, 85 percent of CFS/ME patients have thyroiditis."[5]

Cheney's observations had been made by other scientists, including in a 2008 study published in *Genomics* that found, "statistically significant differences between CFS and control networks determined mainly by remodeling around pituitary and thyroid nodes as well as an emergent immune sub-network" that "align with known mechanisms of chronic inflammation and support possible immune-mediated loss of thyroid function in CFS."[6]

Cheney was also disturbed by the exclusion of patients with borrelia, the spirochete that causes Lyme disease. He also noted that one of the most common antigens found in lyme patients is called P41, which is also an early antigen of HHV-6 and a cleavage product of the SFFV and XMRV envelope precursor.

The exclusion of patients with a "medical or psychiatric illness that might be associated with fatigue" was a giant head-scratcher. "That particularly doesn't make much sense. I suppose it's how hard they hit that." Many chronic illnesses can cause secondary, often transient, grief and depression. "And medical conditions associated with fatigue? Well, this is a medical condition associated with fatigue, so they're trying to exclude the very disease they're supposed to be looking for."[7]

* * *

The overall difficulty Cheney and Mikovits had with the exclusions based on infection by another agent, was that the very thing that led her to suspect an immune-altering retrovirus in ME/CFS was that putting the blood of these patients on a pathogen micro-array made it light up like a Christmas tree.

So again, this gated community-style study was akin to excluding AIDS patients from research because of secondary coinfections, rather than including those with telltale immune markers. The ME/CFS patients were loaded with viruses, bacteria, and parasites.

So the study should examine the reasons why.

As with the other negative studies that seemed to be using crab pots to catch unrelated fish, Lipkin's study appeared to be another survey of the wrong population, those who might never have had ME/CFS to at all.

* * *

"The one point you could make in favor of Lipkin is that he wanted to create a study that was airtight, if it was positive," said Paul Cheney, surveying the CFS/ME landscape from his days as a young physician working with Dan Peterson at Lake Tahoe in 1984 to Columbia University in 2012 and the release of the Lipkin study. "But he runs the risk of not finding the very thing he was supposed to find. And that's the risk he took, and it may have ended up that way, unintended."[8]

For microbiologist, chemist, and CFS/ME patient, Gerwyn Morris, the problems with the Lipkin study started with the group selected for inclusion.[9] "They used unvalidated, home-made selection criteria, not based on any international consensus, which is bad enough, but something they cobbled together at the meeting. From that point on in my view, the study has almost no meaning."

To Morris, the Lipkin investigation was as unscientific as a study could be. "Because you do not know whether the patient group you had in the *Science* study, in terms of underlying etiology or pathophysiology were the same patients that were in the Lipkin study. There is absolutely no way of telling. You can't expect to classify people on the basis of symptom clusters, and expect people to have the same underlying disease. It's just nonsense."[10]

Like Cheney, Morris questioned whether the Lipkin study had been designed to find anything. "If they had been genuinely looking for patients with a possibility of having a retroviral infection, they would have selected patients on the basis of running a very similar cytokine profile that Mikovits and Lombardi produced for so-called XMRV positive patients. It's the sort of cytokine profile you'd expect from a retrovirus infection of this sort of

HTLV-1 type, which is very similar in the way it's replicated with the murine leukemia viruses. The inclusion criteria were rubbish."

When Morris reviewed both the inclusion and exclusion criteria he said it was like, "Looking for HIV, but excluding homosexual males, IV drug users, and those who'd received a blood transfusion."[11]

* * *

For Deckoff-Jones, the former clinical director of the WPI, the problem with the Lipkin study was the failure to look for related viruses.

"They had to put it to bed. The study was an incredible waste of money because they knew the outcome before they started. They should have redesigned it (after the revelations about VP-62 and Silverman's mistake) to look for what was really there. And we do have the technology." Deckoff-Jones felt Lipkin did what was "politically required." "He didn't get to be the world's greatest virus hunter with three floors and 65 people working under him because he doesn't know how to play the system."[12]

* * *

Even with all these problems, Mikovits had participated in the deliberations and it was her duty to see it through to the end.

"We set out to see if XMRV as defined by Silverman was in the population, and it wasn't," Lipkin urged her.

And he was right.

Still, it felt like they were presenting only one sliver of truth; whether the lab contaminant XMRV Silverman was afflicting ME/CFS patients, not whether a retrovirus existed in ME/CFS at all.

The large truth revolved around the question of whether a mouse leukemia related retrovirus, or some chimera of mouse and human virus, was causing disease in a significant number of a well-defined patient population. Lipkin, of course knew this for he had already collected samples with private money that had been donated to a private CFS initiative.

The one issue, which threatened to derail both her participation and that of Frank Ruscetti in the study was whether the ensuing article should deal with the issue of antiretroviral therapy.

Lipkin had written strong language in the draft condemning the use of antiretroviral therapy by the patient population.[13] For Mikovits the use of antiretroviral therapy presented an acceptable risk on a trial basis for these very sick patients. And besides, the study wasn't about the use of antiretroviral therapy, so their disagreement wasn't relevant. If she wasn't supposed to go beyond the boundaries of what had been agreed to, neither should Lipkin.

It all came to a head in August of 2012. Mikovits and Ruscetti had told Lipkin they'd take their names off the paper if any of the antiretroviral comments remained in the manuscript.

Lipkin called Mikovits while she was meeting with an autism doctor and some patients. Mikovits was working for free and Lipkin knew her desperate financial situation brought about by the crushing legal bills she'd had to pay because of the Whittemores and the difficulty of finding a job after having her mug shot plastered around the world in *Science*.

Prior to the call, Frank had sent Lipkin an angry email, pointing out that he and Alter would be fine, both professionally and financially, but Judy was ruined. Frank told Ian that Judy would probably never be allowed to set foot in a lab again and had done nothing wrong.

This claim by Ruscetti seemed to deeply unsettle Lipkin.

When Lipkin called Mikovits she was in still in the autism meeting. She excused herself and stepped outside. Lipkin's voice softened as he learned what she was doing, continuing to try to help people even though she wasn't getting any money for it. "I'm trying to help restore your reputation," he said.

"Frank is right. You can't restore my reputation," she replied. "I'll never be able to work again. Nobody will ever believe a word I say, and I didn't make the mistake. The only mistake I made was saying it was Silverman's XMRV because he claimed he had a full sequence virus and he didn't."

Then they got down to negotiating.

Lipkin would take out the section of the paper which was critical of the use of antiretrovirals in the patient population. Judy would keep in the paper the results of the spleen-focus-forming virus antibody test, showing 6 percent of controls and 6 percent of the patients were positive. The positive numbers for the spleen-focus forming virus antibody test had remained constant among the population from the original *Science* paper, through the Singh and Alter-Lo study, the Blood Working Group, and now the Lipkin study. There was still something unexplained in the blood of ten to twenty million Americans.

Mikovits thought that might be enough to let her start to rebuild her reputation.

She knew she had to give Lipkin something in return and decided it would be the gag sequences they'd recovered from both the patient and control samples. It certainly wasn't a full sequence, but it could be enough to cause an abnormal immune responses in some people. She'd argued that the gag sequences should be in the main body of the paper, but Lipkin thought that would confuse the issue.

In return for Lipkin taking out his comments on the use of anti-retroviral therapy and allowing her to keep the data on the results of the antibody test for the spleen-focus forming virus, she agreed to let the gag sequences be placed in the supplementary materials.

She knew that at the press conference to come she was supposed to be the good little girl, praising the wonderful collaboration between all of the researchers, and the kind and caring staff of the NIH. And maybe, just maybe, when all the dust had settled she could get back into a lab and find out what was really going on with this disease.

To help rebuild her career, Lipkin promised they would collaborate with Mikovits and Ruscetti as coauthors on the next two or three papers he did in ME/CFS.

* * *

On September 16, 2012, two days before the press conference for the Lipkin study, Ian Lipkin sent an email to Judy Mikovits, copying Frank Ruscetti, Martin Enserink of *Science*, Harvey Alter of the NIH, and John Coffin regarding the approach he wanted to take at the press conference. The email appeared to have been in response to a request from Martin Enserink about the release of the upcoming study and was addressed to Judy but was intended for the entire group. Lipkin wrote that it was important for scientists to recognize that their words had an enormous power to influence public opinion. He stated that while he believed XMRV met the definition of a virus, it seemed likely to have been a chimera created in a laboratory, but the last papragraph of the recent Chiu paper still left the door open to implicate XMRV in human disease.[14]

On September 17, there was a very nice dinner for all the participants in the study. During the dinner, Lipkin told Judy he would hire her as a consultant but said it would not be for very much money. Judy said that would be fine, adding that she did not need much money.

With everybody aware of what was to be said at the press conference, the stage was set for the final act of the XMRV in ME/CFS drama.

* * *

"My name is Ian Lipkin and I'm the John Snow Professor of Epidemiology at Columbia University in the Center for Infection and Immunity. This broadcast is coming to you from the campus of Columbia University in the city of New York."[15]

Instead of being attired in his usual jeans and sneakers to race between the floors of his "empire of viruses" like an eager grad student, Lipkin was dressed in a sober suit and tie. His voice was calm, sometimes passionate and even funny, but it was that of a scientist wanting to present his findings to the public.

"I'd like to say it's a beautiful day here in New York," Lipkin began. "There is some rain, but there is some blue peeking through which I think is a wonderful metaphor for the morning."

Lipkin began by describing the disease and previous research. "Chronic fatigue syndrome/myalgic encephalomyelitis is an unexplained, incapacitating syndrome that afflicts at least a million people in the United States. For many years there has been a thought that it could be attributed to an infectious process of some sort because many of the individuals with this syndrome have malaise, night sweats, lymphadenopathy, sore throat, and fever. And there have been a wide range of microbes that have been proposed as causative agents—herpes virus, borna viruses, borrelia—but none has really captured the attention of the scientific community and the population more than retroviruses.[16]

Lipkin noted the papers of Mikovits and the confirmatory study of Harvey Alter of the National Institutes of Health and Shyh-Ching Lo of the Food and Drug Administration. Lipkin also observed that while many commentators had suggested the Lo/Alter study could not be considered a confirmation it was his opinion that: "The viruses discussed were somewhat different, but they really fell into fairly similar classification. It's not important to get into the details of viral taxonomy. It's simply important to say that these were retroviruses that were related and they were distinctive in many ways."[17]

Many but not all groups were unable to replicate these findings, but Lipkin generously noted that none of them had the power to definitively refute the findings, nor did they offer the investigators an opportunity to use their best methods to answer the question.

"Almost two years ago with support from Francis Collins, director of the NIH, Anthony Fauci, director of the NIAID, Harold Varmus, director of the NCI, Peggy Hamburg, commissioner of the FDA, we initiated a study to look into this question in an absolutely clear cut fashion so I'm just going to talk about it very briefly."[18]

Lipkin acknowledged the clinical investigators: Anthony Komaroff out of Boston, Dan Peterson out of Nevada, Nancy Klimas out of Miami, Jose Montoya out of Palo Alto, Susan Levine from New York, and Lucinda Bateman from Salt Lake City:

"They worked tirelessly to define the clinical criteria that would be used in the study. We used Fukada criteria, the Canadian criteria, we tried to hedge the study in every which way so we could favor an infectious hypothesis if one were present."[19]

Mikovits couldn't help but grimace at what she felt to be a misleading statement. If they had wanted to "favor an infectious hypothesis" they obviously wouldn't have kicked out those patients with hepatitis B, hepatitis C, borrelia burgdorferi infection, or those with a "medical or psychiatric illness that might be associated with fatigue." The condition was called "chronic fatigue syndrome," yet they were excluding one of three words in the title, leaving only "chronic" and "syndrome"—a nebulous description at best.

Lipkin continued with his presentation: "And the laboratories, the efforts were led by Shyh-Ching Lo at the FDA, Harvey Alter, Judy Mikovits and Frank Ruscetti here on the podium, Bill Switzer at the Centers for Disease Control, and Maureen Hanson of Cornell University who can't be with us today. The result of the genetics test were chiefly PCR and people used whatever methods they wanted to use, found no evidence of XMRV or related viruses in either the subjects with chronic fatigue syndrome/myalgic encephalomyelitis or the controls."[20]

Then Lipkin made a concession, keeping information in the study that went against that thesis that mouse leukemia viruses weren't causing problems in the human population, although he tried to scuttle past it as quickly as possible: *"There was in fact a finding of some antibody responses, we don't really know what that means, in about six percent of the controls and six percent of the experimental subjects.* I don't mean experimental subjects, excuse me, but subjects with disease. But again, there was no association between the presence of those antibodies and disease. And we really don't know the validity of those particular tests because we don't have positive controls here with which to work" [emphasis added].[21]

When he said of patients, "I don't mean experimental subjects, excuse me, but subjects with disease," it almost seemed like a Freudian slip, since he had excluded so many "subjects with disease" and "experimental subjects" sounded vaguely dehumanizing.

Lipkin had also skated past the troubling information that somewhere around twenty million Americans had abnormal antibodies to proteins, which were closely related to those from a mouse leukemia retrovirus.

Along the way Lipkin would dazzle them with the nice shiny promise of cutting-edge science: "Many in the patient community have written me over the past twelve hours since there have been leaks about the press release and findings of the paper with dismay. With concern that this meant the end of ME/CFS research. Nothing could be further from the truth. Everybody here at this podium, scientists around the world are committed to solving this problem. It is likely to be a constellation of disorders, not necessarily a single agent, be it viral, bacterial, or otherwise.

"The samples that were collected during the course of this project will be an extraordinary resource for addressing questions related to the causes, the treatment, the management, the pathogenesis, the basic science of these disorders.

"And these samples have been collected, they've been stored in ultra-low freezers, they will be released to qualified investigators who make application through the NIH. I'm also informed that there are at least three active grant opportunities that will allow people to study these materials. So this has been a highly leveraged study."[22]

Lipkin next turned to the investigators themselves and it was here that Mikovits was most surprised. He was trying to restore her reputation, and although she doubted it would make even a ripple among the professional community, she was grateful: "The other thing I want to say is that it took extraordinary courage on the part of Drs. Alter, Mikovits, and Ruscetti to participate in this study. To see it through to conclusion. To actually say, we now believe based on the data we obtained that this first report may have been misleading. They too are committed to working on this problem, on this challenge.

"And I can say in all my years of doing science I cannot think of a single instance where an investigator came back, went through this sort of process, and stood up and said, I want to be counted. I want to admit we made an error here, but I'm committed to moving forward and being public and forthright with that science.

"I'm going to read one quote from Judy Mikovits which I think is representative of what many people say, who participated in this study. 'I greatly appreciated the opportunity to fully participate in this unprecedented study, unprecedented because of the level of collaboration, the integrity of the investigators and the commitment of the National Institutes of

Health to provide its considerable resources to the CFS community for this important study. Although I am disappointed we found no association of XMRV/pMLVs with CFS, the silver lining is that our 2009 *Science* report resulted in global awareness of this crippling disease and has sparked new interest in this research. I am dedicated to continuing to work with leaders in the field of pathogen discovery in the effort to determine the etiologic agent for CFS.'"[23]

Lipkin then turned to his left to directly face Mikovits and said, "Judy, I think that was extraordinary. And I applaud you for it." Lipkin began to clap, then looked to the audience and nodded as the cue for them to clap as well. There was a light ripple of applause and Lipkin opened the floor for questions.

* * *

Journalist Hillary Johnson was the first to pipe up. Because she suffered from ME/CFS, she spoke with labored breath: "Dr. Lipkin, these patients have tended to be very ill for many years, even decades, some of them, and there's been a lot of evidence for infection over the years. Might there not potentially be retroviral infection in tissue other than blood, such as spleen, liver, thyroid, lymph nodes, lymphoid tissues, etc? Is that out of the question to be explored in the future?"[24]

Johnson and Mikovits had developed a close relationship over the past few years and Johnson understood better than most outside journalists the unanswered questions of this research. The reason Mikovits had taken multiple patient samples over several months was exactly because of the concern expressed by Hillary. Mikovits thought that evidence of the virus in the blood might wax and wane over time, so it was only by testing several samples from the same individual over a period of time you could positively rule out infection.

Lipkin answered: "In science and medicine, you can never say never. That's part of the nature of what we do. The question of course is when you put a study like this in operation you're testing work that has been previously reported. And in the initial report, the findings were in blood. So, where would one stop?

"You could survey potentially the entire body. And then you could make the argument that this was not one of those individuals, we need to do another hundred and so forth. So what we try and do in designing studies to look at the likelihood of an association with a disease is to examine the first paper, which came out to describe the finding, to make certain to at least replicate it with a fair margin. And that's the approach we took. But you're correct. I cannot say there is no person anywhere in the world who is not infected with a retrovirus in some organ that we have not sampled. But I don't see any way in which we can formally and practically test that hypothesis."[25]

Well, actually, Mikovits had already shown the way to "formally and practically test that hypothesis," by including patients with medical fatigue, thyroid problems, and infections, and by looking in her original study for abnormal natural killer cell presentations, elevated levels of inflammatory markers for cytokines and chemokines, and positive results for antibodies to the spleen-focus forming virus test. The way to "test that hypothesis" was fairly simple: include genuine patients in the study group instead of stand-ins.

Mikovits hadn't excluded patients due to coinfections with hepatits B or C, treponema pallidium, borrelia burgdorferi, whether they had a medical or psychiatric illness that might be associated with fatigue, or abnormal characteristics in their blood, or problems with thyroid functions. She had been prepared to really hunt, taking a targeted and not a scattershot approach.

After Harvey Alter made a few comments, Lipkin had some additional things he wanted to say: "In addition, there are people who will be using immunological methods, this is not work we would do, but others have already proposed it, that would get at questions of serology.

"And whether or not there's previous infection and immune responses that support that hypothesis. There's an independent effort, we have some three hundred samples that will be studied using this approach, a hundred and fifty from patients and a hundred and fifty from controls. A bit less, but that's roughly where we are. That's under the auspices of the Hutchins Family Foundation, which has an even larger cohort. And in addition to doing blood that group is also going to be looking at meta-genomics. It's going to look at fecal flora to see whether there are some changes in the microbiome which might also be associated with disease."[26]

Mikovits thought the last elements were good and even believed that looking at the microbiome of the digestive tract of patients with CFS was a reasonable investigation. In fact, Maureen Hanson had written a beautiful grant proposal to study the gut microbiome with Ruth Ley in 2010, which was rejected twice, but ultimately funded. If a mouse leukemia retrovirus was causing disturbances in the immune system, the damage could logically originate in the gastrointestinal system. A large part of what happened in the immune system was determined by what took place in the gut so it was a sensible place to investigate.

The next question came from Deborah Warnoff. At one point she'd noted that due to the decreased lifespan of ME/CFS sufferers according to the government tables, she had about forty-five days left to live.

After that grim statement, she asked: "Just to follow up in Hillary's question and what you just mentioned about the CFS initiative which will do flora. I'm just wondering if there's someplace where one can't get in, and you've done so much work compiling great cohorts for the CFA initiative and for this project. And with HIV a good deal of work was done with lymph glands and so forth. I just think it would be very tempting since you could aspirate fairly readily, to go back and get lymph for all these people.

"I'm a little surprised, although I'm not used to the confines of the scientific method, that you could resist the temptation to extract a bit of lymph as you were going along. But since that was used in HIV, I just wonder if that wouldn't be a very tempting target?"[27]

Mikovits thought Warnoff was right on the money. The lymph glands were a reservoir for HIV and samples were fairly easy to obtain compared to other organ tissues.

Lipkin responded: "Well, doing lymph node dissection is far more invasive than phlebotomy, and once again, the objective here is to follow up on studies that were based on analysis of blood. And the findings in those studies were so robust, that when we used, I think, was it Mady, three times as many subjects as we thought we needed to confirm the results in that study, we felt that we were powering it sufficiently, and adequately, and rigorously test the validity of those first findings. Again, this is what I do for a living, we discover viruses. So, we frequently use tissue samples and other samples and so forth. But the point of this study was to test the original findings."[28]

In other words, Lipkin offered a big fat *no* to the woman who'd said she might have a month and a half to live. They weren't going to be testing lymph tissue for retroviruses, even though such an approach in AIDS had shown that lymph was a reservoir for HIV (and lymph was relatively easy tissue to extract), so Deborah Warnoff had better pull out her bucket list for her remaining forty-five days, because nobody was going to aspirate her nodes.

It fell to Harvey Alter to pull the curtain away and reveal what seemed to be the opinion of most of the research community: "One has to resist the temptation to keep the murine retroviral hypothesis alive. What you're suggesting is to say, well, it all came out negative, but maybe it's still true by doing something else. And I think rather you have to say, really, this study was quite definitive. And has basically excluded these agents as the cause of CFS. And yet, this is the impetus for moving on to more extensive, deeper studies to find the actual cause."[29]

Quite definitive? Every time the door was cracked open around retroviral theories for ME/CFS, patients felt the foreshadowing of the same door slamming shut.

Even Elaine DeFreitas, the researcher who also published evidence of a retrovirus in ME/CFS in the early 1990s, told Hillary Johnson, "I could see myself twenty years from now, when I'm a high school biology teacher and someone calls and says, 'Hey they just found a retrovirus in CFS.' And maybe that's how it will happen. And I know how I'll feel—I'll feel great."[30] But did there have to be decades of treading old waters in between? Couldn't someone aspirate a lymph node?

Lipkin spent some additional time defending the study, but then made remarkable claims: "Our first foray, if you call it that, into chronic fatigue syndrome was back in the middle 1990s when a Japanese team said 50 percent of CFS in Japan was due to Borna disease virus, a virus that we'd identified several years earlier. And when we went through this and were unable to find any link between Borna virus and this syndrome, the one thing that did impress me, was that there was an enormous amount of immune-reactivity that appeared to be non-specific in these individuals.

"So at a time when people were saying this was a psychosomatic disorder, I said, two-thirds to three-quarters of these individuals whom we've studied have polyclonal B cell activation. They're sick. We don't know why. But they're sick. That's really the take-home point."[31]

Mikovits did have to agree the finding of polyclonal B cell activation was rock-solid. The immune system of these individuals was way out of whack and needed to be further studied. Mikovits didn't care if they made the link now or twenty years from now with a mouse leukemia related retrovirus, but patients like Deborah Warnoff and Hillary Johnson needed therapies now.

Yet it was precisely at junctures like this that Lipkin's momentarily supportive words seemed to carry the least weight with his fellow scientists. The research community had no trouble with his efforts to "un-discover" previous findings, but when he pointed out facts that might lead to progress against these diseases he was talking to a pretty rigid wall.

Lipkin continued in this vein for several minutes and some of the other investigators offered their opinions before the ideas expressed were summarized by Mady Hornig, the director of Translational Medicine at Columbia: "There are clearly a host of questions that remain that we're dedicated to pursuing along with other investigators. We know, as Ian alluded, that there appears to be at least in a portion of patients with CFS, an immune-reactivity, differential immune profiles.

"We don't know what those mean, why they exist, we don't know which subsets of patients may be most likely to have those sorts of indices. But with these sorts of study design, we've done such exquisite characterization of patients, we've controlled for so many factors that could alter the infectious agents in the catchment area, you know, age and sex, and all of those other factors, so we might begin to pursue what the indices might be that may lead us to some of the etiologic factors."[32]

The next question came again from Hillary Johnson and addressed sexism in disease investigation: "The origination of the discovery of XMRV was by, as you know, Joe DeRisi and Silverman, and also a number of papers published in 2008–2009 relating XMRVs to prostate cancers as well. Or associating it, not saying it was the cause of it. Nor did the Mikovits or Ruscetti paper say it was the cause either.

"My question is, why so much emphasis, so much urgency to retract the connection between XMRV and CFS and there hasn't been any calls for the retraction of the prostate cancer papers where XMRV and prostate cancer were associated?"[33]

Lipkin didn't appear to want to discuss the issue. Hillary knew from her previous conversations that Lipkin was appalled at the way Silverman, DeRisi, and Gannon had stuck their heads in the sand over this issue, but he was avoiding public debate.

Lipkin replied: "I think that's an interesting question. I don't see any delicate way for us to address it, and it's off-point. So, I'll be happy to talk with you about that off-line, but not on-line. Other questions?"[34]

The next question was from Deborah Warnoff again, and it concerned how the patient community could speed up the pace of research for progress within her lifetime (despite the forty-five day statistical prognosis, she was hoping for a few more years). Lipkin's reply sounded like a call to arms, but because of the very nature of their disease, political activism to these patients was like embarking on a space shuttle flight.

Lipkin was calling for a crusade he knew would be difficult, if not impossible, to mount (as if patients had not already tried): "I've been involved with a number of organizations that have been successful in promoting research. The most successful of course was the HIV/AIDS community and thereafter the autism community. And you know there have been others that have been more successful and less successful.

"There's no question but that your political leadership will respond to pressure. So, this may not be a very popular thing to say with everyone because at the National Institutes of Health we like to focus on basic science and applied science without talking about specific diseases. But the fact remains that if you say there's going to be a certain amount of work that's going to be focused on a particular area, and you can focus it in that way, then people will work in that area."

"So I would encourage you to try and motivate your colleagues, people with this disease, related diseases, and the families and loved ones, to request additional support in this area. This is what happened with autism and it made a huge difference in the investment at the level of the NIH and even in the DOD.

"Again, this is very difficult for people who have a chronic illness, which is debilitating. It's difficult to be aggressive in pursuing this level of support, but I think it's what you can do."[35]

Even Harvey Alter who had just seemed suddenly so intent about ditching the mouse leukemia retrovirus theory seemed to get excited by the question of what to do next in CFS research. Alter said: "I think what's going to accelerate things at this point is that there's been a shift in thinking. To date it's been looking at organisms, mostly viruses, as causes of this disease. And every one has sort of petered out with further investigation.

"But now I think we're thinking of the uniqueness of the host. And why do certain people respond to common viruses in a different way? Why is there a hyper-response, to probably cytokine production, other things, released in certain patients in response to viruses that most people handle differently?"[36]

Mikovits, sitting quietly for the most part, thought the immune response did seem to be key. While the evidence from the spleen-focus forming antibody virus test showed that something similar to a mouse leukemia retrovirus was present in three to six percent of the US population, the majority of these people were controlling it. But for how long, Mikovits wondered?

Were these millions of people a ticking time-bomb of chronic disease as their immune systems underwent the changes associated with aging or other environmental challenges? Was the scientific research community acting like the guy in the joke who jumped off the Empire State building and as he passed each floor called out, "Everything's good so far!"[37]

There was some further discussion of how the immune system was acting abnormally in these patients and then a question came in over the internet from Martin Enserink, one of the writers from the journal *Science*, who had penned the article *False Positive* with Jon Cohen.

Enserink asked: "Can you explain why the many failed attempts to replicate your 2009 study, the Coffin-Pathak paper about XMRV's origins, and the Blood Working Group's study published in 2009 failed to convince you that your 2009 study was wrong? And what sets this new study apart?"[38]

The question seemed to annoy Lipkin and he spoke with a clipped tone in his voice. "I think we've already been over that ad nauseum. We've talked about the fact that the first study was not appropriately powered. We've talked about the fact that the other studies did not allow them to do exactly what they wanted to do and the way they wanted to do it. I don't know that there's anything else to add to that."[39]

Lipkin asked if Ruscetti or Mikovits had anything to add. They had nothing else to say to *Science*.

* * *

The final question in the press conference came from Hillary Johnson: "What were the good things that came from this work? And we've seen that XMRV is a chimera as it's been described, and isn't this somewhat worrisome in that we've seen how rapidly XMRV can replicate in human cells? We saw what XMRV did in the Rhesus macaques, and we're still using MLV (mouse leukemia retroviruses) and gammaretrovirus research in creating recombinants and vaccine therapy. Has this caused people at NIH (National Institutes of Health) to be somewhat concerned and even alarmed at the fact that these recombinants can occur in a lab? That there can be man-made chimera viruses?"[40]

Maybe it was because of the lateness of the meeting, or the fact that those with ME/CFS who were present had exhausted their meager reserves of energy, but Dr. Lipkin's response could have been the opening scene of a science fiction movie where the best intentions of scientists go horribly wrong: "There's a term that's used to describe this. It's called gain a function research. And it's come up most recently in the context of H5N1, the work that was done in Rotterdam and Wisconsin and Japan, and generating recombinant influenza viruses that appear to have more virulence, more transmissibility in ferret models.

"And this is an issue that's being addressed by the scientific community. There is no clear pathway as of yet. It's being developed at very high levels in fact within the World Health Organization. So this is something because whenever you're creating new viruses that have potential to infect humans there is an issue of how that should be addressed and how that should be contained.

"I can tell you with respect to H5N1 in the same journal in which our work has appeared, *MBIO*, there will be a series of reports at the end of this month that tackle this issue. And there will be people speaking for H5N1 recombinant research at various levels, people speaking vociferously against it, and people who are in the middle who are talking about doing work of that sort under high level containment.

"And all these different perspectives will be addressed, debated in *MBIO*, and I encourage you to wait for that. It will be coming up in about a week's time. And there will be a press release and I will probably be participating in that as well, but it won't be here.

"But in any event that is with a disease that is acute. It's something that causes disease very rapidly. But I think the same arguments, the same principles are at play, whether you're talking about something that causes acute or chronic illness.

"And this is something that the scientific community and the community at large is going to have to wrestle with and make decisions about.

"Very good point."[41]

Johnson was asking if the public at large should be worried about man-made chimera viruses that could replicate rapidly in human cells.

And all he could say was, very good point?

But while Lipkin admitted that this fear of "man-made chimera viruses" was a "very good question," he was avoiding the central question at the heart of the ME/CFS epidemic: what had started it all, beginning with the outbreak at the Los Angeles County Hospital in 1934–1935, and why was nobody interested in the question?

* * *

The Lipkin multi-center study would claim one more victim, Max Pfost, Mikovits's loyal research assistant from the Whittemore Peterson Institute. The young man who'd loved science so much, he'd had tattoos of DNA proteins being translated, veins blowing up with viruses coming out of them, immune cells and B cells, as well as a Latin inscription which read *res firma mitescere nescit* ("a firm resolve is not easily broken"), would pay a heavy price.

In the aftermath of her termination, Mikovits had made sure Pfost successfully transferred to the University of Alberta, where he would be a PhD student under the mentorship of of Dr. Andy Mason, a researcher who had also discovered a mouse related virus which had jumped into humans, a betaretrovirus which was linked to an autoimmune liver disease, primary biliary cirrhosis.

Prior to starting work in Alberta, Pfost said he "talked with Andy when I got up there and I said I know you accepted me before all this legal stuff happened and this affidavit showed up and if you're not comfortable with me being here, please tell me. But Mason said he'd be fine with it."[42]

Pfost was generally satisfied with his time in Canada from January of 2012 until the Lipkin study came out in September of that year, but then things changed. It was his birthday week and he'd missed a couple days of school, but "the week prior to that I was speaking with Andy about working on some type of scholarship proposal for a study. He said you should look into some of these things and we can work on that. And I'll help get you set up, get you some more money for school. So I was working on that at home while I was sick and he called me at home for a meeting, like a week after that."[43]

When Pfost arrived, his key card didn't work, just as on that morning he'd arrived at the WPI after Mikovits had been fired. When he got in, he found that the "meeting" was a serious meeting, and they'd brought in a faculty advisor who told Pfost he'd made no progress and that at this rate it would take him twelve years to graduate. "And I found that really weird because I didn't have a lot of results, but I did have results, but they were negative results. [Pfost had been researching subversive viruses.] So for me I was really surprised because they said I had nothing to show for it."[44]

Pfost said, "They explained to me I needed to leave the program, so we're going to offer you $3,500, we'll give you a couple days to think about it. If you don't take it, we're going to do everything in our power to make it miserable for you here and you'll have to leave. I was blown away because I'm in a foreign country, it's Canada, and I don't have a support system in place, and I can't just pick up a job as I found out. You need a work permit. So I didn't know what to do. I'd already had to deal with this whole legal battle with the Whittemores, and watched how that played out, and to me I just didn't want to deal with going up against the wall anymore. So I just said, okay fine, I'll take the money and go home."[45]

At the time Pfost was contacted in March of 2014 he was working for a boat company as a delivery parts driver, and sometimes being allowed to do electrical work. Of his current coworkers he says, "It's funny to listen to them be stressed out about things. They just have no idea. This is the most stress-free job I've ever had in my life."[46]

When asked what Pfost thought he'd learned from these events, he was both philosophical and pessimistic. "I always felt that in the end, science prevails. You have data, you put that data together, and if it's irrefutable, then that's what you go with. And I'm learning that's not the case. You can have good data, and now you've got all these big people at the top and they say 'we don't want that for our journal.' Or, 'they probably shouldn't get any funding because I don't like the project they're working on,' or 'it doesn't make sense.' So now your funding is cut and you can't work. It's tough. I want to get back into it. It's probably taken me a while to realize that. But it feels like science doesn't want me back."[47]

CHAPTER TWENTY-ONE

The Rediscovery of XMRV?

*The results suggest that xenograft approaches commonly used
in the study of human cancer promote the evolution of novel
retroviruses with pathogenic properties.[1]*

—Dr. Vinay Pathak, in a paper released on March 27, 2013,
showing that proteins from the envelope of a mouse
leukemia virus create larger cancerous tumors.

*We found retroviruses in 85 percent of the sample pools. Again, it is
very difficult to know whether or not this is clinically significant or
not. And given the previous experience with retroviruses in chronic
fatigue, I am going to be very clear in telling you, although I am
reporting them in Professor Montoya's samples, neither he, nor we,
have concluded that there is a relationship to disease.[2]*

—Dr. Ian Lipkin in a public conference call with the Centers
for Disease Control on September 10, 2013

In 2009, Mikovits and her team reported finding evidence of XMRV in 67
percent of her ME/CFS patient cohort who also had diminished natural

killer cell activity and other markers of impaired immunity. In 2010 Drs. Shyh-Ching Lo and Harvey Alter reported 83 percent of their cohort of ME/CFS patients showed evidence of mouse leukemia retroviruses. In 2011, Drs. John Coffin and Vinay Pathak, suggested that XMRV or something very similar had arisen through the passage of prostate cancer samples through mouse tissue between 1992–1994 at Case Western University in Ohio.

By 2012, the first two papers had been retracted or their conclusions altered. What had caused the scientific switchback? The studies from the Blood Working Group and Lipkin could not replicate Mikovits's team's original findings. In 2013, Ruscetti re-derived the B cells lines from uncultured frozen materials and they now contained no XMRV.

* * *

In 2012 Ian Lipkin, had to acknowledge the danger recombinant viruses created in labs might pose. The mouse leukemia viruses and the theory of the possible connection to human disease was now considered dead, or as the journal *Nature* proclaimed of Lipkin in an article also released on September 18, 2012, entitled, "The Scientist Who Put the Nail in XMRV's Coffin,"[3] the situation had been "contained."

"Contained" seemed a more appropriate word for the lingering mystery. Scientists did not know what drove the abnormal immune response that even Lipkin had admitted witnessing in ME/CFS. The three positive studies reports using Mikovits's antibody test for spleen-focus forming virus still needed to be explained.

Judy Mikovits might escape the "burning at the stake," but the research community would do the next best thing and try to forget she had ever existed, effectively shunning her. Even the powerful support of Ian Lipkin could do little to bring her back to the field to which she had devoted her life. Lipkin's powers seemed to be restricted to un-discovery of research, not the resurrection of researchers.

The millions who suffered would have to make do with the promise of bright, shiny science in a hazy but amazing future to come, the same life raft they had clung to for decades, rebranded in new packaging. Their salvation might arrive in the form of ultra-low freezers to maintain their samples, deep-sequencing, genomics, RNA profiles, immune system modulators, cytokine and chemokine tests, and investigating the micro-biome of their digestive tracts.

These advances held promise, but where was the urgency, the fast-tracking, seen in other diseases? Nobody would look for the truth of what had happened in 1934–1935 at Los Angeles County General Hospital and sickened nearly two hundred doctors and nurses. Nobody would look for the truth of what happened to start the modern epidemic of ME/CFS which began in 1984–1985 in Incline Village, Nevada, where it was first observed by Dan Peterson and Paul Cheney; and nobody would ask why 1 in 50 children in the United States had autism, a condition that was unknown before its description in 1943 by Leo Kanner, the former Rockefeller Foundation fellow.

* * *

On November 27, 2012, a glimmering curiosity of a paper was submitted to the journal *Retrovirology*. It was unusual for a couple reasons.

First, it was odd because of the subject matter, XMRV, when it was well-known that the editor in chief of *Retrovirology*, Kuan-Teh Jeang, did not believe in the value of research involving the XMRV retrovirus, which his journal had lead a vigorous campaign against for three years since he accepted it was simply a lab creation and had shouted his virulent dismissal at the Leuvan conference in Belgium.

Two months had passed since the publication of the Lipkin study and *Nature* magazine's identification of Lipkin as "The Man Who Put the Nail in XMRV's Coffin."[4] But it seemed this pesky little mouse-human retrovirus wouldn't stay dead and buried: instead it reappeared with church-mouse-quiet in *Retrovirology*, heralding a rebirth. But then, in the awful style of confounding cover-ups, a tragedy also unfolded.

Kuan-Teh Jeang had recently delivered the October 24, 2012, George Khoury Lecture at the NIH on cellular transformation by the human T cell leukemia virus (HTLV-1).[5] On January 27, 2013, a Sunday night, Teh, as he was called by his friends, chief of the Molecular Virology Section of the National Institute of Infection and Allergic Diseases Laboratory of Molecular Microbiology, would go to his office and inexplicably kill himself.

It is almost hard to reconcile his prior dismissal of XMRV with what several of his Asian-American scientific colleagues wrote in an admiring obituary when they said that Teh "always kept his curiosity alive and was ready to embrace serendipity."[6] He sounded, in that description, like another potential ally stolen away.

Whatever the cause of his suicide, the field of retrivology looked more and more like a crime scene, from the willful neglect of patients to institutional neglect and malfeasance, to mug shots of innocent scientists in respected journals. The overt blood-throwing days of ACT-UP had gotten quieter, yet more insidious.

The second curious thing about this *Retrovirology* paper was one of its authors, Vinay Pathak, who along with John Coffin had first promulgated the idea that XMRV was an inadvertent lab creation made sometime between 1992–1994 at Case Western University and thus couldn't be the cause of the modern outbreak of ME/CFS that began in 1984–1985. If XMRV was simply a lab creation, what was the interest in studying it? But Pathak and his colleagues seemed very interested.

As a Christian, Mikovits felt in her soul that God had placed her in the middle of this controversy for a reason, and her faith directed that if she held onto her faith and acted with integrity, she might also understand why. Harvey's actions may have been a long con, but she too felt that she was in the middle of a longer, greater plan, and it would ultimately shake down to its pure meaning.

The title of the *Retrovirology* paper was "Xenotropic MLV Envelope Proteins Induce Tumor Cells to Secrete Factors that Promote the Formation of Immature Blood Vessels," and it was published on March 27, 2013, two months to the day after Jeang's suicide. In the background section of their paper the authors wrote:

> XMRV had first been identified in tissue from prostate cancer tumors and later in the blood of patients with ME/CFS. Those findings were now in dispute, but most parties agreed that the virus itself had evolved as the result of "retroviral combination events in human tumor cell lines established through murine xenograft events."

Simply stated, when human tissue passed through mouse tissue, some mouse viruses could recombine with human viruses to form new viruses.

In the results section of their paper, Pathak and his colleagues observed that when XMRV was injected into mice that had existing tumors, the virus suppressed the differentiation of blood vessels, leading to larger tumors. In the conclusions section they wrote:

> Although it is highly unlikely that either XMRV or B4Rv themselves infect humans and are pathogenic, the results suggest that xenograft approaches commonly used in the study of human cancer promote the evolution of novel retroviruses with pathogenic properties.[7]

Mikovits read the line, "promote the evolution of novel retroviruses with pathogenic properties," and wanted to cry out, "that's what I've been saying!"

It brought her immediately back to the November 2006 Human Herpes Virus 6 (HHV-6) meeting in Barcelona, Spain, where she'd rushed the stage after the presentation of Dan Peterson showing his ME/CFS with cancers had an abnormal T-cells populations.

To Mikovits, the data looked like the footprint of an immune-altering retrovirus just as she said to Gary Owens in Cleveland November 10, 2009, when he dared call B4rv, XMRV2. Indeed, the 2013 paper in *Retrovirology* (that Dr. Owens was as a coauthor on the Pathak paper) reported, that "Phylogenetic tree analyses demonstrated that while B4rv and XMRV share greater similarity to one another than other MLVs, B4rv represents a distinct sequence from that of XMRV (Figure 1D). Together, these data show that B4rv is a xenotropic MLV found in a human cell line derived through xenograft experiments in nude mice that is distinct in sequence and proviral origin from XMRV."[8]

Mikovits had entered the ME/CFS field because of the elevated rate of certain rare types of cancer among the patients, one a doctor who treated ME/CFS had died of just such a highly vascularized and aggressive tumor shortly after she detected high levels of env protein in his blood. Now Pathak and Owens showed the envelope proteins of that very virus and its related family members could indeed promote cancer formation.

The mouse leukemia related viruses and the diseases they might cause among the human population were not quite dead. Vasculitis in mice could be caused by XMRV or B4RV but somehow—despite this fairly damning finding—they concluded that vaculitis in humans could not be?

The place from which these viruses had mutated, the course of their evolution and the path, which had led them from mice to humans, were still obviously unmapped regions that needed to be explored.

* * *

An article published in the journal *PLOS ONE* on March 13, 2013, from researchers at the University of California, San Diego School of Medicine about the reversal of autism in a mouse model of the disorder prompted new theories about what might be behind autism and other neuroimmune diseases, including ME/CFS.

The approach was called *antipurinergic therapy*, or APT, and it was based on the theory that cells got stuck in a defensive mode as they suspected some type of invader. In this immune-activated state the cells failed to communicate with each other. The lead researcher was Robert Naviaux, MD, PhD, professor of medicine and co-director of the Mitochondrial and Metabolic Disease Center at UC San Diego. Mikovits wondered if he would shortly find himself in trouble or whether he might more successfully navigate the treacherous waters of neuroimmune diseases.

From an article on *Science Daily* on March 13, 2013, discussing the findings:

> "When cells are exposed to classical forms of danger, such as a virus, infection or toxic environmental substances, a defense mechanism is activated," Naviaux explained. "This results in changes to metabolism and gene expression, and reduces the communication between neighboring cells. Simply put, when cells stop talking to each other, children stop talking . . ."[9]

When Mikovits read the article she smiled. She had worked with suramin back in the late 1980s in prostate cancer, looking for treatment for her step-father.

It was not effective against prostate cancer, but maybe that was because the process was too far along. Could suramin be effective earlier, used at lower doses to prevent the development of cancer? She also had concerns about long-term use of the drug, but there were other compounds that might do the same thing.

Mikovits noted the elegant, smooth, and politically inoffensive way the lead researcher, Naviaux had framed the issue, "When cells are exposed to classical forms of danger, such as a virus, infection or toxic environmental substances, a defense mechanism is activated."[10]

That covered a lot of ground.

It also allowed people to skip over the messy middle of what had activated this defense mechanism (such as a retrovirus) and concentrate on bringing relief to patients.[11]

* * *

Ian Lipkin held a public conference call on September 10, 2013, with the Centers for Disease Control. The subject was his latest research on ME/CFS patients. Lipkin said:

> So, again, there seems to be a significant difference in profiles obtained from people who have had the disease for less than three years and people who have had the disease for more than three years.

> In what I would call the "Early Group," which is less than three years, there seems to be a number of markers suggestive of some sort of allergy aspect. They have increased numbers, typically, of eosinophils in blood, and when we examine their markers from areas, cytokines—we find there are definite differences. I'm not going to talk a great deal about the differences between before three years and after three years, because we are still working on this.

Lipkin then entered into what he hoped would be the main point of his talk and the take-away from his findings, namely that they had identified a pattern of inflammatory markers that were unique for ME/CFS.

> So, I will confine my discussion to a number of cytokines and chemokines that are up or down in people who have disease, and again, this is important because I think that we may find that there are drugs which can be used to modulate the levels of cytokines; and while these may not get at the cause of this disease, the primary cause which I still believe is likely to be an infectious agent, it may give us some insight in ways in which we can manage and decrease some of the disabilities associated with Chronic Fatigue Syndrome.[12]

It was an abnormal pattern of cytokines and chemokines and her dedication to development of therapeutic strategies to modulate cytokines that had first led Mikovits to conclude that a retrovirus might lie at the heart of the disorder.

And then Lipkin reported a retroviral finding which was just as robust as the ones Mikovits and the Lo/Alter group had described, but it didn't seem he wanted to spend much time discussing the implications:

> We found retroviruses in 85 percent of the sample pools. Again, it is very difficult at this point to know whether or not this is clinically significant or not, and given the previous experience with retroviruses in chronic fatigue, I am going to be very clear in telling you, although I am reporting this at present in Professor Montoya's samples, neither he nor we have concluded that there is a relationship to disease.[13]

When Ruscetti heard this he was apoplectic: "What gives Lipkin the right to decide that retroviral sequences which Coffin had been finding since the 1980s had no relationship to disease?" Ruscetti screamed. It was nothing but political expediency.

The "stealth" nature of retroviruses was well known, so if they were really doing any kind of virus hunting, they wouldn't be looking only for buffalo on an open plain. A study published in November 2013 in *Nature* found, interestingly, that HIV uses an "invisibility cloak" made up of the body's own cells to evade detection.

As reported by Andrew Naughtie in *The Conversation*, "HIV uses molecules inside host cells in an infected person to avoid alerting the body's innate immune system (IIS)—cells and mechanisms that form the first line of defense in our bodies—to its presence. By using the host's own cells to hide, the virus is able to replicate in the body undisturbed.

The team was able to identify how the HIV virus used this cloaking tactic, and was then able to look at how the viral invader might be uncloaked."[14]

Ruscetti wondered what it would take for someone to "uncloak" the potential role of retroviruses in neuroimmune disease. Just about everyone agreed that taming the cytokine/chemokine storm in ME/CFS patients was probably an important piece of treating the disorder, just as 2010 was dubbed "the year of inflammation" by researchers at the *Conference on Retroviruses and Opportunistic Infections* that focused heavily on the role of inflammation in AIDS.[15]

But why were they running from the retrovirus in ME/CFS, if a "year of inflammation" could swirl around AIDS and inflammatory markers were clearly noted in ME/CFS? Something was causing the cytokine/chemokine storm. Didn't they want to know what? They were like meteorologists not wanting to figure out why tornadoes spin.

* * *

When Mikovits later reflected on all that had taken place in the XMRV investigation she couldn't help but think that her troubles stemmed not from being the first to consider new ideas, but the first to take them to public forums and discuss their broader public health implications. There was clear evidence that others had pondered the same ideas: chief among them were her most prominent detractors, John Coffin and Jonathan Stoye.

In 1995 the two of them had in fact collaborated on a Letter to the Editor of *Nature Medicine* entitled "The Dangers of Xenotransplantation."[16] Xenotransplantation refers to the growing of tissue of human organ tissue in an animal, then transplanting that organ into a human being. Both Coffin and Stoye were deeply concerned about the possible dangers of such a procedure being done on a regular basis.

Their concerns about the risks of xenotransplantation would mirror the fears Mikovits would later express about whether the passaging of human tissue through mice and possibly other animals in the early days of vaccine development had resulted in just such a catastrophe.

In the 1995 letter, Coffin and Stoye wrote that despite the recent enthusiasm for xenotransplantation, they had significant concerns. What about viruses, which might be contained in these tissues? They wrote:

> The pathogenicity of these viruses has not been studied. However, since growth of related viruses in immunosuppressed hosts can lead to tumor development, this prospect is somewhat alarming.
>
> We suggest that strenuous efforts be made to identify or breed virus-negative donors and that all protocols for xenotransplantation include specific steps to monitor human recipients for possible infection by retrovirus of donor origin.[17]

Science does have the full capacity for reflection, wisdom, and discretion, and in their campaign against xenograft transplants Coffin and Stoye displayed admirable insight. Viruses, which had adapted themselves to live in mammals might become especially virulent when transplanted into humans.

The fact that the organ transplant recipients would in all likelihood be immune-suppressed as a result of the operation itself, and the immunosuppressant drugs typically given after a transplant, raised the possibility that even relatively harmless viruses might cause extensive harm to such individuals.

If Coffin and Stoye were correct in their concerns about the potential for unknown viruses to cross into humans as a result of human organs grown in animal tissue, it's difficult to see much of a distinction between these notions of organ transplants, and vaccines prepared in animal cultures (such as mice), and then injected into people.

Infants would likely be at special risk for such viruses as their immune systems are still developing and might react in unexpected way to such a challenge. The same might be said of the ME/CFS patients, some of whom worked in high-pressure jobs or were extreme athletes and could have put their bodies through novel stressors, others who may have had prior immunosuppression from another factor.

But the question remains whether science is willing to look at its own past practices and exhume those that people like Coffin have tried to bury. The question is in many ways a philosophical one, whether any powerful group should be "self-policing" without significant oversight from independent bodies. From those who have endured the abuse of members of the clergy, to the reckless behavior of large financial institutions, it seems that groups whose ambitions and plans are not periodically challenged often find themselves committing mistakes of devastating proportions. And while industrialized nations no longer have the epidemics of polio or yellow fever ravaging their populations, there are other, quieter plagues.

Conservative estimates of ME/CFS in our country are more than a million individuals, and autism affects 1 in 50 children. At its height polio affected no more than 1 in 2,000 children, most of whom ultimately recovered, and for that there was a widespread social outcry for a cure. ME/CFS, autism, and other neuroimmune diseases are lingering afflictions, persisting for decades and lifetimes with incurable devastation.

These are not ancient diseases, but relatively new denizens of our planet. We have not seen similar outbreaks before. One wonders if the physical plague of these desperately sick individuals is not matched by an equally deadly spiritual plague among the scientific community that dissuades scientists from having a direct, empathic relationship with patients.

The larger question is whether the plague feared by Coffin and Stoye has already arrived, whether we fail to see the clever virus which does not kill its host, but has learned to live with it. When a disease takes so much from a patient, but stops just short of death, how does the medical community respond?

And will the scientific community have the courage to answer the question of whether these diseases might be the result of their own creation? Or will those who suffer have to do what one brave patient ME/CFS did, setting up a mattress in a public space and lying on it in protest, until someone adequately responds?

The final stunning paper in the possible resurrection of the xenotropic murine viruses playing roles in human disease was published by Coffin and Pathak in November 2013. It is entitled "Generation of Multiple Replication-Competent Retroviruses through Recombination between PreXMRV-1 and PreXMRV-2." The abstract read like the nightmare scenario of some science fiction movie, showing that what scientists had believed to be benign, carried terrifying risks. In the experiment they put two endogenous mouse leukemia proviruses, preXMRV-1 and PreXMRV-2 in culture, and waited to see what happened. They found:

> the presence of reverse transcriptase activity at 10 days postinfection indicated the presence of RCRs [retroviruses that can replicate]. Population sequencing of proviral DNA indicated that all RCRs contained the gag and 5' half of

pol from PreXMRV-2 and the long terminal repeat, 3' half of pol (including integrase), and env from PreXMRV-1.

It had taken 10 days to create a new retrovirus in those conditions. And what about all the *other* xenotropic murine leukemia virus related replication-competent retroviruses that might have been generated by this recombination experiment, beyond the dismissed XMRV?

Could they cause a natural history of infection in humans? Could they relate to Lo and Alter's findings of variant MLVs? There is no evidence of a natural history of infection in humans by these viruses as of yet. However, it would have taken and will still take only one. Mikovits's question to Coffin in Ottawa takes on more prescient meaning "How many new recombinant retroviruses have we created?"

Ruscetti thinks that this recombination experiment should be done in human cells expressing human endogenous viruses like HERV K.

XMRV can cause pathogenic changes in mice. Lipkin found retroviral sequences in 85 percent of ME/CFS patients; and Coffin generated novel replication competent retroviruses related to XMRV in ten days.

In February of 2014 the American Society for Microbiology published an article on the effectiveness of alcohol-based disinfectants to prevent XMRV contamination. The concluding sentence of their abstract states, "Their applications will help to prevent unintended XMRV contamination of cell cultures in laboratories and *minimize the risk for laboratory personnel and healthcare workers to become infected with this biosafety level 2 organism.*"[18] (italics added) In the article's concern about the risk to healthcare workers, one can almost hear the echoes of the original 1934–1935 outbreak of ME/CFS among the doctors and nurses of Los Angeles County General Hospital.

As for Mikovits, she is like her former research assistant, Max Pfost, a person that science does not seem to want. The legal battles with the Whittemores left her financially devastated, and the research grants, which would have enabled her to continue her investigations at another institution were taken away from her. She still has not received copies of her notebooks or those of the team she led at the WPI for nearly five years. It's as if Darwin, having sailed around the world for five years on the *Beagle*, collecting his samples and making his notes, published his theory of evolution, only to have the Church of England swoop in and take all of his materials away, then feign surprise that he couldn't keep working on his theory. One can only speculate why those in positions of power in science have not demanded that her notebooks be returned.

And yet Mikovits continues to go to conferences and follow the latest scientific research. She has formed a company with Frank Ruscetti, MAR Consulting Inc, so that they can consult with physicans, patients, companies, and academic institutions. For Mikovits, it has always been and will always be about the patients.

Mikovits wonders whether science will continue to dismiss and dehumanize the patients, severing those unmistakable connections, and not following hopeful leads. Or will the NIH and other government agencies finally welcome the great insights of patients and researchers?

The question, which Mikovits continues to ask, the question, which urgently needs to be answered, is whether the plague feared by Coffin and Stoye has already arrived, but we do not recognize it. As Mikovits and Ruscetti have often said to each other over the years: they see what they want to see, and that is the real plague.

Acknowledgments

Kent Heckenlively

I'd like to thank my lovely wife and partner in life, Linda, my son and first reader, Ben, and my daughter, Jacqueline, who shows me on a daily basis the meaning of courage.

I'd also like to thank my father, who has always supported my efforts and my dear deceased mother who taught me that if a cause is just you fight no matter the size of your enemy. I'd like to thank my brother, Jay, who has been my champion and best friend since the day I was born.

I'd like to thank some of the great teachers in my life, my seventh grade science teacher, Paul Rago, high school teachers Ed Balsdon and Brother Richard Orona, college professors Clinton Bond, David Alvarez, and Carol Lashoff, writing teachers James N. Frey and Donna Levin, and in law school, the always entertaining Bernie Segal.

I've been blessed to have some of the greatest friends anybody could ever want who have stood by me through both success and failure. Thank you John Wible, John Henry, Chris Sweeney, Pete Klenow, Eric Holm, Bill Wright, and Max Swafford for your constant friendship.

I want to thank Judy Mikovits for her constant patience with me as I struggled with this very important story. I also want to thank the many scientists who freely gave of their time to better help me understand the complexities of the subject matter.

I'd like to thank the patients with ME/CFS who shared their stories with me, even when it was physically difficult for them. At many times during this odyssey I have felt like I have been bearing witness to a terrible tragedy. I am humbled by the task and hope that I have justified your trust in me to tell this tale.

I also want to thank Dr. Amy Yasko for her dedication to children with autism, the hours she spent answering my questions, and the comments she made to me regarding yellow fever and the earliest autism cases.

And last I'd like to thank the wonderful staff at Skyhorse Publishing for taking a chance on this project and steering it through some very troubled waters. I am forever in your debt.

Judy Mikovits
I would like to thank my family, those who have guided my long career in science, and those patients and scientists who remain committed to solving the problems of chronic diseases.

I'd like to give special thanks to Robin Moulton, Rivka Solomon, and Peggy Munson for their invaluable support and guidance to help me better understand the patient perspective, as well as their thoughtful and incisive comments on earlier drafts of this book.

I'd like to thank all of my great friends at PBYC, CPC, and cancer support groups for their spiritual counsel in my times of darkness and despair.

And last, I'd like to thank you, the readers of this book, who will determine whether the issues raised in this book provide a blueprint of hope for millions who suffer from chronic disease.

Notes

PROLOGUE

[1] Jon Cohen and Martin Enserink, "False Positive" *Science*, Vol.333, (September 23, 2011), 1694–1701.

[2] Vincent C. Lombardi, Francis W. Ruscetti, Jaydip Dad Gupta, Max Pfost, Kathryn S. Hagen, Daniel L. Peterson, Sandra K. Ruscetti, Cari Petrow-Sandowski, Bert Gold, Michael Dean, Robert H. Silverman, Judy A. Mikovits. "XMRV, in Blood Cells of Patients with Chronic Fatigue Syndrome", *Science*, Vol. 326, (October 23, 2009), 585–588.

[3] Vincent C. Lombardi, Francis W. Ruscetti, Jaydip Das Gupta, Max Pfost, Kathryn S. Hagen, Daniel L. Peterson, Sandra K. Ruscetti, Cari Petrow-Sandowski, Bert Gold, Michael Dean, Robert H. Silverman, Judy A. Mikovits. Vincent C. Lombardi, "Detection of an Infectious Retrovirus, XMRV, in Blood Cells of Patients with Chronic Fatigue Syndrome." (Partial Retraction). *Science*, Vol.334. (October 14, 2011), 334.

[4] Dr. William Reeves, CDC Chief of Viral Diseases Branch, Press Conference on ME/CFS, (2006).

[5] Regan Harris, (Volunteer for the Whittemore-Peterson Institute), Telephone Interview by Kent Heckenlively, August 15, 2013.

[6] Amy Dockser Marcus, "Scientist Who Led XMRV Research Team Let Go," *Wall Street Journal*, October 3, 2011.

[7] Annette Whittemore, Termination Letter to Dr. Judy Mikovits, September 30, 2011.

[8] Judy Mikovits, Response to Termination Letter, October 1, 2011.

[9] Press Release, "Whittemore Peterson Institute Announces Availability of Updated XMRV Testing – UK Study Publication Does Not Impact XMRV Research", 2010, http://webarchive.org/web/2011100604405/http://www.wpiinstitute.org/news/docs/WPI pressrel_O_11410.pdf

[10] Dr. Jamie Deckoff-Jones, "Square One," *X Rx Blog*, October 3, 2011.

[11] Graham Simmon, Michael Busch et al, "Failure to Confirm XMRV/MLVs in the Blood of Patients with Chronic Fatigue Syndrome; A Multi-Laboratory Study," *Science*, Vol. 334, (November 11, 2011), 814–817.

[12] Judy Mikovits, Telephone Interview by Kent Heckenlively, March 20, 2013.

[13] Ian Lipkin, Email to Frank Ruscetti and Judy Mikovits, November 14, 2011.

[14] Judy Mikovits, Telephone Interview by Kent Heckenlively, May 15, 2012.

[15] Chuck Neubauer and Richard Cooper, "Desert Connections," *Los Angeles Times,* August 20, 2006.

[16] Chuck Neubauer and Richard Cooper, "Desert Connections," *Los Angeles Times,* August 20, 2006.

[17] Chuck Neubauer and Richard Cooper, "Desert Connections," *Los Angeles Times,* August 20, 2006.

[18] Chuck Neubauer and Richard Cooper, "Desert Connections," *Los Angeles Times,* August 20, 2006.

[19] Richard Disney, "Harvey Whittemore and the Politics of Pull in Nevada," *The Nevada Observer,* September 1, 2006.

[20] Richard Disney, "The Nevada 'Power Rangers': Pete Ernaut and the Harry Reid Land Deal." *The Nevada Observer,* October 15, 2006.

[21] Martha Bellisle, "Seeno Construction Family Made Mark in Reno," *Reno Gazette Journal,* February 12, 2012.

[22] Dennis Neal Jones, (Attorney for Judy Mikovits) Telephone Interview by Kent Heckenlively, July 29, 2013.

[23] David Follin, (Attorney for Judy Mikovits) Telephone Interview by Kent Heckenlively, May 5, 2014.

[24] David Follin, (Attorney for Judy Mikovits) Telephone Interview by Kent Heckenlively, May 5, 2014.

[25] Sam Shad, "Interview with Annette Whittemore and Judy Mikovits," *Nevada Newsmakers,* October 14, 2009.

[26] Vincent C. Lombardi, Francis W. Ruscetti, Jaydip Das Gupta, Max Pfost, Kathryn S. Hagen, Daniel L. Peterson, Sandra K. Ruscetti, Cari Petrow-Sandowski, Bert Gold, Michael Dean, Robert H. Silverman, Judy A. Mikovits, "Vincent C. Lombardi Detection of an Infectious Retrovirus, XMRV, in Blood Cells of Patients with Chronic Fatigue Syndrome", *Science,* Vol. 334, (October 14,2011), 334.

[27] Frederick Hecht and Annie Luetkemeyer, "Immunizations and HIV," University of California, San Francisco HIV web-site, Accessed May 3, 2013 http://hivinsite.ucsf.edu/InSite?page=kb-00&doc=kb-03-01-08.

[28] Timothy Grady, "The Golf Course at the End of the World," *Las Vegas Seven,* January 6, 2011.

[29] Timothy Grady, "The Golf Course at the End of the World," *Las Vegas Seven,* January 6, 2011.

[30] Timothy Grady, "The Golf Course at the End of the World," *Las Vegas Seven,* January 6, 2011.

[31] Timothy Grady, "The Golf Course at the End of the World," *Las Vegas Seven,* January 6, 2011.

[32] Timothy Grady, "The Golf Course at the End of the World," *Las Vegas Seven.* January 6, 2011.

[33] Bill Burns, Telephone Interview with Kent Heckenlively, November 30, 2012.

[34] Jamie Deckoff-Jones, Telephone Interview with Kent Heckenlively, August 14, 2013.

[35] Jamie Deckoff-Jones, Telephone Interview with Kent Heckenlively, August 14, 2013.

[36] Jamie Deckoff-Jones, Telephone Interview with Kent Heckenlively, August 14, 2013.

[37] Jamie Deckoff-Jones, Telephone Interview with Kent Heckenlively, August 14, 2013.

[38] Jamie Deckoff-Jones, Telephone Interview with Kent Heckenlively, August 14, 2013.

39 John M. Coffin, and Jonathon P. Stoye, "A New Virus for Old Diseases," *Science* Vol. 326 (October 23, 2009), 530–531.

CHAPTER ONE

1   Robert Gallo, "Second Oral History Interview with Dr. Robert Gallo of the National Cancer Institute about the History of AIDS," Interviewed by Dr. Victoria A. Harden and Dennis Rodrigues (November 4, 1994) 19.

2   Robert Gallo, Keynote Speech, 5th International Conference on HHV-6 &7, Barcelona, Spain, May 1, 2006.

3   Salahuddin, SZ, Ablashi, DV, Markham PD, Josephs SF, Sturzenegger, S, Kaplan, M, Biberfeld, P, Wong-Staal, F, Kramasy, B, et al, "Isolation of a new Virus, HBLV, in Patients with Lymphoproliferative Disorders", *Science*, Vol. 234, (October 31, 1986), 596–601.

4   World Health Organization, "Global Health Observatory on HIV/AIDS," Accessed 5 May 2013, http://www.who.int/gho/hiv/en/

5   About Dr. Gallo, *Institute of Human Virology, University of Maryland*, Accessed May 5, 2013, http://www.ihv.org/about/robert_gallo.html/

6   About Dr. Gallo, *Institute of Human Virology, University of Maryland*, Accessed May 5, 2013, http://www.ihv.org/about/robert_gallo.html/

7   Lawrence Altman, "Discoverers of AIDS and Cancer Viruses Win Nobel," *The New York Times*, October 7, 2008.

8   John Crewdson, "The Great AIDS Quest", *Chicago Tribune*, November 19, 1989.

9   Douglas Kneeland, "The Gallo Case: A Three-Year Odyssey in Search of the Truth." *Chicago Tribune*, December 6, 1992.

10 Phillip Hilts, "Federal Inquiry Finds Misconduct by a Discoverer of the AIDS Virus", *The New York Times*, December 31, 1992/

11 John Horgan, "Autopsy of a Medical Breakthrough," *The New York Times*, March 3, 2002.

12 John Crewdson, *Science Fictions – A Scientific Mystery, a Massive Cover-up, and the Dark Legacy of Robert Gallo*, (New York: Little, Brown and Company, 2002).

13 Lawrence Altman, "Discoverers of AIDS and Cancer Viruses Win Nobel," *The New York Times*, October 7, 2008.

14 Lawrence Altman, "Discoverers of AIDS and Cancer Viruses Win Nobel," *The New York Times*, October 7, 2008.

15 Declan Butler, "Nobel Fight over African HIV Centre," *Nature*, Vol. 486, issue 7403 (June 19, 2012), 301–302, doi:10.1038/486301a.

16 Declan Butler, "Nobel Fight over African HIV Centre," *Nature*, Vol. 486, issue 7403 (June 19, 2012), 301–302, doi:10.1038/486301a.

17 Kendall Smith, "The Discovery of the Interleukin 2 Molecule." (From Dr. Kendall Smith's Immunology Research Site. Dr. Smith is the Rochelle Butler Professor of Medicine and Immunology at Cornell's Weil Medical College and Graduate School of Biomedical Sciences), Accessed May 5, 2013, http://www.kendallasmith.com/molecule.html.

18 Judy Mikovits, Telephone Interview with Kent Heckenlively, July 3, 2013.

19 Seth Roberts, "What AIDS Researcher Dr. Robert Gallo Did in Pursuit of the Nobel Prize." *Spy*, July 1990.

[20] Seth Roberts, "What AIDS Researcher Dr. Robert Gallo Did in Pursuit of the Nobel Prize." *Spy*, July 1990.

[21] Frank Ruscetti, Telephone Interview with Kent Heckenlively, September 4, 2013.

[22] Frank Ruscetti, Telephone Interview with Kent Heckenlively, September 4, 2013.

[23] Frank Ruscetti, Telephone Interview with Kent Heckenlively, September 4, 2013.

[24] Frank Ruscetti, Telephone Interview with Kent Heckenlively, September 4, 2013.

[25] Judy Mikovits, Written Recollection, May 13, 2012.

[26] Judy Mikovits, Written Recollection, May 13, 2012.

[27] Judy Mikovits, Written Recollection, May 13, 2012.

[28] Judy Mikovits, Written Recollection, May 13, 2012.

[29] Frank Ruscetti, Telephone Interview with Kent Heckenlively, June 15, 2012.

[30] Salahuddin, SZ, Ablashi, DV, Markham PD, Josephs SF, Sturzenegger, S, Kaplan, M, Biberfeld, P, Wong-Staal, F, Kramasy, B, et al, "Isolation of a new Virus, HBLV, in Patients with Lymphoproliferative Disorders", *Science*, Vol. 234, October 31, 1986, 596–601.

[31] Nicholas Regush, *The Virus Within – A Coming Epidemic – How Medical Detectives are tracking a Terrifying Virus that Hides in Almost All of Us*, (New York, Dutton Group, 2000).

[32] Judy Mikovits, Telephone Interview with Kent Heckenlively, June 9, 2012.

[33] Ken Richards, Telephone Interview with Kent Heckenlively, September 1, 2013.

[34] Ken Richards, Telephone Interview with Kent Heckenlively, September 1, 2013.

[35] Ken Richards, Email Communication to Kent Heckenlively, September 6, 2013.

[36] Mikovits, J. A., Raziuddin, Gonda, M., Ruta, M. Lohrey, N. Kung, H-F. and Ruscetti, F. "Negative Regulation of HIV Replication in Monocytes: Distinctions between Restricted and latent Expression in THP-1 Cells, *Journal of Experimental Medicine*, Vol. 171, 1705–1720, (1990).

[37] Judy Mikovits, Telephone Interview with Kent Heckenlively, June 9, 2012.

[38] John Crewdson, *Science Fictions – A Scientific Mystery, a Massive Cover-up, and the Dark Legacy of Robert Gallo*, (New York: Little, Brown and Company, 2002).

[39] Preliminary Agenda, 5th International Conference on HHV-6 & 7, April 30th-May 3, 2006, Barcelona, Spain.

[40] Judy Mikovits, Telephone Interview with Kent Heckenlively, June 9, 2012.

Chapter Two

[1] Ed Vogel, "Nevada Power Broker Whittemore Now a Pariah," *Las Vegas Review Journal*, February 26, 2012.

[2] Judy Mikovits, Telephone Interview with Kent Heckenlively, June 9, 2012.

[3] Victor Williams, "Reno-Sparks Red Hawk Takes Wing into a Bright Future," *The Examiner*, June 1, 2009.

[4] "Lakes Course," Red Hawk Golf and Resort, Sparks, Nevada, Accessed June 21, 2012, http://www.resortatredhawk.com/lakes-course.html.

[5] Anne McMillian, "Gregory Pari, Ph.D., Named Chair of School of Medicine's Department of Microbiology and Immunolgy," *Nevada Today*, July 12, 2010.

[6] Judy Mikovits, Telephone Interview with Kent Heckenlively, June 9, 2012.

[7] Scientific Advisory Board, *Whittemore Peterson Institute*, Accessed, 21 June 21, 2012, http://www.wpinstitute.org/research/research_advisoryBoard.html.

8   Annette Whittemore, "The Whittemore Peterson Institute: Building Bridges through Private and Public Sector Collaboration," *Molecular Interventions*, Vol. 10, Issue 3, June 2010, 120–126.

9   Andrea Whittemore-Goad, "My Name is Andrea Whittemore." Facebook, November 9 2009.

10 Annette Whittemore, "The Whittemore Peterson Institute: Building Bridges through Private and Public Sector Collaboration," *Molecular Interventions*, June 2010. Volume 10, Issue 3. pp.120–126.

11 Annette Whittemore, "The Whittemore Peterson Institute: Building Bridges through Private and Public Sector Collaboration," *Molecular Interventions*, June 2010, Volume 10, Issue 3, pp.120–126.

12 Andrea Whittemore-Goad, "My Name is Andrea Whittemore," Facebook, November 9, 2009.

13 Judy Mikovits, Telephone Interview with Kent Heckenlively, June 9, 2012.

14 Judy Mikovits, Telephone Interview with Kent Heckenlively, June 9, 2012.

15 RB Schwartz, AL Komaroff, BM Garada, M Gleit, TH Doolittle, DW Bates, RG Vasile, BL Holman, "SPECT Imaging of the Brain: A Comparison of Patients with Chronic fatigue Syndrome, AIDS Dementia Complex, and Major Unipolar Depression," *American Journal of Roentolology*, Vol. 164 (April 1994), 943–951.

16 "Tuberculosis: The Connection between TB and HIV (the AIDS Virus), Centers for Disease Control, http://www.cdc.gov/tb/publications/pamphlets/tbandhiv_eng.htm.

17 Brian Gazzard and Jens Lundgren, "British HIV Association and British Infection Association Guidelines for the Treatment of opportunistic Infection in HIV-seropositive Individuals 2011," *HIV Medicine* 12.2, (September 2011).

18 Judy Mikovits, Telephone Interview with Kent Heckenlively, June 9, 2012.

19 Judy Mikovits, Telephone Interview with Kent Heckenlively, June 9, 2012.

20 Judy Mikovits, Telephone Interview with Kent Heckenlively, June 9, 2012.

21 Frank Ruscetti, Telephone Interview with Kent Heckenlively, June 14, 2012.

CHAPTER THREE

1   Judy Mikovits, Telephone Interview with Kent Heckenlively, June 27, 2012.

2   Visiting Information, "Ventura County Sheriff's Office – Todd Road Visiting Hours." Accessed June 7, 2012, http://www.vcsd.org/sub-visiting-information.php.

3   Judy Mikovits, Telephone Interview with Kent Heckenlively, June 27, 2012.

4   Judy Mikovits, Telephone Interview with Kent Heckenlively, June 27, 2012.

CHAPTER FOUR

1   Joseph DeRisi, "Hunting the Next Killer Virus," TED Talks, Monterey, California, January 29, 2006.

2   Judy Mikovits, Telephone Interview with Kent Heckenlively, July 10, 2012.

3   Paul Cheney, Telephone Interview with Kent Heckenlively, July 25, 2013.

4   Hillary Johnson, *Osler's Web: Inside the Labyrinth of the Chronic Fatigue Syndrome Epidemic*, (New York: Penguin Group, 1996), 18.

5   Hillary Johnson, *Osler's Web: Inside the Labyrinth of the Chronic Fatigue Syndrome Epidemic*, (New York: Penguin Group, 1996), 19.

6   Hillary Johnson, *Osler's Web: Inside the Labyrinth of the Chronic Fatigue Syndrome Epidemic*, (New York: Penguin Group, 1996), 13.

7   Hillary Johnson, *Osler's Web: Inside the Labyrinth of the Chronic Fatigue Syndrome Epidemic*, (New York: Penguin Group, 1996), 26.

8   Hillary Johnson, *Osler's Web: Inside the Labyrinth of the Chronic Fatigue Syndrome Epidemic*, (New York: Penguin Group, 1996), 28.

9   Hillary Johnson, *Osler's Web: Inside the Labyrinth of the Chronic Fatigue Syndrome Epidemic*, (New York: Penguin Group, 1996), 16.

10 Hillary Johnson, *Osler's Web: Inside the Labyrinth of the Chronic Fatigue Syndrome Epidemic*, (New York: Penguin Group, 1996), 28

11 Hillary Johnson, *Osler's Web: Inside the Labyrinth of the Chronic Fatigue Syndrome Epidemic*, (New York: Penguin Group, 1996), 29.

12 Paul Cheney, Telephone Interview with Kent Heckenlively, July 25, 2013.

13 Paul Cheney, Telephone Interview with Kent Heckenlively, July 25, 2013.

14 Hillary Johnson, *Osler's Web: Inside the Labyrinth of the Chronic Fatigue Syndrome Epidemic*, (New York: Penguin Group, 1996), 94.

15 Hillary Johnson, *Osler's Web: Inside the Labyrinth of the Chronic Fatigue Syndrome Epidemic*, (New York: Penguin Group, 1996), 94

16 Hillary Johnson, *Osler's Web: Inside the Labyrinth of the Chronic Fatigue Syndrome Epidemic*, (New York: Penguin Group, 1996), 94.

17 Hillary Johnson, *Osler's Web: Inside the Labyrinth of the Chronic Fatigue Syndrome Epidemic*, (New York: Penguin Group, 1996), 94.

18 Hillary Johnson, *Osler's Web: Inside the Labyrinth of the Chronic Fatigue Syndrome Epidemic*, (New York: Penguin Group, 1996), 94.

19 Hillary Johnson, *Osler's Web: Inside the Labyrinth of the Chronic Fatigue Syndrome Epidemic*, (New York: Penguin Group, 1996), 94.

20 Hillary Johnson, *Osler's Web: Inside the Labyrinth of the Chronic Fatigue Syndrome Epidemic*, (New York: Penguin Group, 1996), 95.

21 Hillary Johnson, *Osler's Web: Inside the Labyrinth of the Chronic Fatigue Syndrome Epidemic*, (New York: Penguin Group, 1996), 95.

22 Hillary Johnson, *Osler's Web: Inside the Labyrinth of the Chronic Fatigue Syndrome Epidemic*, (New York: Penguin Group, 1996), 95.

23 Hillary Johnson, *Osler's Web: Inside the Labyrinth of the Chronic Fatigue Syndrome Epidemic*, (New York: Penguin Group, 1996), 99.

24 Hillary Johnson, *Osler's Web: Inside the Labyrinth of the Chronic Fatigue Syndrome Epidemic*, (New York: Penguin Group, 1996), 164–165.

25 Paul Cheney, Telephone Interview with Kent Heckenlively, July 25, 2013.

26 Paul Cheney, Telephone Interview with Kent Heckenlively, July 25, 2013.

27 Paul Cheney, Telephone Interview with Kent Heckenlively, July 25, 2013.

28 Paul Cheney, Telephone Interview with Kent Heckenlively, July 25, 2013.

29 Paul Cheney, Telephone Interview with Kent Heckenlively, July 25, 2013.

30 Steven E Schutzer, Thomas E Angel, Tao Liu, Athena A Schepmoes, Therese R Clauss, Joshua N Adkins, David G Camp, Bart K Holland, Jonas Bergquist, Patricia K Coyle, Richard D Smith, Brian A Fallon, Benjamin H Natelson, "Distinct Cerebrospinal Fluid Proteomes Differentiate Post-Treatment Lyme Disease from Chronic Fatigue Syndrome," *PLoS ONE*, 23 Feb 2011, DOI:10.1371/journal.pone.0017287.

31 Dr. Bryan Rosner and Tami Duncan, *The Lyme-Autism Connection*, (South Lake Tahoe: BioMed Publishing Group, 2008).

32 Steven E Schutzer, Thomas E Angel, Tao Liu, Athena A Schepmoes, Therese R Clauss, Joshua N Adkins, David G Camp, Bart K Holland, Jonas Bergquist, Patricia K Coyle, Richard D Smith, Brian A Fallon, Benjamin H Natelson, "Distinct Cerebrospinal Fluid Proteomes Differentiate Post-Treatment Lyme Disease from Chronic Fatigue Syndrome," *PLoS ONE*, Feb 23, 2011, DOI:10.1371/journal.pone.0017287.

33 Paul Cheney, Telephone Interview with Kent Heckenlively, July 25, 2013.

34 Paul Cheney, Telephone Interview with Kent Heckenlively, July 25, 2013.

35 Judy Mikovits, Telephone Interview with Kent Heckenlively, July 10, 2012.

36 Joseph L. DeRisi, Anatoly Ursiman, Ross J. Molinaro, Nicole Fischer, Sarah J. Plummer, Graham Casey, Eric A. Klein, Malathi Krishnamurthy, Christina Magi-Galuzzi, Raymond R. Tubbs, Don Ganem, Robert H. Silverman, Joseph L. DeRisi, "Identification of a Novel Gammaretrovirus in Prosatte Tumors of Patients Homozygous for R462Q RNase L Variant," *PLOSpathogens*, March 31, 2006.

37 Judy Mikovits, Telephone Interview with Kent Heckenlively, July 10, 2012.

38 Joseph L. DeRisi, Anatoly Ursiman, Ross J. Molinaro, Nicole Fischer, Sarah J. Plummer, Graham Casey, Eric A. Klein, Malathi Krishnamurthy, Christina Magi-Galuzzi, Raymond R. Tubbs, Don Ganem, Robert H. Silverman, Joseph L. DeRisi, "Identification of a Novel Gammaretrovirus in Prosatte Tumors of Patients Homozygous for R462Q RNase L Variant," *PLOSpathogens*, March 31, 2006.

39 David Biello, "Scientists Identify Gene Differences Between Humans and Chimps," *Scientific American*, August 17, 2006.

40 Vin LoPresti, "Stalking the AIDS Virus – A Killer with Many Faces," *Los Alamos Research Quarterly*, Fall 2009.

41 Judy Mikovits, Telephone Interview with Kent Heckenlively, July 10, 2012.

42 Judy Mikovits, Telephone Interview with Kent Heckenlively, July 10, 2012.

43 Whittemore Peterson Institute for Neuro-Immune Disease, 2011 Annual Report.

44 Whittemore Peterson Institute for Neuro-Immune Disease, 2011 Annual Report.

45 Nevada State Senate, Senate Bill No. 105 – Committee on Finance, February 21, 2005.

46 Whittemore Peterson Institute for Neuro-Immune Disease, 2011 Annual Report.

47 Annette Whittemore, Personal Communication to Dr. Judy Mikovits, January 14, 2008.

48 J. Niis, K. De Merlier, "Impairments of the 2-5A Synthenase/RNase L Pathway in Chronic Fatigue Syndrome," *In Vivo*. Vol. 6, (November-December 2005) 1013–1021.

49 Judy Mikovits, Telephone Interview with Kent Heckenlively, July 10, 2012.

50 Judy Mikovits, Telephone Interview with Kent Heckenlively, July 10, 2012.

51 Judy Mikovits, Telephone Interview with Kent Heckenlively, July 10, 2012.

52 Judy Mikovits, Telephone Interview with Kent Heckenlively, July 10, 2012.

53 Hillary Johnson, *Osler's Web: Inside the Labyrinth of the Chronic Fatigue Syndrome Epidemic*, (New York: Penguin Group, 1996).

54 Max Pfost, Telephone Interview with Kent Heckenlively, March 13, 2014.

55 Max Pfost, Telephone Interview with Kent Heckenlively, March 13, 2014.

56 Max Pfost, Telephone Interview with Kent Heckenlively, March 13, 2014.

57 Max Pfost, Telephone Interview with Kent Heckenlively, March 13, 2014.

58 Judy Mikovits, Telephone Interview with Kent Heckenlively, July 10, 2012.

59 Judy Mikovits, Telephone Interview with Kent Heckenlively, July 10, 2012.

60 Judy Mikovits, Telephone Interview with Kent Heckenlively, July 10, 2012.

61 Judy Mikovits, Telephone Interview with Kent Heckenlively, July 10, 2012.
62 Judy Mikovits, Telephone Interview with Kent Heckenlively, July 10, 2012.
63 Confidential Disclosure Agreement between the National Cancer Institute, the Whittemore Peterson Institute for Neuro-Immune Disorders, and the Cleveland Clinic, (January 21, 2009).
64 Confidential Disclosure Agreement between the National Cancer Institute, the Whittemore Peterson Institute for Neuro-Immune Disorders, and the Cleveland Clinic, (January 21, 2009).

CHAPTER FIVE

1   Maurice Brodie, "Attempts to Produce Poliomyelitis in Refractory Lab Animals." *Experimental Biology and Medicine*, (March 1, 1935), 832–836, doi: 10.3181/00379727-32-7876.
2   John F. Enders, Frederick C. Robbins, Thomas H. Weller, "The Cultivation of Poliomyelitis Viruses in Tissue Culture," Nobel Lecture Presentation, (December 11).
3   Randy Shilts, *And the Band Played On – Politics, People and the AIDS Epidemic*, (New York: St. Martin's Press, 1987), 20.
4   Sam Whiting, "Where History Was Made: A tour of 41 points of gay interest all across the city." *San Francisco Chronicle*, June 23, 2000. Also: Weissman, David, *We Were Here*, A Documentary, 2010.
5   Randy Shilts, *And the Band Played On – Politics, People and the AIDS Epidemic*, (New York: St. Martin's Press, 1987), 21–23.
6   "Disease Study Experts Here," *Los Angeles Times*, June 15, 1934, 14.
7   John R. Paul, *A History of Poliomyelitis,* (New Haven and London: Yale University Press, 1971), 215.
8   John R. Paul, *A History of Poliomyelitis,* (New Haven and London: Yale University Press, 1971), 215.
9   "Disease Study Experts Here," *Los Angeles Times*, June 15, 1934, 14.
10 Paul Offit, *The Cutter Incident – How America's First Polio Vaccine Led to the Growing Vaccine Crisis*, (New Haven and London: Yale University Press, 2005), 16.
11 "Specter of Paralysis Stalks Carolina," *The Literary Digest*, July 20, 1935.
12 Maurice Brodie, "Attempts to Produce Poliomyelitis in Refractory Lab Animals," *Experimental Biology and Medicine*, March 1, 1935.
13 Paul Offit, *The Cutter Incident – How America's First Polio Vaccine Led to the Growing Vaccine Crisis*, (New Haven and London: Yale University Press, 2005), 17.
14 Maurice Brodie and William Park, "Active Immunization Against Poliomyelitis," *American Journal of Public Health,* (February 1936), 119–125.
15 John R. Paul, *A History of Poliomyelitis*, (New Haven and London: Yale University Press, 1971), 221.
16 John R. Paul, *A History of Poliomyelitis*, (New Haven and London: Yale University Press, 1971), 222.
17 John R. Paul, *A History of Poliomyelitis*, (New Haven and London: Yale University Press, 1971), 224.
18 W. A. Sawyer, S. F. Kitchen, Lloyd Wray, "Vaccination Against Yellow Fever with Immune Serum and Virus Fixed for Mice," *Journal of Experimental Medicine*, (May 31, 1932), 945–969.

19 W. A. Sawyer, S. F. Kitchen, Lloyd Wray, "Vaccination Against Yellow Fever with Immune Serum and Virus Fixed for Mice." *Journal of Experimental Medicine*, (May 31, 1932), 945.

20 Carl Zimmer, "Using a Virus's Knack for Mutating to Wipe It Out," *New York Times*, January 4, 2010.

21 W. A. Sawyer, S. F. Kitchen, Lloyd Wray, "Vaccination Against Yellow Fever with Immune Serum and Virus Fixed for Mice," *Journal of Experimental Medicine*, (May 31, 1932), 948.

22 W. A. Sawyer, S. F. Kitchen, Lloyd Wray, "Vaccination Against Yellow Fever with Immune Serum and Virus Fixed for Mice," *Journal of Experimental Medicine*, (May 31, 1932), 947.

23 W. A. Sawyer, S. F. Kitchen, Lloyd Wray, "Vaccination Against Yellow Fever with Immune Serum and Virus Fixed for Mice," *Journal of Experimental Medicine*, (May 31, 1932), 954–956.

24 G. Stuart, "The Problem of Mass Vaccination Against Yellow Fever." World Health Organization – Expert Committee on Yellow Fever, 14–19 September 1953, Kampala, Uganda, Presentation.

25 G. Stuart, "The Problem of Mass Vaccination Against Yellow Fever," World Health Organization – Expert Committee on Yellow Fever. 14–19 September 1953, Kampala, Uganda, Presentation.

26 Paul Offit, *The Cutter Incident – How America's First Polio Vaccine Led to the Growing Vaccine Crisis*, (New Haven and London: Yale University Press, 2005), 17.

27 Paul Offit, *The Cutter Incident – How America's First Polio Vaccine Led to the Growing Vaccine Crisis*, (New Haven and London: Yale University Press, 2005), 18.

28 "Paralysis Serum Kills Doctor's Boy," *The New York Times*, July 25, 1934.

29 "Foundation Denies Issuing Vaccine Ban," *The New York Times*, December 28, 1935.

30 "Foundation Denies Issuing Vaccine Ban," *The New York Times,* December 28, 1935.

31 Paul Offit, *The Cutter Incident – How America's First Polio Vaccine Led to the Growing Vaccine Crisis*, (New Haven and London: Yale University Press, 2005), 18.

32 John F. Enders, Frederick C. Robbins, Thomas H. Weller, "The Cultivation of Poliomyelitis Viruses in Tissue Culture," Nobel Lecture Presentation, December 11, 1954.

33 A. G. Gilliam, R. H. Onstott, "Result of Field Studies with a Poliomyelitis Vaccine," *American Journal of Public Health*, (February 1936), 113–118.

34 A. G. Gilliam, "Epidemiological Study of an Epidemic, Diagnosed as Poliomyelitis, occurring Among the Personnel of the Los Angeles County General Hospital during the Summer of 1934," *Public Health Bulletin No. 240*, (April 1938).

35 A. G. Gilliam, "Epidemiological Study of an Epidemic, Diagnosed as Poliomyelitis, occurring Among the Personnel of the Los Angeles County General Hospital during the Summer of 1934," *Public Health Bulletin No. 240*, (April 1938).

36 "Chronic Fatigue Syndrome: Managing Activities and Exercise." *Centers for Disease Control and Prevention website*, Page last reviewed: 14 February 2013, http://www.cdc.gov/cfs/management/managing-activities.html.

37 A. G. Gilliam, "Epidemiological Study of an Epidemic, Diagnosed as Poliomyelitis, occurring Among the Personnel of the Los Angeles County General Hospital during the Summer of 1934," *Public Health Bulletin No. 240*, (April 1938).

38 Hillary Johnson, *Osler's Web: Inside the Labyrinth of the Chronic Fatigue Syndrome Epidemic*, (New York: Penguin Group, 1996), 200.

39 Dr. Byron Marshall Hyde, "The Clinical and Scientific Basis of Myalgic Encephalomyelitis/ Chronic Fatigue Syndrome," (Ottawa, Ontario, Canada, Nightingale Research Foundation, 1992), 119.

40 Dr. Byron Marshall Hyde, "The Clinical and Scientific Basis of Myalgic Encephalomyelitis/ Chronic Fatigue Syndrome," (Ottawa, Ontario, Canada, Nightingale Research Foundation, 1992), 120–121.

41 Dr. Byron Marshall Hyde, "The Clinical and Scientific Basis of Myalgic Encephalomyelitis/ Chronic Fatigue Syndrome," (Ottawa, Ontario, Canada, Nightingale Research Foundation, 1992), 120–121.

42 Dr. Byron Marshall Hyde, Telephone Interview with Kent Heckenlively, January 21, 2013.

43 Leo Kanner, "Autistic Disturbances of Affective Contact," Nervous Child Vol. 2, (1943) 217–250.

44 Dan Olmsted, Mark Blaxill, The Age of Autism, (New York: Thomas Dunne Books-St. Martin's Press 2010), 163–199.

45 Dan Olmsted, Mark Blaxill, The Age of Autism, (New York: Thomas Dunne Books-St. Martin's Press 2010), 168.

46 The Rockefeller Foundation Annual Report (1933), 39.

47 John F. Kessel, Anson S. Hoyt, and Roy T. Fisk. "Use of Serum and the Routine and Experimental Laboratory Findings in the 1934 Poliomyelitis Epidemic." American Journal of Public Health and the Nations Health, Vol. 24, No. 12, (December 1934), 1215–1223. doi: 10.2105/AJPH.24.12.1215. One might say that this was the moment when the mouse (as in a mouse retrovirus) met mercury.

48 David M. Oshinsky, Polio – An American Story, (New York. Oxford University Press 2006), 56–57.

49 The Rockefeller Foundation Annual Report (1934).

50 The Rockefeller Foundation Annual Report (1934) pp. 34–35.

51 The Rockefeller Foundation Annual Report (1934) pp.83.

52 The Rockefeller Foundation Annual Report (1935).

53 The Rockefeller Foundation Annual Report (1936) (1937) (1938) (1939).

54 Antoinette Cornelia Van Der Kuyl, Marion Comelissen, Ben Berkhout, "Of Mice and Men: on the origin of XMRV," Frontiers in Microbiology, Vol. 1, Article 147, (January 17, 2011).

55 Antoinette Cornelia Van Der Kuyl, Marion Comelissen, Ben Berkhout, "Of Mice and Men: on the origin of XMRV," Frontiers in Microbiology, Vol. 1, Article 147, (January 17, 2011), 4–5.

CHAPTER SIX

1 Judy Mikovits, Telephone Interview with Kent Heckenlively, August 1, 2012.

2 Denise Grady, "Is a Virus the Cause of Chronic Fatigue Syndrome?" The New York Times, October 12, 2009.

3 Jodi Bassett, "ME—The Shocking Disease." The Hummingbirds' Foundation for M.E. (HFME), Copyright March 2010, revised June 2011, www.hfme.org.

4 Mark Fuerst, "Infectious diseases may play a role in heart disease in general, not just in heart failure," Heartwire, 20 September 2000, http://www.medscape.com/viewarticle/786881.

5   WA Schwartzman, M. Patnaik, FJ Angulo, BR Visscher, EN Miller, JB Peter, "Bartonella (Rochalimaea) antibodies, dementia, and cat ownership among men infected with human immunodeficiency virus," Clin Infect Dis. (October 1995); 21(4) 954–9.

6   Michael Specter, "Darwin's surprise: why are evolutionary biologists bringing back extinct deadly viruses?" *New Yorker,* December 3, 2007.

7   Michael Specter, "Darwin's surprise: why are evolutionary biologists bringing back extinct deadly viruses?" *New Yorker,* December 3, 2007.

CHAPTER SEVEN

1   Judy Mikovits, Telephone Interview with Kent Heckenlively, August 8, 2012.

2   Michael Specter, "Darwin's surprise: why are evolutionary biologists bringing back extinct deadly viruses?" *New Yorker,* December 3, 2007.

3   "Readers Ask: A Virus Linked to Chronic Fatigue Syndrome," 15 October 2009, *The New York Times Health Blog.* http://consults.blogs.nytimes.com/2009/10/15/readers-ask-a-virus-linked-to-chronic-fatigue-syndrome/?_r=0.

4   Judy Mikovits, Telephone Interview with Kent Heckenlively, August 8, 2012.

5   Confidential Disclosure Agreement "Relating to Novel Assays to Detect Xenotropic Murine Leukemia Virus-related Virus (XMRV) Infection in Humans," between Whittemore Peterson Institute, National Cancer Institute, and Cleveland Clinic, 21 January 2009.

6   Judy Mikovits, Telephone Interview with Kent Heckenlively, August 8, 2012.

7   Center for Cancer Research, *Our Science – Sandra K. Ruscetti,* Accessed May 18, 2013.

8   Kazuo Nishigaki, Charlotte Hanson, Takashi Ohashi Spadaccini, Sandra Ruscetti, "Erythroblast Transformation by the Friend Spleen Focus-Forming Virus is Associated with a Block in Erythropoietin-Induced STAT1 Phosphorylation and the DNA Binding and Correlates with High Expression of the Hematopoietic Phosphatase SHP-1," *Journal of Virology,* (June 2006), 5678–5685.

9   Gopi Mohan, Wenfang Li, Richard Compans, Chingiai Yang, "Antigenic Subversion: A Novel Mechanism of Host Immune Evasion by Ebola Virus." *PLOSpathogens,* (December 13, 2012).

10 9th International Association for Chronic Fatigue Syndrome/ME, 12–15 March 2009.

11 9th International Association for Chronic Fatigue Syndrome/ME, 12–15 March 2009.

12 Maureen Hanson, Telephone Interview with Kent Heckenlively, July 16, 2013.

13 Maureen Hanson, Telephone Interview with Kent Heckenlively, July 16, 2013.

14 Judy Mikovits, Telephone Interview with Kent Heckenlively, August 8, 2012.

15 Judy Mikovits, Telephone Interview with Kent Heckenlively, August 8, 2012.

16 Judy Mikovits, Telephone Interview with Kent Heckenlively, August 8, 2012.

17 Judy Mikovits, Telephone Interview with Kent Heckenlively, August 20, 2012.

18 Seeno vs. Whittemore, Filed on January 27, 2012.

19 Seeno vs. Whittemore, Filed on January 27, 2012.

20 Seeno vs. Whittemore, Filed on January 27, 2012.

21 Hillary Johnson, "The Why," Delivered at the *Invest in ME Conference,* London, England, May 28, 2009.

22 Email from Paula Kiberstis to Dr. Judy Mikovits, June 4, 2009.

23 Personal Communication from Dr. Judy Mikovits to Dr. Paula Kiberstis, July 21, 2009.

CHAPTER EIGHT

[1] Frank Ruscetti, Telephone Interview with Kent Heckenlively, June 14, 2012.

[2] Letter from Dr. Stuart Le Grice and Dr. John Coffin to Dr. Frank Ruscetti, June 12, 2009.

[3] Robert Schalberg, Daniel Choe, Harsh Taker, Ila Singh, "Prevalence and Distribuition of XMRV, a Novel Retrovirus, in Human Prostate Cancers," Cold Spring Harbor Meeting on Retroviruses, May 18–23, 2009.

[4] Robert Schalberg, Daniel Choe, Harsh Taker, Ila Singh, "Prevalence and Distribuition of XMRV, a Novel Retrovirus, in Human Prostate Cancers," Cold Spring Harbor Meeting on Retroviruses, May 18–23, 2009.

[5] Rika Furuta, Takayuki Myazawa, Takei Sugiyama, Takatumi Kimura, Fumiya Hirayama, Yoshihiko Tani, and Hirotoshi Shibata, "The Prevalence of Xenotropic Murine Leukemia Virus-Related Virus in Healthy Blood Donors in Japan," Cold Spring Harbor Meeting on Retroviruses, May 18–23, 2009.

[6] Rika Furuta, Takayuki Myazawa, Takei Sugiyama, Takatumi Kimura, Fumiya Hirayama, Yoshihiko Tani, and Hirotoshi Shibata, "The Prevalence of Xenotropic Murine Leukemia Virus-Related Virus in Healthy Blood Donors in Japan," Cold Spring Harbor Meeting on Retroviruses, May 18–23, 2009.

[7] "Public Health Implications of XMRV Infection", Center for Cancer Research & Center of Excellence in HIV/AIDS and Cancer Virology – Workshop, July 22, 2009.

[8] National Cancer Institute – HIV Drug Resistance program – John M. Coffin, Ph.D. Accessed September 13, 2012, http://home.ncifcrf.gov/hivdrp/Coffin.html.

[9] Frank Ruscetti, Telephone Interview with Kent Heckenlively, June 15, 2012.

[10] "Public Health Implications of XMRV Infection", Center for Cancer Research & Center of Excellence in HIV/AIDS and Cancer Virology – Workshop, July 22, 2009.

[11] "Public Health Implications of XMRV Infection", Center for Cancer Research & Center of Excellence in HIV/AIDS and Cancer Virology – Workshop, July 22, 2009.

[12] "Public Health Implications of XMRV Infection", Center for Cancer Research & Center of Excellence in HIV/AIDS and Cancer Virology – Workshop, July 22, 2009.

[13] Martin Enserink, "Retraction of First Paper on XMRV Takes Authors by Surprise." Science Now, September 19, 2012.

[14] "Public Health Implications of XMRV Infection", Center for Cancer Research & Center of Excellence in HIV/AIDS and Cancer Virology – Workshop, July 22, 2009.

[15] Yu-An Zhang, Maitra Anirban, Adi Gazdar, et al, "Frequent Detection of Xenotropic Murine Leukemia Virus (XMLV) in Human Cultures Established from Mouse Xenografts." Cancer Biology and Therapy, (October 1, 2011), 617–628.

[16] Debbie Bookchin and Jim Schumacher, The Virus and the Vaccine – Contaminated Vaccine, Deadly Cancers, and Government Neglect, (New York; St. Martin's Griffin, 2004).

[17] Yu-An Zhang, Maitra Anirban, Adi Gazdar, et al, "Frequent Detection of Xenotropic Murine Leukemia Virus (XMLV) in Human Cultures Established from Mouse Xenografts," Cancer Biology and Therapy, (October 1, 2011), 617–628.

[18] Yu-An Zhang, Maitra Anirban, Adi Gazdar, et al, "Frequent Detection of Xenotropic Murine Leukemia Virus (XMLV) in Human Cultures Established from Mouse Xenografts," Cancer Biology and Therapy, (October 1, 2011), 617–628.

[19] Frank Ryan, "I, Virus: Why You're Only Half-Human," New Scientist, 29 January 29, 2010.

[20] Zhang, Yu-an, Anirban, Maitra, Gazdar, Adi, et al, "Frequent Detection of Xenotropic Murine Leukemia Virus (XMLV) in Human Cultures Established from Mouse Xenografts," Cancer Biology and Therapy, (October 1, 2011), p. 617–628.

21 Yu-An Zhang, Maitra Anirban, Adi Gazdar, et al, "Frequent Detection of Xenotropic Murine Leukemia Virus (XMLV) in Human Cultures Established from Mouse Xenografts," *Cancer Biology and Therapy*, (October 1, 2011), 617–628.

22 "Public Health Implications of XMRV Infection", Center for Cancer Research & Center of Excellence in HIV/AIDS and Cancer Virology – Workshop, July 22, 2009.

23 "Public Health Implications of XMRV Infection", Center for Cancer Research & Center of Excellence in HIV/AIDS and Cancer Virology – Workshop, July 22, 2009.

24 "Public Health Implications of XMRV Infection", Center for Cancer Research & Center of Excellence in HIV/AIDS and Cancer Virology – Workshop, July 22, 2009.

25 "Public Health Implications of XMRV Infection", Center for Cancer Research & Center of Excellence in HIV/AIDS and Cancer Virology – Workshop, July 22, 2009.

26 "Public Health Implications of XMRV Infection", Center for Cancer Research & Center of Excellence in HIV/AIDS and Cancer Virology – Workshop, July 22, 2009.

27 "Public Health Implications of XMRV Infection", Center for Cancer Research & Center of Excellence in HIV/AIDS and Cancer Virology – Workshop, July 22, 2009.

28 Judy Mikovits, Telephone Interview with Kent Heckenlively, August 8, 2012.

29 Frank Ruscetti, Telephone Interview with Kent Heckenlively, September 4, 2013.

30 "Public Health Implications of XMRV Infection", Center for Cancer Research & Center of Excellence in HIV/AIDS and Cancer Virology – Workshop, July 22, 2009.

31 "Public Health Implications of XMRV Infection", Center for Cancer Research & Center of Excellence in HIV/AIDS and Cancer Virology – Workshop, July 22, 2009.

32 "Public Health Implications of XMRV Infection", Center for Cancer Research & Center of Excellence in HIV/AIDS and Cancer Virology – Workshop, July 22, 2009

33 "Public Health Implications of XMRV Infection", Center for Cancer Research & Center of Excellence in HIV/AIDS and Cancer Virology – Workshop, July 22, 2009.

34 "Public Health Implications of XMRV Infection", Center for Cancer Research & Center of Excellence in HIV/AIDS and Cancer Virology – Workshop, July 22, 2009.

35 "Public Health Implications of XMRV Infection", Center for Cancer Research & Center of Excellence in HIV/AIDS and Cancer Virology – Workshop, July 22, 2009.

36 "Public Health Implications of XMRV Infection", Center for Cancer Research & Center of Excellence in HIV/AIDS and Cancer Virology – Workshop, July 22, 2009.

37 "Public Health Implications of XMRV Infection", Center for Cancer Research & Center of Excellence in HIV/AIDS and Cancer Virology – Workshop, July 22, 2009.

38 "Public Health Implications of XMRV Infection", Center for Cancer Research & Center of Excellence in HIV/AIDS and Cancer Virology – Workshop, July 22, 2009.

39 "Public Health Implications of XMRV Infection", Center for Cancer Research & Center of Excellence in HIV/AIDS and Cancer Virology – Workshop, July 22, 2009.

40 "Public Health Implications of XMRV Infection", Center for Cancer Research & Center of Excellence in HIV/AIDS and Cancer Virology – Workshop, July 22, 2009.

41 "Public Health Implications of XMRV Infection", Center for Cancer Research & Center of Excellence in HIV/AIDS and Cancer Virology – Workshop, July 22, 2009.

42 "Public Health Implications of XMRV Infection", Center for Cancer Research & Center of Excellence in HIV/AIDS and Cancer Virology – Workshop, July 22, 2009.

43 "Public Health Implications of XMRV Infection", Center for Cancer Research & Center of Excellence in HIV/AIDS and Cancer Virology – Workshop, July 22, 2009.

44 "Public Health Implications of XMRV Infection", Center for Cancer Research & Center of Excellence in HIV/AIDS and Cancer Virology – Workshop, July 22, 2009.

45   "Public Health Implications of XMRV Infection", Center for Cancer Research & Center of Excellence in HIV/AIDS and Cancer Virology – Workshop, July 22, 2009.
46   Judy Mikovits, Telephone Interview with Kent Heckenlively, August 8, 2012.

CHAPTER NINE

1   Max Pfost, Telephone Interview with Kent Heckenlively, March 13, 2014.
2   Judy Mikovits, Telephone Interview with Kent Heckenlively, September 11, 2012.
3   Ewen Callaway, "The Scientist who Put the Nail in XMRV's Coffin," *Nature,* September 18, 2012.
4   Declaration of Frank Ruscetti, May 17, 2012.
5   Ewen Callaway, "The Scientist who Put the Nail in XMRV's Coffin," *Nature,* September 18, 2012.
6   Judy Mikovits, Telephone Interview with Kent Heckenlively, September 11, 2012.
7   Max Pfost, Telephone Interview with Kent Heckenlively, March 13, 2014.
8   Judy Mikovits, Telephone Interview with Kent Heckenlively, September 11, 2012.
9   Max Pfost, Telephone Interview with Kent Heckenlively, March 13, 2014.
10   Max Pfost, Telephone Interview with Kent Heckenlively, March 13, 2014.
11   Max Pfost, Telephone Interview with Kent Heckenlively, March 13, 2014.
12   Max Pfost, Telephone Interview with Kent Heckenlively, March 13, 2014.
13   Max Pfost, Telephone Interview with Kent Heckenlively, March 13, 2014.
14   Judy Mikovits, Telephone Interview with Kent Heckenlively, September 11, 2012.
15   Judy Mikovits, Telephone Interview with Kent Heckenlively, September 11, 2012.
16   Max Pfost, Telephone Interview with Kent Heckenlively, March 13, 2014.
17   Max Pfost, Telephone Interview with Kent Heckenlively, March 13, 2014.
18   Judy Mikovits, Telephone Interview with Kent Heckenlively, September 11, 2012.
19   Judy Mikovits, Telephone Interview with Kent Heckenlively, September 11, 2012.
20   Judy Mikovits, Telephone Interview with Kent Heckenlively, September 11, 2012.
21   Judy Mikovits, Telephone Interview with Kent Heckenlively, September 11, 2012.
22   Judy Mikovits, Telephone Interview with Kent Heckenlively, September 11, 2012.
23   Judy Mikovits, Telephone Interview with Kent Heckenlively, September 11, 2012.
24   Max Pfost, Telephone Interview with Kent Heckenlively, March 13, 2014.
25   Max Pfost, Telephone Interview with Kent Heckenlively, March 13, 2014.
26   Max Pfost, Telephone Interview with Kent Heckenlively, March 13, 2014.
27   Max Pfost, Telephone Interview with Kent Heckenlively, March 13, 2014.
28   Max Pfost, Telephone Interview with Kent Heckenlively, March 13, 2014.
29   Max Pfost, Telephone Interview with Kent Heckenlively, March 13, 2014.
30   Max Pfost, Telephone Interview with Kent Heckenlively, March 13, 2014.
31   Max Pfost, Telephone Interview with Kent Heckenlively, March 13, 2014.
32   Affidavit of Harvey Whittemore, November 7, 2011.
33   Declaration of Frank Ruscetti, May 17, 2012.
34   Max Pfost, Telephone Interview with Kent Heckenlively, March 13, 2014.
35   Max Pfost, Telephone Interview with Kent Heckenlively, March 13, 2014.
36   Max Pfost, Telephone Interview with Kent Heckenlively, March 13, 2014.
37   Affidavit of Harvey Whittemore, Whittemore Peterson Institute for Neuro-Immune Disease v. Judy A. Mikovits, Second Judicial District Court for the State of Nevada in and for the County of Washoe, November, 7, 2011, Case No. CV11-03232.

CHAPTER TEN

[1] Jon Cohen and Martin Enserink, "False Positive," *Science*, Vol.333, (September 23, 2011),1694–1701.

[2] *Nevada Newsmakers* with Sam Shad, Interview with Annette Whittemore and Dr. Judy Mikovits, (October 8, 2009), TV broadcast.

[3] A. M. Enstron, L. Lit, Judy Van De Water, et al, "Altered Gene Expression and Function of Peripheral Blood Natural Killer Cells in Children with Autism," *Journal of Brain and Behavioral Immunity*, August 14, 2008.

[4] M. Caliguri, C. Murray, D. Buchwald, H. Levine, P. Cheney, D. Peterson, AL Komaroff, J. Ritz, "Phenotypic and functional deficiency of natural killer cells in patients with chronic fatigue syndrome," *Journal of Immunology*, Volume 139(10), (November 15, 1987), 3306–13.

[5] "Natural Killer Cell" *Science Daily*, Accessed November 3, 2012.

[6] A. M. Enstron, L. Lit, Judy Van De Water, et al, "Altered Gene Expression and Function of Peripheral Blood Natural Killer Cells in Children with Autism," *Journal of Brain and Behavioral Immunity*, August 14, 2008.

[7] Judy Mikovits, Telephone Interview with Kent Heckenlively, October 21, 2012.

[8] Paula Goines, Lori Haapenen, Judy Van De Water, et al., "Auto-Antibodies to Cerebellum in Children with Autism Associate with Behavior," *Journal of Brain and Behavioral Immunity*, December 4, 2010.

[9] Ethel Cesarman, Yuan Chang, Daniel Knowles, et al., "Kaposi's Sarcoma-Associated Herpesvirus-Like DNA Sequences in AIDS-Related Body-Cavity-Based-Lymphomas," *New England Journal of Medicine*, (May 4, 1995), 1186–1191.

[10] Judy Mikovits, Telephone Interview with Kent Heckenlively, October 21, 2012.

[11] Judy Mikovits, Telephone Interview with Kent Heckenlively, October 21, 2012.

[12] Paul Levine, "Summary and Perspective: Epidemiology of Chronic Fatigue Syndrome," *Journal of Clinical and Infectious Diseases*, Vol. 18 (January 1994), S57-S60.

[13] Judy Mikovits, Telephone Interview with Kent Heckenlively, October 21, 2012.

[14] Judy Mikovits, Telephone Interview with Kent Heckenlively, October 21, 2012.

[15] Interview: Larry Kramer, "The Age of AIDS", Frontline PBS.

[16] Craig Maupin, "The CFS Program at the NIH – Past, present, and future," *The CFS Report*, (September 2005).

[17] Craig Maupin, "The CFS Program at the NIH – Past, present, and future," *The CFS Report*, (September 2005).

[18] Judy Mikovits, Telephone Interview with Kent Heckenlively, October 21, 2012.

[19] *Nevada Newsmakers* with Sam Shad, Interview with Annette Whittemore and Dr. Judy Mikovits, (October 8, 2009), TV broadcast.

[20] *Nevada Newsmakers* with Sam Shad, Interview with Annette Whittemore and Dr. Judy Mikovits, (October 8, 2009), TV broadcast.

[21] *Nevada Newsmakers* with Sam Shad, Interview with Annette Whittemore and Dr. Judy Mikovits, (October 8, 2009), TV broadcast.

[22] *Nevada Newsmakers* with Sam Shad, Interview with Annette Whittemore and Dr. Judy Mikovits, (October 8, 2009), TV broadcast.

[23] *Nevada Newsmakers* with Sam Shad, Interview with Annette Whittemore and Dr. Judy Mikovits, (October 8, 2009), TV broadcast.

[24] *Nevada Newsmakers* with Sam Shad, Interview with Annette Whittemore and Dr. Judy Mikovits, (October 8, 2009), TV broadcast.

[25] Judy Mikovits, Telephone Interview with Kent Heckenlively, October 21, 2012.
[26] Judy Mikovits, Telephone Interview with Kent Heckenlively, October 21, 2012.
[27] Judy Mikovits, Telephone Interview with Kent Heckenlively, October 21, 2012.
[28] Judy Mikovits, Telephone Interview with Kent Heckenlively, October 21, 2012.
[29] Senior Scientist at the National Cancer Institute who wished to remain Annonymous (scientist had worked extensively with envelope proteins of murine viruses), Telephone Interview with Kent Heckenlively, September 11, 2013.
[30] Email exchange from Dr. Judy Mikovits to Dr. Robert Silverman, November 13, 2009.
[31] Email exchange from Dr. Robert Silverman to Dr. Judy Mikovits, November 13, 2009.
[32] Email exchange from Dr. Judy Mikovits to Dr. Robert Silverman, 13 November 2009.
[33] Judy Mikovits, Telephone Interview with Kent Heckenlively, October 21, 2012.
[34] Frank Ruscetti, Telephone Interview with Kent Heckenlively, 2012.

CHAPTER ELEVEN

[1] Original Submission, Vincent C. Lombardi, Francis W. Ruscetti, Jaydip Das Gupta, Max Pfost, Kathryn S. Hagen, Daniel L. Peterson, Sandra K. Ruscetti, Cari Petrow-Sandowski, Bert Gold, Michael Dean, Robert H. Silverman, Judy A. Mikovits, "Partial Retraction, Detection of an Infectious Retrovirus, XMRV, in Blood Cells of Patients with Chronic Fatigue Syndrome", *Science*, Vol.334 (October 14, 2011).

[2] Original Submission, Vincent C. Lombardi, Francis W. Ruscetti, Jaydip Das Gupta, Max Pfost, Kathryn S. Hagen, Daniel L. Peterson, Sandra K. Ruscetti, Cari Petrow-Sandowski, Bert Gold, Michael Dean, Robert H. Silverman, Judy A. Mikovits, "Partial Retraction, Detection of an Infectious Retrovirus, XMRV, in Blood Cells of Patients with Chronic Fatigue Syndrome", *Science*, Vol.334 (October 14, 2011).

[3] Original Submission, Vincent C. Lombardi, Francis W. Ruscetti, Jaydip Das Gupta, Max Pfost, Kathryn S. Hagen, Daniel L. Peterson, Sandra K. Ruscetti, Cari Petrow-Sandowski, Bert Gold, Michael Dean, Robert H. Silverman, Judy A. Mikovits, "Partial Retraction, Detection of an Infectious Retrovirus, XMRV, in Blood Cells of Patients with Chronic Fatigue Syndrome", *Science*, Vol.334 (October 14, 2011).

[4] Original Submission, Vincent C. Lombardi, Francis W. Ruscetti, Jaydip Das Gupta, Max Pfost, Kathryn S. Hagen, Daniel L. Peterson, Sandra K. Ruscetti, Cari Petrow-Sandowski, Bert Gold, Michael Dean, Robert H. Silverman, Judy A. Mikovits, "Partial Retraction, Detection of an Infectious Retrovirus, XMRV, in Blood Cells of Patients with Chronic Fatigue Syndrome", *Science*, Vol.334 (October 14, 2011).

[5] Original Submission, Vincent C. Lombardi, Francis W. Ruscetti, Jaydip Das Gupta, Max Pfost, Kathryn S. Hagen, Daniel L. Peterson, Sandra K. Ruscetti, Cari Petrow-Sandowski, Bert Gold, Michael Dean, Robert H. Silverman, Judy A. Mikovits, "Partial Retraction, Detection of an Infectious Retrovirus, XMRV, in Blood Cells of Patients with Chronic Fatigue Syndrome", *Science*, Vol.334 (October 14, 2011).

[6] Original Submission, Vincent C. Lombardi, Francis W. Ruscetti, Jaydip Das Gupta, Max Pfost, Kathryn S. Hagen, Daniel L. Peterson, Sandra K. Ruscetti, Cari Petrow-Sandowski, Bert Gold, Michael Dean, Robert H. Silverman, Judy A. Mikovits, "Partial Retraction, Detection of an Infectious Retrovirus, XMRV, in Blood Cells of Patients with Chronic Fatigue Syndrome", *Science*, Vol.334 (October 14, 2011).

[7] Original Submission, Vincent C. Lombardi, Francis W. Ruscetti, Jaydip Das Gupta, Max Pfost, Kathryn S. Hagen, Daniel L. Peterson, Sandra K. Ruscetti, Cari Petrow-

Sandowski, Bert Gold, Michael Dean, Robert H. Silverman, Judy A. Mikovits, "Partial Retraction, Detection of an Infectious Retrovirus, XMRV, in Blood Cells of Patients with Chronic Fatigue Syndrome", *Science*, Vol.334 (October 14, 2011).

8   Original Submission, Vincent C. Lombardi, Francis W. Ruscetti, Jaydip Das Gupta, Max Pfost, Kathryn S. Hagen, Daniel L. Peterson, Sandra K. Ruscetti, Cari Petrow-Sandowski, Bert Gold, Michael Dean, Robert H. Silverman, Judy A. Mikovits, "Partial Retraction, Detection of an Infectious Retrovirus, XMRV, in Blood Cells of Patients with Chronic Fatigue Syndrome", *Science*, Vol.334 (October 14, 2011).

9   Judy Mikovits, Telephone Interview with Kent Heckenlively, 21 October 21, 2012.

10 John Coffin, Jonathon Stoye, "A New Virus for Old Diseases?" *Science*, October 8, 2009.

11 John Coffin, Jonathon Stoye, "A New Virus for Old Diseases?" *Science*, October 8, 2009.

12 John Coffin, Jonathon Stoye, "A New Virus for Old Diseases?" *Science*, October 8, 2009.

13 John Coffin, Jonathon Stoye, "A New Virus for Old Diseases?" *Science*, October 8, 2009.

14 John Coffin, Jonathon Stoye, "A New Virus for Old Diseases?" *Science*, October 8, 2009.

15 John Coffin, Jonathon Stoye, "A New Virus for Old Diseases?" *Science*, October 8, 2009.

16 John Coffin, Jonathon Stoye, "A New Virus for Old Diseases?" *Science*, October 8, 2009.

17 John Coffin, Jonathon Stoye, "A New Virus for Old Diseases?" *Science*, October 8, 2009.

18 Judy Mikovits, Telephone Interview with Kent Heckenlively, October 21, 2012.

19 "Children With Autism Have Mitochondrial Dysfunction, Study Finds," *Science Daily*, November 30, 2010.

20 Sarah Myhill, Norman E. Booth, John McLaren-Howard, "Chronic Fatigue Syndrome and Mitochondrial Dysfunction," International Journal of Clinical Experimental Medicine, 2(1): 1–16, (January 15, 2009).

21 John Coffin, Jonathon Stoye, "A New Virus for Old Diseases?" *Science*, October 8, 2009.

22 John Coffin, Jonathon Stoye, "A New Virus for Old Diseases?" *Science*, October 8, 2009.

CHAPTER TWELVE

1   Frank Ruscetti, "Declaration of Frank Ruscetti, May 17, 2012," Filed with Motion: Civil Court, Dennis Jones, Attorney for Dr. Judy Mikovits.

2   Judy Mikovits, Telephone Interview with Kent Heckenlively, November 21, 2012.

3   Jon Cohen, "Inmate Mikovits Meets Judge," *Science Insider*, November 22, 2011.

4   Jon Cohen, "Inmate Mikovits Meets Judge," *Science Insider*, November 22, 2011.

5   Frank Ruscetti, "Declaration of Frank Ruscetti, May 17, 2012," Filed with Motion: Civil Court, Dennis Jones, Attorney for Dr. Judy Mikovits.

6   Frank Ruscetti, "Declaration of Frank Ruscetti, May 17, 2012," Filed with Motion: Civil Court, Dennis Jones, Attorney for Dr. Judy Mikovits.

7   Frank Ruscetti, Email to Harvey Whittemore, November 21, 2011.

8   Harvey Whittemore, Email to Frank Ruscetti, November 21, 2011.

9   Frank Ruscetti, Declaration of Frank Ruscetti, May 17, 2012.

10 Adam Garcia (Employee of the University of Nevada, Reno), email to Washoe County District Attorney, Richard Gammick, November 21, 2011.

11 Harvey Whittemore, email to Frank Ruscetti, November 22, 2011.

12 Frank Ruscetti, "Declaration of Frank Ruscetti, May 17, 2012," Filed with Motion: Civil Court, Dennis Jones, Attorney for Dr. Judy Mikovits.

13 Frank Ruscetti, "Declaration of Frank Ruscetti, May 17, 2012," Filed with Motion: Civil Court, Dennis Jones, Attorney for Dr. Judy Mikovits.

14 Frank Ruscetti, "Declaration of Frank Ruscetti, May 17, 2012," Filed with Motion: Civil Court, Dennis Jones, Attorney for Dr. Judy Mikovits.

15 Frank Ruscetti, "Declaration of Frank Ruscetti, May 17, 2012," Filed with Motion: Civil Court, Dennis Jones, Attorney for Dr. Judy Mikovits.

16 Judy Mikovits, email received, communication of calls/texts between Harvey Whittemore and Frank Ruscetti, Proposed Apology for Dr. Judy Mikovits, November 22, 2011.

17 Frank Ruscetti, "Declaration of Frank Ruscetti, May 17, 2012," Filed with Motion: Civil Court, Dennis Jones, Attorney for Dr. Judy Mikovits.

18 Judy Mikovits, email to Ian Lipkin, November 15, 2011.

19 Ian Lipkin, email to Harold Varmus (Head of the NCI), December 1, 2011.

20 Frank Ruscetti, email to Kent Heckenlively, July 8, 2014.

21 Bill Burns, Telephone Interview with Kent Heckenlively, November 30, 2012.

22 Jon Cohen, "Dispute over Lab Notebooks Lands Judy Mikovits in Jail," *Science,* December 2, 2011.

23 Mary Kearney, Jonathon Spindler, Ann Wiegand, Wei Shao, Elizabeth Anderson, Frank Maldarelli, Francis Ruscetti, John Mellors, Steve Hughes, Stuart Le Grice, John Coffin, "Multiple Sources of Contamination in Samples from Patients Reported to Have XMRV Infection", PLOS One, February 20, 2012, DOI:10.1371/journal.pone.0030889.

24 Frank Ruscetti, email to Kent Heckenlively, Written Recollection, July 8, 2014.

25 Frank Ruscetti, email to Kent Heckenlively, Written Recollection, July 8, 2014.

CHAPTER THIRTEEN

1   Denise Grady, "Is a Virus the Cause of Fatigue Syndrome?" *The New York Times*, October 12, 2009.

2   Urs M. Nater, Jame F. Jones, Jim M. Lin, Elizabeth Maloney, William C. Reeves, Christine Heim, "Personality Features and Personality Disorders in Chronic Fatigue Syndrome: A Population-Based Study," *Psychotherapy and Psychosomatics*, July 28, 2010, 79(5):312-8, doi: 10.1159/000319312.

3   Claudia Dreifus, "A Conversation with: Stephen Straus; Separating Remedies from Snake Oil," The *New York Times*, April 3, 2001.

4   Lizzie Buchen, "Virus Linked to Chronic Fatigue Syndrome" *Nature,* October 8, 2009.

5   Lizzie Buchen, "Virus Linked to Chronic Fatigue Syndrome" *Nature,* October 8, 2009.

6   Judy Mikovits, Telephone Interview with Kent Heckenlively, December 11, 2012.

7   Nicky Campbell, Shelagh Fogarty, BBC Radio 5 Live, November 9, 2009.

8   Simon Wesselley, Chaichana Nimnuan, Michael Sharpe, "Functional Somatic Syndromes: One or Many?" *Lancet,* September 11, 1999, Vol. 354, 936–9.

9   Judy Mikovits, Telephone Interview with Kent Heckenlively, December 11, 2012.

10 Judy Mikovits, Letter from Paula Kiberstis, January 5, 2010.

11 Judy Mikovits, "Response to Lloyd, Sudlow, and Van Der Meer and their Co-Authors," *Science,* January 2010.

12 Judy Mikovits, email Communication from Dr. Carl Ware, January 5, 2010.

13 Otto Erlwein, Steve Kaye, Myra McClure, Jonathon Weber, Gillian Wills, David Collier, Simon Wessely, Athony Cleare, "Failure to Detect the Novel Retrovirus XMRV in Chronic Fatigue Syndrome, *PLOSone*. January 6, 2010, Vol. 5, Issue 1, doi: 10.1371/journal.pone.0008519.

[14] Frank Van Kuppeveld, Arjan De Jong, Kjerstin Lanke, Gerald Verhaegh, Willem Melchers, Caroline Swanink, Gijs Bleijenberg, Mahia Netea, Jochem Galama "Prevalence of Xenotropic Murine Leukemia Virus-Related Virus in Patients with Chronic Fatigue Syndrome in the Netherlands" Retrospective Analysis of Samples from an Established Cohort," *British Medical Journal*, February 25, 2010, doi: 10.1136/bmj.c108.

[15] Frank Van Kuppeveld, Arjan De Jong, Kjerstin Lanke, Gerald Verhaegh, Willem Melchers, Caroline Swanink, Gijs Bleijenberg, Mahia Netea, Jochem Galama "Prevalence of Xenotropic Murine Leukemia Virus-Related Virus in Patients with Chronic Fatigue Syndrome in the Netherlands" Retrospective Analysis of Samples from an Established Cohort." *British Medical Journal*, February 25, 2010, doi: 10.1136/bmj.c108.

[16] Judy Mikovits, Ying Huang, Max Pfost, Vincent Lombardi, Daniel Bertolette, Kathryn Hagen, Frank Ruscetti, Frank, "Distribution of Xenotropic Murine Leukemia Virus-Related Virus (XMRV) Infection in Chronic Fatigue Syndrome and Prostate Cancer," *AIDS Review*, July-September 2010, 12(3) 149–152.

[17] B. M. Carruthers, et al, "Myalgic Encephalomyelitis: International Consensus Criteria" Journal of Internal Medicine, Vol. 270, issue 4, October 2011, 327–338.

[18] Nicole Fischer, Claudia Schulz, Kristin Stieler, Oliver Hohn, Cristoph Lange, Christian Drosten, Martin Aepfelbacher, "Xenotropic Murine Leukemia Virus-Related Gammaretrovirus in Respiratory Tract", *Journal of Emerging Infectious Diseases*, Vol. 16, Number 6, June 2010, doi: 10.3201/eid1606.100066.

[19] Judy Mikovits, Telephone Interview with Kent Heckenlively, December 11, 2012.

[20] K. Patrick Ober, *Mark Twain and Medicine: Any Mummery Will Cure*, (University of Missouri Press, Reprint Edition, 2011).

[21] Conference Programme 2010: The 5th Invest in ME International ME/CFS Conference 2010, May 19, 2010

[22] Geoffrey Cowley and Mary Hager. "A Chronic Fatigue Cover-Up? A New Book Says the Health Establishment Has Ignored an AIDS-Like Epidemic, but it's a Hard Case to Sustain" *Newsweek*, April 22, 1996.

[23] Trine Tsouderous, "Healthcare Media Director, Golin Harris", *Chicago Tribune*, September 18, 2012.

[24] Trine Tsouderous, "Hope Outrunning Science on Illness", *Chicago Tribune*, June 8, 2010.

[25] Judy Mikovits, Telephone Interview with Kent Heckenlively, December 11, 2012.

[26] Jill Neimark, 11 December 2012. *Discover* magazine, January/February 2010 p. 56.

[27] Alexander Grant, "Vaccine-Phobia Becomes a Public Health Threat," *Discover*, January/February 2010, 18–19.

CHAPTER FOURTEEN

[1] Press Release, *ORTHO*, June 22, 2010.

[2] Email from Dr. Shyh-Ching Lo to Dr. Frank Ruscetti, July 12, 2010.

[3] Email from Dr. Shyh-Ching Lo to Dr. Frank Ruscetti, July 12, 2010.

[4] Judy Mikovits, Telephone Interview with Kent Heckenlively, December 30, 2012.

[5] Blood Systems Research Institute, "Michael P. Busch, MD, PhD," Blood Systems Research Institute (BSRI) website, accessed August 5, 2014, www.bsrisf.org/i-mbusch.html.

[6] Judy Mikovits, Telephone Interview with Kent Heckenlively, July 5, 2013.

[7] Judy Mikovits, Telephone Interview with Kent Heckenlively, December 30, 2012.

[8] Email from Dr. Judy Mikovits to members of the Blood Working Group, June 27, 2010.

[9] Ian Lipkin, "Microbe Hunting," *Microbiology and Molecular Biology Reviews*, September 2010 Vol. 74. No. 3, 363–377.

[10] American Association of Blood Banks, "AABB Statement on XMRV," February 10, 2011, www.aabb.org/advocacy/statements/Pages/statement021011.aspx.

[11] Harvey Klein, Roger Dodd, et al., "Promising Outcomes," Transfusion, vol. 51, March 2011, 654–661.

[12] Email from Dr. Shyh-Ching Lo to Dr. Frank Ruscetti, June 10, 2010.

[13] Cristina Luiggi, "Why I Delayed the XMRV Paper," *The Scientist*, August 23, 2010.

[14] Cristina Luiggi, "Why I Delayed the XMRV Paper," *The Scientist*, August 23, 2010.

[15] Email from Dr. Judy Mikovits to members of the Blood Working Group, 28 June 2010.

[16] Amy Dockser Marcus, " Papers Held from Publication", *The Wall Street Journal*, June 30, 2010.

[17] William, M. Switzer, Hongwei Jia, Oliver Hohn, HaoQiang Zheng, Shaohua Tang, Anupama Shankar, Norbert Bannert, Graham Simmons, R. Michael Hendry, Virginia R Falkenberg, William C. Reeves, and Walid Heneine, "Absense of Evidence of Xenotropic Murine Leulemia Virus-Related Virus Infection in Persons with Chronic Fatigue Syndrome and Healthy Controls in the United States," *Retrovirology*, July 1, 2010, doi:10.1186/1742-4690-7-57.

[18] Judy Mikovits, Telephone Interview with Kent Heckenlively, December 30, 2012.

[19] Press Release, Whittemore Peterson Institute for Neuro-Immune Disease, August 16, 2010.

[20] Shyh-Ching Lo, Natalia Pripuzova, Bingjie Li, Anthony L. Komaroff, Guo-Chiuan Hung, Richard Wang, Harvey J. Alter, "Detection of MLV-related Virus Gene Sequences in Blood of Patients with Chronic Fatigue Syndrome and Healthy Blood Donors," *Proceedings of the National Academy of Sciences*, August 23, 2010, doi:10.1073/pnas.1006901107.

CHAPTER FIFTEEN

[1] Dennis Jones, Telephone Interview with Kent Heckenlively, August 3, 2010.

[2] Judy Mikovits, Judy, Telephone Interview with Kent Heckenlively, December 30, 2012.

[3] Whittemore Peterson Institute for Neuro-Immune v. Judy A Mikovits, Order to Show Cause, Second Judicial District Court of the State of Nevada in and for the County of Washoe, The Honorable Brent Adams, District Judge, Case No. CV11-03232, December 19, 2011, Reno, Nevada.

[4] Whittemore Peterson Institute for Neuro-Immune v. Judy A Mikovits, Order to Show Cause, Second Judicial District Court of the State of Nevada in and for the County of Washoe, The Honorable Brent Adams, District Judge, Case No. CV11-03232, December 19, 2011, Reno, Nevada.

[5] Whittemore Peterson Institute for Neuro-Immune v. Judy A Mikovits, Order to Show Cause, Second Judicial District Court of the State of Nevada in and for the County of Washoe, The Honorable Brent Adams, District Judge, Case No. CV11-03232, December 19, 2011, Reno, Nevada.

[6] Whittemore Peterson Institute for Neuro-Immune v. Judy A Mikovits, Order to Show Cause, Second Judicial District Court of the State of Nevada in and for the County

of Washoe, The Honorable Brent Adams, District Judge, Case No. CV11-03232, December 19, 2011, Reno, Nevada.

7   Whittemore Peterson Institute for Neuro-Immune v. Judy A Mikovits, Order to Show Cause, Second Judicial District Court of the State of Nevada in and for the County of Washoe, The Honorable Brent Adams, District Judge, Case No. CV11-03232, December 19, 2011, Reno, Nevada.

8   Whittemore Peterson Institute for Neuro-Immune v. Judy A Mikovits, Order to Show Cause, Second Judicial District Court of the State of Nevada in and for the County of Washoe, The Honorable Brent Adams, District Judge, Case No. CV11-03232, December 19, 2011, Reno, Nevada.

9   Whittemore Peterson Institute for Neuro-Immune v. Judy A Mikovits, Order to Show Cause, Second Judicial District Court of the State of Nevada in and for the County of Washoe, The Honorable Brent Adams, District Judge, Case No. CV11-03232, December 19, 2011, Reno, Nevada.

10 Whittemore Peterson Institute for Neuro-Immune v. Judy A Mikovits, Order to Show Cause, Second Judicial District Court of the State of Nevada in and for the County of Washoe, The Honorable Brent Adams, District Judge, Case No. CV11-03232, December 19, 2011, Reno, Nevada.

11 Whittemore Peterson Institute for Neuro-Immune v. Judy A Mikovits, Order to Show Cause, Second Judicial District Court of the State of Nevada in and for the County of Washoe, The Honorable Brent Adams, District Judge, Case No. CV11-03232, December 19, 2011, Reno, Nevada.

12 Judy Mikovits, Telephone Interview with Kent Heckenlively, December 30, 2012.

CHAPTER SIXTEEN

1   Grant Delin, "Discover Interview: The World's Most Celebrated Virus Hunter, Ian Lipkin," Discover, April 2012.

2   Judy Mikovits, Telephone Interview with Kent Heckenlively, January 20, 2013.

3   Jonathan R. Kerr, Robert Petty, Beverly Burke, John Gough, David fear, Lindsey I. Sinclair, Derek Mattey, Selwyn C. M. Richards, Jane Montgomery, Don A. Baldwin, Paul Kellem, Tim J. Harrison, George E. Griffin, Janice Main, Derek Enlander, David J. Nutt, and Stephen T. Holgate, "Gene Expression Subtypes in Patients with Chronic Fatigue Syndrome/Myalgic Encephalomyelitis," Journal of Infectious Diseases, April 15, 2008, 197:1171–84, doi:10.1086/533453.

4   Jonathan R. Kerr, Robert Petty, Beverly Burke, John Gough, David fear, Lindsey I. Sinclair, Derek Mattey, Selwyn C. M. Richards, Jane Montgomery, Don A. Baldwin, Paul Kellem, Tim J. Harrison, George E. Griffin, Janice Main, Derek Enlander, David J. Nutt, and Stephen T. Holgate, "Gene Expression Subtypes in Patients with Chronic Fatigue Syndrome/Myalgic Encephalomyelitis," Journal of Infectious Diseases, April 15, 2008, 197:1171–84, doi:10.1086/533453.

5   Chris Cairns, CFS Patient Advocate, June 9, 2010.

6   Chris Cairns, CFS Patient Advocate, June 9, 2010.

7   Chris Cairns, CFS Patient Advocate, June 9, 2010.

8   Gerwyn Morris, Telephone Interview with Kent Heckenlively, August 24, 2013.

9   Gerwyn Morris, Telephone Interview with Kent Heckenlively, August 24, 2013.

10 Gerwyn Morris, Telephone Interview with Kent Heckenlively, August 24, 2013.

[11] Gerwyn Morris, Telephone Interview with Kent Heckenlively, August 24, 2013.

[12] Gerwyn Morris, Telephone Interview with Kent Heckenlively, August 24, 2013.

[13] Gerwyn Morris, Telephone Interview with Kent Heckenlively, August 24, 2013.

[14] Gerwyn Morris, Telephone Interview with Kent Heckenlively, August 24, 2013.

[15] Gerwyn Morris, Telephone Interview with Kent Heckenlively, August 24, 2013.

[16] Gerwyn Morris, Telephone Interview with Kent Heckenlively, August 24, 2013.

[17] J. A.Petros, R. S. Arnold, C. Plattner, L. Yue, N. V. Makarova, J. L. Blackwell, E. Hunter, "Variant XMRVs in Clinical Prostate Cancer," First International Conference on XMRV, Abstract, September 7, 2010.

[18] Tzong-Hae Lee, Elona Gusho, Jaydip Das Gupta, Eric A. Klein, Robert H. Silverman, "XMRV Infection induces Host Genes that Regulate Inflammation and Cellular Physiology," First International Conference on XMRV, Abstract, September 7, 2010.

[19] Paul Cheney, "XMRV Detection in a National Practice Specializing in Chronic Fatigue Syndrome (CFS)," First International Conference on XMRV, Abstract, September 7, 2010.

[20] Paul Cheney, "XMRV Detection in a National Practice Specializing in Chronic Fatigue Syndrome (CFS)," First International Conference on XMRV, Abstract, September 7, 2010.

[21] Vincent Raccinello, *This Week in Virology*, February 17, 2011.

[22] Shyh-Ching Lo, Natalia Pripuzova, Bingjie Li, Anthony L. Komaroff, Guo-Chiuan Hung, Richard Wang, Harvey J. Alter (NIH). "Detection of MLV-Related Virus Gene Sequences in Blood of Patients with Chronic Fatigue Syndrome and Healthy Blood Donors," First International Conference on XMRV, Abstract, September 7, 2010.

[23] Judy A. Mikovits, Vincent C. Lombardi, and Francis W. Ruscetti, "Detection of an Infectious Retrovirus, XMRV, in Blood Cells of Patients with Chronic Fatigue Syndrome," First International Conference on XMRV, Abstract, September 7, 2010.

[24] Judy A. Mikovits, Vincent C. Lombardi, and Francis W. Ruscetti, "Detection of an Infectious Retrovirus, XMRV, in Blood Cells of Patients with Chronic Fatigue Syndrome," First International Conference on XMRV, Abstract, September 7, 2010.

[25] Valerie Courgnaud, Jean-Luc Battini, Marc Sitbon, Andrew Mason, "Mouse Retroviruses and Chronic Fatigue Syndrome: Does X (or P) Mark the Spot?" *Proceedings of the National Academy of Sciences*, September 7, 2010, Vol. 107, 15666–15667, doi: 10.1073/pnas.1007944107.

[26] Grant Delin, "Discover Interview: The World's Most Celebrated Virus Hunter, Ian Lipkin," Discover, April 2012.

[27] Judy Mikovits, Telephone Interview with Kent Heckenlively, January 20, 2013.

[28] David Tuller, "Study Links Chronic Fatigue to Virus Class," *The New York Times*, August 23, 2010.

CHAPTER SEVENTEEN

[1] Judy Mikovits, Telephone Interview with Kent Heckenlively, February 5, 2013.

[2] John M. Coffin, Howard M. Temin, "Comparison of Rous Sarcoma Virus-Specific Deoxyribonucleic Acid Polymerases in Virions of Rous Sarcoma Virus and in Rous Sarcoma Virus-Infected Chicken Cells," *Journal of Virology*, May 1971; 7(5): 625–634.

[3] John Coffin, Jonathon Stoye, "A New Virus for Old Diseases?" *Science*, October 23, 2009.

[4] Judy Mikovits, Telephone Interview with Kent Heckenlively, February 5, 2013.

[5] Judy Mikovits, Telephone Interview with Kent Heckenlively, February 5, 2013.

6   Tobias Paprotka, Krista Delviks-Frankenberry, Oya Cingoz, Anthony Martinez, Hsing-Jien Kung, Clifford G. Tepper, Wei-Shau Hu, Matthew J. Fivash Jr., John M. Coffin, Vinay Pathak, "Recombinant Origin of the Retrovirus XMRV," *Science*, July 1, 2011, Volume 333, 97–101, doi: 10.1126/science.1205292.

7   Judy Mikovits, Telephone Interview with Kent Heckenlively, February 5, 2013.

8   DeRisi, Joseph, "Hunting the Next Killer Virus," TED Talks. Monterey, California, February 5, 2006.

9   Judy Mikovits, Telephone Interview with Kent Heckenlively, February 5, 2013.

10   Harvey Alter, Discussion, State of the Knowledge Conference, April 7, 2011.

11   Harvey Alter, Discussion, State of the Knowledge Conference, April 7, 2011.

12   Harvey Alter, Discussion, State of the Knowledge Conference, April 7, 2011.

13   Harvey Alter, Discussion, State of the Knowledge Conference, April 7, 2011.

14   Harvey Alter, Discussion, State of the Knowledge Conference, April 7, 2011.

15   Harvey Alter, Discussion, State of the Knowledge Conference, April 7, 2011.

16   Cort Johnson, "NIH Steps Up to the Plate—Pathogen Ace Picked to Lead Major XMRV/CFS Study," *Phoenix Rising*, September 8, 2010.

17   Lombardi VC1, Hagen KS, Hunter KW, Diamond JW, Smith-Gagen J, Yang W, Mikovits JA, Xenotropic murine leukemia virus-related virus-associated chronic fatigue syndrome reveals a distinct inflammatory signature. In Vivo. 2011 May-Jun; 25(3): 307–14.

18   Judy Mikovits, Discussion, State of the Knowledge Conference, April 7, 2011.

19   Tobias Paprotka, Krista Delviks-Frankenberry, Oya Cingoz, Anthony Martinez, Hsing-Jien Kung, Clifford G. Tepper, Wei-Shau Hu, Matthew J. Fivash Jr., John M. Coffin, Vinay Pathak, "Recombinant Origin of the Retrovirus XMRV," *Science*, July 1, 2011, Volume 333, 97–101, doi: 10.1126/science.1205292.

20   Tobias Paprotka, Krista Delviks-Frankenberry, Oya Cingoz, Anthony Martinez, Hsing-Jien Kung, Clifford G. Tepper, Wei-Shau Hu, Matthew J. Fivash Jr., John M. Coffin, Vinay Pathak, "Recombinant Origin of the Retrovirus XMRV," *Science*, July 1, 2011, Volume 333, 97–101, doi: 10.1126/science.1205292.

21   Judy Mikovits, Telephone Interview with Kent Heckenlively, February 5, 2013.

CHAPTER EIGHTEEN

1   Martha Bellisie, "Wit and Work Made Lobbyist, Harvey Whittemore, 'An Institution,'" *Reno Gazette Journal*, February 12, 2012.

2   David McGrath Schwartz, "Federal Grand Jury Indicts Former Nevada Power Broker, Harvey Whittemore," *Las Vegas Sun*, June 6, 2012.

3   Judy Mikovits, Telephone Interview with Kent Heckenlively, February 18, 2013.

4   Wingfield Nevada Group Holding Company LLC, Tuffy Ranch Properties LLC, The Foothills at Wingfield LLC vs. F. Harvey Whittemore, Annette Whittemore, The Lakeshore House Limited Partnership. District Court, Clark County, Nevada. Case No. A-12-655426-B, Complaint filed January 27, 2012.

5   Wingfield Nevada Group Holding Company LLC, Tuffy Ranch Properties LLC, The Foothills at Wingfield LLC vs. F. Harvey Whittemore, Annette Whittemore, The Lakeshore House Limited Partnership, District Court, Clark County, Nevada, Case No. A-12-655426-B, Complaint filed January 27, 2012.

6   Wingfield Nevada Group Holding Company LLC, Tuffy Ranch Properties LLC, The Foothills at Wingfield LLC vs. F. Harvey Whittemore, Annette Whittemore, The

Lakeshore House Limited Partnership, District Court, Clark County, Nevada, Case No. A-12-655426-B, Complaint filed January 27, 2012.

7 Wingfield Nevada Group Holding Company LLC, Tuffy Ranch Properties LLC, The Foothills at Wingfield LLC vs. F. Harvey Whittemore, Annette Whittemore, The Lakeshore House Limited Partnership, District Court, Clark County, Nevada, Case No. A-12-655426-B, Complaint filed January 27, 2012.

8 Wingfield Nevada Group Holding Company LLC, Tuffy Ranch Properties LLC, The Foothills at Wingfield LLC vs. F. Harvey Whittemore, Annette Whittemore, The Lakeshore House Limited Partnership, District Court, Clark County, Nevada, Case No. A-12-655426-B, Complaint filed January 27, 2012.

9 Wingfield Nevada Group Holding Company LLC, Tuffy Ranch Properties LLC, The Foothills at Wingfield LLC vs. F. Harvey Whittemore, Annette Whittemore, The Lakeshore House Limited Partnership, District Court, Clark County, Nevada, Case No. A-12-655426-B, Complaint filed January 27, 2012.

10 F. Harvey Whittemore, Annette Whittemore vs. Thomas A Seeno, Albert D. Seeno, Jr., and Albert Seeno III. Complaint. United States District Court in and for the District of Nevada, Case 3:1 2-cv-00063-LRH, Complaint filed February 1, 2012.

11 Mathias Gafni, "Seeno Family Settles Contentious Nevada Lawsuits," *Contra Costa Times*, February 7, 2013.

12 Judy Mikovits, Telephone Interview with Kent Heckenlively, February 18, 2013.

13 F. Harvey Whittemore, Annette Whittemore vs. Thomas A Seeno, Albert D. Seeno, Jr., and Albert Seeno III, Complaint, United States District Court in and for the District of Nevada, Case 3:1 2-cv-00063-LRH, Complaint filed February 1, 2012.

14 Judy Mikovits, Telephone Interview with Kent Heckenlively, February 18, 2013.

15 F. Harvey Whittemore, Annette Whittemore vs. Thomas A Seeno, Albert D. Seeno, Jr., and Albert Seeno III, Complaint, United States District Court in and for the District of Nevada, Case 3:1 2-cv-00063-LRH, Complaint filed February 1, 2012.

16 Martha Bellisile, Whittemore Manager Claims Seenos Violated Environmental Permits, *Reno Gazette Journal*, February 25, 2012.

17 Martha Bellisile, Whittemore Manager Claims Seenos Violated Environmental Permits, *Reno Gazette Journal*, February 25, 2012.

18 Jeff German and Francis McCabe, FBI Investigating Whittemore Activities, '07 Campaign Contributions, *Las Vegas Review Journal*, February 10, 2012.

19 Jeff German and Francis McCabe, FBI Investigating Whittemore Activities, '07 Campaign Contributions, *Las Vegas Review Journal*, February 10, 2012.

20 Jeff German, "FBI Subpoenas in Whittemore Probe Aimed to Surprise," *Las Vegas Review Journal*, February 20, 2012.

21 Judy Mikovits, Telephone Interview with Kent Heckenlively, December 23, 2012.

22 Jeff German, "FBI Subpoenas in Whittemore Probe Aimed to Surprise," *Las Vegas Review Journal*, February 20, 2012.

23 Jeff German, "FBI Subpoenas in Whittemore Probe Aimed to Surprise," *Las Vegas Review Journal*, February 20, 2012.

24 Jeff German, "FBI Subpoenas in Whittemore Probe Aimed to Surprise," *Las Vegas Review Journal*, February 20, 2012.

25 Martha Bellisle, "Fired Whittemore-Peterson Institute Researcher Claims Justice System Flawed," *Reno Gazette Journal*, April 24, 2012.

26 California Code of Judicial Ethics, 2013, http://www.courts.ca.gov/documents/ca_code_judicial_ethics.pdf.

27 State of Nevada Commission on Judicial Discipline, "Nevada Code of Judicial Conduct," adopted January 19, 2010, http://www.dorothyforjudge.com/files/NevadaCodeofJudicial Conduct.pdf.

28 California Code of Judicial Ethics, 2013, Canon 2(E)(2)(b)(i), www.courts.ca.gov/documents/ca_code_judicial_ethics.pdf

29 California Code of Civil Procedure, Section 170.1, http://www.leginfo.ca.gov/cgi-bin/displaycode?section=ccp&group=00001-01000&file=170-170.9

30 Nevada Code of Judicial Conduct, Preamble (2), Adopted January 19, 2010, http://www.dorothyforjudge.com/files/NevadaCodeofJudicialConduct.pdf.

31 Judy Mikovits, Telephone Interview with Kent Heckenlively, February 18, 2013.

32 Ann Hall, email to Dennis Jones, March 7, 2012.

33 Judy Mikovits, email to Dennis Jones. 8 March 2012.

34 Dennis Jones, email to Judy Mikovits. 9 March 2012.

35 Annette Whittemore, email to Judy Mikovits, August 2, 2011.

36 Judy Mikovits, Telephone Interview with Kent Heckenlively, February 18, 2013.

37 Judy Mikovits, Telephone Interview with Kent Heckenlively, February 18, 2013.

38 Matthias Gafni, "Seeno Family Settles Contentious Nevada Lawsuits." Contra Costa Times, February 7, 2013.

39 WPI vs. Dr. Judy Mikovits, Motion for Reconsideration, Second Judicial District Court of the State of Nevada in and for the County of Washoe, Case No. CV11-03232, May 19, 2012.

40 WPI vs. Dr. Judy Mikovits, Motion for Reconsideration, Second Judicial District Court of the State of Nevada in and for the County of Washoe, Case No. CV11-03232, May 19, 2012.

41 WPI vs. Dr. Judy Mikovits, Motion for Reconsideration, Second Judicial District Court of the State of Nevada in and for the County of Washoe, Case No. CV11-03232, May 19, 2012.

42 Anjeanette Damon, "Indictment Caps Lobbyist Harvey Whittemore's Dramatic Fall from Grace," Las Vegas Sun, June 10, 2012.

43 Anjeanette Damon, "Indictment Caps Lobbyist Harvey Whittemore's Dramatic Fall from Grace," Las Vegas Sun, June 10, 2012.

44 Anjeanette Damon, "Indictment Caps Lobbyist Harvey Whittemore's Dramatic Fall from Grace," Las Vegas Sun, June 10, 2012.

45 Anjeanette Damon, "Indictment Caps Lobbyist Harvey Whittemore's Dramatic Fall from Grace," Las Vegas Sun, June 10, 2012.

46 Anjeanette Damon, "Indictment Caps Lobbyist Harvey Whittemore's Dramatic Fall from Grace," Las Vegas Sun, June 10, 2012.

47 Jon Cohen, email to Dr. Judy Mikovits, June 13, 2012.

48 Judy Mikovits, email to Jon Cohen, June 13, 2012.

49 Jon Cohen, email to Dr. Judy Mikovits, June 13, 2012.

50 Judy Mikovits, email to Jon Cohen, June 13, 2012.

51 Jon Cohen, email to Dr. Judy Mikovits, June 13, 2012.

52 Jon Cohen, "Criminal Charges Dropped Against Chronic Fatigue Syndrome Researcher Judy Mikovits," Science Insider, June 13, 2012.

53   Judy Mikovits, Telephone Interview with Kent Heckenlively, February 18, 2013.
54   Press Release, "Former Nevada lobbyist Harvey Whittemore Convicted of making Unlawful Senate Camapign Contributions," Federal Bureau of Investigation, May 29, 2013.
55   Press Release, "Former Nevada Lobbyist Harvey Whittemore Convicted of Making Unlawful Senate Camapign Contributions," Federal Bureau of Investigation. May 29, 2013.
56   Government's Sentencing memorandum, US v. F. Harvey Whittemore, September 23, 2013.
57   US v. F. Harvey Whittemore, United States District Court, District of Nevada, Transcript of imposition of Sentence, Case 3:12-cr-00058-LRH-WGC, September 30, 2013, 101–102.
58   US v. F. Harvey Whittemore, United States District Court, District of Nevada, Transcript of imposition of Sentence, Case 3:12-cr-00058-LRH-WGC, September 30, 2013, 101–102.
59   US v. F. Harvey Whittemore, United States District Court, District of Nevada, Transcript of imposition of Sentence, Case 3:12-cr-00058-LRH-WGC, September 30, 2013, 101–102.
60   US v. F. Harvey Whittemore, United States District Court, District of Nevada, Transcript of imposition of Sentence, Case 3:12-cr-00058-LRH-WGC, September 30, 2013, 101–102.
61   US v. F. Harvey Whittemore, United States District Court, District of Nevada, Transcript of imposition of Sentence, Case 3:12-cr-00058-LRH-WGC, September 30, 2013, 101–102.
62   US v. F. Harvey Whittemore, United States District Court, District of Nevada, Transcript of imposition of Sentence, Case 3:12-cr-00058-LRH-WGC, September 30, 2013, 101–102.
63   US v. F. Harvey Whittemore, United States District Court, District of Nevada, "Partial Transcript of Jury Trial (Day Nine), Closing Arguments of Counsel", Case 3:12-cr-00058-LRH-WGC, May 28, 2013, 89.
64   US v. F. Harvey Whittemore, United States District Court, District of Nevada, "Partial Transcript of Jury Trial (Day Nine), Closing Arguments of Counsel", Case 3:12-cr-00058-LRH-WGC, May 28, 2013, 89.
65   US v. F. Harvey Whittemore, United States District Court, District of Nevada, Transcript of imposition of Sentence, Case 3:12-cr-00058-LRH-WGC, September 30, 2013, 101–102.
66   Jeff German, "Harvey Whittemore Ordered to Surrender to Federal Prison Authorities," Las Vegas Review, June 5, 2014.

CHAPTER NINETEEN

1   Hillary Johnson, "Hunting the Shadow Virus," Discover, March 2013.
2   Howard Hughes Medical Institute, "Joseph DeRisi Receives Heinz Award," September 10, 2008, www.hhmi.org/news/joseph-derisi-receives-heinz-award.
3   Deanna Lee, Jaydip Das Gupta, Christina Gaughan, Imke Steffen, Ning Tang, Ka-Ceung Luk, Anatoly Urisman, Nicole Fisher, Ross Molinaro, Miranda Broz, Gerald Schochetman, Eric A. Klein, Don Ganem, Joseph L. DeRisi, John Hackett Jr., Robert H. Silverman, Charles Y. Chiu, "In-Depth Investigation of Archival and Prospectively Collected Samples Reveals no Evidence for XMRV Infection in Prostate Cancer," PLOSOne, September 18, 2012, doi: 10.1371/journal.pone.0044954.
4   Anatoly Urisman, Ross Molinaro, Nicole Fischer, SJ Plummer, G Case, Eric A. Klein, K Malathi, C Magi-Galluzzi, RR Tubbs, Don Ganem, Robert Silverman, Joseph DeRisi, "Identification of a novel Gammaretrovirus in prostate tumors of patients homozygous for R462Q RNase L variant," PLoS Pathogen, Epub, March 31, 2006.

5   Erin Podolak, "Template-Blocking PCR for Genome Walking," *BioTechniques*, March 23, 2010, www.biotechniques.com/news/Template-blocking-PCR-for-genome-walking/biotechniques-217395.html.

6   "Science Reference: Dolly the Sheep." Undated. *ScienceDaily*. Web: http://www.sciencedaily.com/articles/d/dolly_the_sheep.htm.

7   Judy Mikovits, Telephone Interview with Kent Heckenlively, March 3, 2013.

8   Judy Mikovits, Telephone Interview with Kent Heckenlively, March 3, 2013.

9   Rebecca S. Arnold, Natalia V. Makarova, Abedoye O. Osunkoya, Suganthi Suppiah, Tkara A Scott, Nicole A. Johnson, Sushma M. Bhosle, Dennis Liotta, Eric Hunter, Fray F. Marshall, Hinh Ly, Ross J. Molinaro, Jerry L. Blackwell, John A. Petros, "XMRV Infection in Patients with Prostate Cancer: Novel Serologic Assay and Correlation with PCR and FISH," *Urology*, April, 2010;75(4):755-61. doi: 10.1016/j.urology.2010.01.038.

10 Session Title: Endogenous Retroviruses, Foamy Viruses, and XMRV, 15th International Conference on Human Retrovirology: HTLV and Related Retroviruses, June 5–8, 2011

11 Robert Silverman, "Human Infection or Lab Artifact: Will the Real XMRV Please Stand up?" *Retrovirology*, June 6, 2011, 8(Suppl 1): A241, doi: 10.1186/1742-4690-8-S1-A241.

12 Judy Mikovits, Telephone Interview with Kent Heckenlively, March 3, 2013.

13 Francis Ruscetti, Vincent C. Lombardi, Michael Snyderman, Dan Bertolette, Kathryn S. Jones, and Judy A. Mikovits, "Development of XMRV Producing B Cell Lines from Lymphomas from Patients with Chronic Fatigue Syndrome," *Retrovirology*, June 6, 2011, 8 (Suppl 1):A230 doi:10.1186/1742-4690-8-S1-A230.

14 Francis Ruscetti, Vincent C. Lombardi, Michael Snyderman, Dan Bertolette, Kathryn S. Jones, and Judy A. Mikovits, "Development of XMRV Producing B Cell Lines from Lymphomas from Patients with Chronic Fatigue Syndrome," *Retrovirology*, June 6, 2011, 8 (Suppl 1): A230 doi:10.1186/1742-4690-8-S1-A230.

15 Francis Ruscetti, Vincent C. Lombardi, Michael Snyderman, Dan Bertolette, Kathryn S. Jones, and Judy A. Mikovits, "Development of XMRV Producing B Cell Lines from Lymphomas from Patients with Chronic Fatigue Syndrome," *Retrovirology*, June 6, 2011, 8 (Suppl 1):A230 doi:10.1186/1742-4690-8-S1-A230.

16 Robert Silverman, email to Judy Mikovits and her co-authors, July 6, 2011.

17 Frank Rucsetti, email to Robert Silverman, July 7, 2011.

18 Robert Silverman, email to Dr. Frank Ruscetti, July 7, 2011.

19 Frank Rucsetti, email to Robert Silverman, July 7, 2011.

20 Frank Rucsetti, email to Robert Silverman, July 7, 2011.

21 Hillary Johnson, "Hunting the Shadow Virus," Discover, March 2013.

22 Judy Mikovits, Telephone Interview with Kent Heckenlively, March 3, 2013.

23 Liz Highleyman, "Inflammation, Immune Activation and HIV," The Body Pro from The San Francisco AIDS Foundation, Winter/Spring 2010.

24 Liz Highleyman, "Inflammation, Immune Activation and HIV," The Body Pro from The San Francisco AIDS Foundation, Winter/Spring 2010.

25 Frank Ruscetti, "The Scientific Deception of Annette Whittemore," Written Recollection, May 2013.

26 Frank Ruscetti, "The Scientific Deception of Annette Whittemore," Written Recollection, May 2013.

27 Frank Ruscetti, "The Scientific Deception of Annette Whittemore," Written Recollection, May 2013.

[28] Frank Ruscetti, "The Scientific Deception of Annette Whittemore," Written Recollection, May 2013.

[29] General Session Agenda-International Association for Chronic Fatigue Syndrome/ME, September 22–25, 2011.

[30] General Session Agenda-International Association for Chronic Fatigue Syndrome/ME, September 22–25, 2011.

[31] Judy Mikovits, Agenda-IACFS/ME 10th International Research & Clinical Conference, September 22–25, 2011, Ottawa, Ontario, Canada.

[32] Francois Villinger, Jaydip Das Gupta, Nattawat Onlamoon, Ross Molinaro, Suganthi Suppiah, Prachi Sharma, Kenneth Rogers, Christina Gaughan, Eric Klein, Xiaoxing Qiu, Gerald Schochetman, John Hackett, and Robert H Silverman, "XMRV replicates preferentially in mucosal sites in vivo: Relevance to XMRV transmission?" *Retrovirology*, June 6, 2011, 8(Suppl 1):A219 doi:10.1186/1742-4690-8-S1-A219.

[33] Judy Mikovits, Agenda-IACFS/ME 10th International Research & Clinical Conference, September 22–25, 2011, Ottawa, Ontario, Canada.

[34] Judy Mikovits, "The Case for Human Gamma Retroviruses (HGRV) in CFS/ME," International Association for Chronic Fatigue Syndrome/ME, September 22–25, 2011, Ottawa, Ontario, Canada.

[35] Judy Mikovits, Telephone Interview with Kent Heckenlively, March 3, 2013.

[36] Judy Mikovits, Telephone Interview with Kent Heckenlively, March 3, 2013.

[37] Jon Cohen and Martin Enserink, "False Positive," *Science*, Vol.333, September 23, 2011, 1694–1701.

[38] Jon Cohen and Martin Enserink, "False Positive," *Science*, Vol.333, September 23, 2011, 1694–1701.

[39] Jon Cohen and Martin Enserink, "False Positive," *Science*, Vol.333, September 23, 2011, 1694–1701.

[40] Jon Cohen and Martin Enserink, "False Positive," *Science*, Vol.333, September 23, 2011, 1694–1701.

[41] Jon Cohen and Martin Enserink, "False Positive," *Science*, Vol.333, September 23, 2011, 1694–1701.

CHAPTER TWENTY

[1] Ewen Calloway, "The Man Who Put the Nail in XMRV's Coffin," *Nature*, September 18, 2012.

[2] Grant Delin, "Discover Interview: The World's Most Celebrated Virus Hunter, Ian Lipkin," *Discover*, April 2012.

[3] Grant Delin, "Discover Interview: The World's Most Celebrated Virus Hunter, Ian Lipkin," *Discover*, April 2012.

[4] Judy Mikovits, Telephone Interview with Kent Heckenlively, March 20, 2013.

[5] Paul Cheney, Telephone Interview with Kent Heckenlively, July 25, 2013.

[6] Jim Fuite, Suzanne D. Vernon, Gordon Broderick, "Neuroendocrine and Immune Network Re-modeling in Chronic Fatigue Syndrome: An Exploratory Analysis," Genomics, December 2008, 92(6):393-9. doi: 10.1016/j.ygeno.2008.08.008.

[7] Paul Cheney, Telephone Interview with Kent Heckenlively, July 25, 2013.

[8] Paul Cheney, Telephone Interview with Kent Heckenlively, July 25, 2013.

[9] Gerwyn Morris, Telephone Interview with Kent Heckenlively, August 23, 2013.

10 Gerwyn Morris, Telephone Interview with Kent Heckenlively, August 23, 2013.
11 Gerwyn Morris, Telephone Interview with Kent Heckenlively, August 23, 2013.
12 Jamie Deckoff-Jones, Telephone Interview with Kent Heckenlively, August 14, 2013.
13 Judy Mikovits, Telephone Interview with Kent Heckenlively, March 20, 2013.
14 Email from Dr. Ian Lipkin to Dr. Judy Mikovits, September 16, 2012.
15 Ian Lipkin, Press Conference on Multi-Center Study at Columbia University, September 18, 2012.
16 Ian Lipkin, Press Conference on Multi-Center Study at Columbia University, September 18, 2012.
17 Ian Lipkin, Press Conference on Multi-Center Study at Columbia University, September 18, 2012.
18 Ian Lipkin, Press Conference on Multi-Center Study at Columbia University, September 18, 2012.
19 Ian Lipkin, Press Conference on Multi-Center Study at Columbia University, September 18, 2012.
20 Ian Lipkin, Press Conference on Multi-Center Study at Columbia University, September 18, 2012.
21 Ian Lipkin, Press Conference on Multi-Center Study at Columbia University, September 18, 2012.
22 Ian Lipkin, Press Conference on Multi-Center Study at Columbia University, September 18, 2012.
23 Ian Lipkin, Press Conference on Multi-Center Study at Columbia University, September 18, 2012.
24 Hillary Johnson, Press Conference on Multi-Center Study at Columbia University, September 18, 2012.
25 Ian Lipkin, Press Conference on Multi-Center Study at Columbia University, September 18, 2012.
26 Ian Lipkin, Press Conference on Multi-Center Study at Columbia University, September 18, 2012.
27 Deborah Warnoff, Press Conference on Multi-Center Study at Columbia University, September 18, 2012.
28 Ian Lipkin, Press Conference on Multi-Center Study at Columbia University, September 18, 2012.
29 Harvey Alter, Press Conference on Multi-Center Study at Columbia University, September 18, 2012.
30 Hillary Johnson, Press Conference on Multi-Center Study at Columbia University, September 18, 2012.
31 Ian Lipkin, Press Conference on Multi-Center Study at Columbia University, September 18, 2012.
32 Mady Hornig, Press Conference on Multi-Center Study at Columbia University, September 18, 2012.
33 Hillary Johnson, Press Conference on Multi-Center Study at Columbia University, September 18, 2012.
34 Ian Lipkin, Press Conference on Multi-Center Study at Columbia University, September 18, 2012.
35 Ian Lipkin, Press Conference on Multi-Center Study at Columbia University, September 18, 2012.

36 Harvey Alter, Press Conference on Multi-Center Study at Columbia University, September 18, 2012.

37 Martin Enserink, Press Conference on Multi-Center Study at Columbia University, September 18, 2012.

38 Martin Enserink, Press Conference on Multi-Center Study at Columbia University, September 18, 2012.

39 Ian Lipkin, Press Conference on Multi-Center Study at Columbia University, September 18, 2012.

40 Hillary Johnson, Press Conference on Multi-Center Study at Columbia University, September 18, 2012.

41 Ian Lipkin, Press Conference on Multi-Center Study at Columbia University, September 18, 2012.

42 Max Pfost, Telephone Interview with Kent Heckenlively, March 13, 2014.

43 Max Pfost, Telephone Interview with Kent Heckenlively, March 13, 2014.

44 Max Pfost, Telephone Interview with Kent Heckenlively, March 13, 2014.

45 Max Pfost, Telephone Interview with Kent Heckenlively, March 13, 2014.

46 Max Pfost, Telephone Interview with Kent Heckenlively, March 13, 2014.

47 Max Pfost, Telephone Interview with Kent Heckenlively, March 13, 2014.

CHAPTER TWENTY-ONE

1 Meera Murgai, James Thomas, Olga Cherepanova, Krista Delviks-Frankenberry, Paul Deeble, Vinay K Pathak, David Rekosh, and Gary Owens, "Xenotropic MLV Envelope Proteins Induce Tumor Cells to Secrete Factors That Promote the Formation of Immature Blood Vessels," *Retrovirology*, March 27, 2013. doi: 10.1186/1742-4690-10-34. PMCID: PMC3681559

2 Ian Lipkin, Public Conference Call with the Centers for Disease Control, September 10, 2013. Transcript by ME/CFS Forums.com/wiki/Lipkin.

3 Ewen Calloway, "The Man Who Put the Nail in XMRV's Coffin," *Nature*, September 18, 2012.

4 Ewen Calloway, "The Man Who Put the Nail in XMRV's Coffin," *Nature*, September 18, 2012.

5 Ben Berkhout et al., "Obituary: Kuan-Teh Jeang," *Retrovirology*, 2013, 10:28.

6 Dong-Yan Jin, Yun-Bo Shi, and T-C Wu. "In Memorium: Kuan-The Jeang, MD PhD (1958–2013)," Editorial in *Cell & Bioscience*, 2013, 3:13.

7 Meera Murgai, James Thomas, Olga Cherepanova, Krista Delviks-Frankenberry, Paul Deeble, Vinay K Pathak, David Rekosh, and Gary Owens, "Xenotropic MLV Envelope Proteins Induce Tumor Cells to Secrete Factors That Promote the Formation of Immature Blood Vessels," *Retrovirology*, March 27, 2013. doi: 10.1186/1742-4690-10–34. PMCID: PMC3681559

8 Meera Murgai, James Thomas, Olga Cherepanova, Krista Delviks-Frankenberry, Paul Deeble, Vinay K Pathak, David Rekosh, and Gary Owens, "Xenotropic MLV Envelope Proteins Induce Tumor Cells to Secrete Factors That Promote the Formation of Immature Blood Vessels," *Retrovirology*, March 27, 2013, doi: 10.1186/1742-4690-10–34. PMCID: PMC3681559

9 Robert Naviaux, Zarazuela Zolkipli, Lin Wang, Tomohiro Nakayama, Jane C. Naviaux, Thuy P Le, Michael A.Schuchbauer, Mihael Rogac, Qingbo Tang, Laura L. Dugan, Susan

B. Powell, "Antipurinergic Therapy Corrects the Autism-Like Features in the Poly(IC) Mouse Model," *PLoS One*, March 13, 2013, doi: 10.1371/journal.pone.0057380.

[10] Robert Naviaux, Zarazuela Zolkipli, Lin Wang, Tomohiro Nakayama, Jane C. Naviaux, Thuy P Le, Michael A.Schuchbauer, Mihael Rogac, Qingbo Tang, Laura L. Dugan, Susan B. Powell, "Antipurinergic Therapy Corrects the Autism-Like Features in the Poly(IC) Mouse Model," *PLoS One*, March 13, 2013, doi: 10.1371/journal.pone.0057380.

[11] "Rain Mouse–R"ecent Experiments Give a Glimmer of Hope for a treatment for Autism," *The Economist*, June 21, 2014, www.economist.com/node/21604533/print.

[12] Ian Lipkin, Public Conference Call with the Centers for Disease Control, September 10, 2013, Transcript by ME/CFS Forums.com/wiki/Lipkin.

[13] Ian Lipkin, Public Conference Call with the Centers for Disease Control, September 10, 2013. Transcript by ME/CFS Forums.com/wiki/Lipkin.

[14] Andrew Naughtie, "HIV 'invisibility cloak' allows virus to evade immune system." The Conversation. November 8, 2013. Web: http://theconversation.com/hiv-invisibility-cloak-allows-virus-to-evade-immune-system-19918.

[15] Liz Highleyman, "Inflammation, Immune Activation and HIV," The Body Pro from the San Francisco AIDS Foundation, Winter/Spring 2010.

[16] Jonathan Stoye and John Coffin, "The Dangers of Xenotransplantation," *Nature Medicine* 1, November 1995, 1(11):1100.

[17] Jonathan Stoye and John Coffin, "The Dangers of Xenotransplantation," *Nature Medicine* 1, November 1995, 1(11):1100.

[18] David Palesch, Mohammad Khalid, Christina Sturzel, and Jan Munch, "How Can Xenotropic Murine Leukemia Virus-Related Virus (XMRV) Contamination be Prevented? Susceptibility of XMRV to Alcohol-Based Disinfectants and its Environmental Stability," *Applied and Environmental Microbiology*, February 14, 2014, doi: 10.1128/AEM.04064-13.

Made in the USA
Middletown, DE
05 May 2020